FUNDAMENTALS OF

INFORMATION SYSTEMS

Ralph M. Stair
Florida State University

George W. Reynolds
The University of Cincinnati

**COURSE
TECHNOLOGY**
—★—
THOMSON LEARNING

Australia • Canada • Mexico • Singapore • Spain • United Kingdom • United States

COURSE TECHNOLOGY
THOMSON LEARNING ™

Fundamentals of Information Systems
by Ralph M. Stair, George W. Reynolds

Managing Editor:
Jennifer Locke

Senior Vice President, Publisher:
Kristen Duerr

Project Management and Development:
Elm Street Publishing Services, Inc.

Senior Product Manager:
Jennifer Muroff

Associate Product Manager:
Matthew Van Kirk

Editorial Assistant:
Janet Aras

Marketing Manager:
Toby Shelton

Text Design:
Elm Street Publishing Services, Inc.

Cover Design:
Efrat Reis

Composition House:
GEX Publishing Services

Photo Researcher:
Abby Reip

Disclaimer
Course Technology reserves the right to revise this publication and make changes from time to time in its content without notice.

Microsoft, Windows 95, and Windows 98 are registered trademarks of Microsoft Corporation.

Some of the product names and company names used in this book have been used for identification purposes only and may be trademarks or registered trademarks of their manufacturers and sellers.

SAP, R/3, and other SAP product/services referenced herein are trademarks of SAP Aktiengesellschaft, Systems, Applications and Products in Data Processing, Neurottstasse 16, 69190 Walldorf, Germany. The publisher gratefully acknowledges SAP's kind permission to use these trademarks in this publication. SAP AG is not the publisher of this book and is not responsible for it under any aspect of press law.

Library of Congress Cataloging-in-Publication Data

Stair, Ralph M.

Fundamentals of information systems/Ralph M. Stair, George W. Reynolds.—1st ed., p. cm.

Includes bibliographical references and index.

ISBN 0-619-03416-5 (alk. paper)

1. Management information systems. J. Reynolds, George Walter, 1944–.II. Title.
T58.6 .S717 2001
658' .0546—dc21

2001028166

ISBN 0-619-03416-5

PREFACE

We are proud to introduce the first edition of an exciting new text, *Fundamentals of Information Systems*. Based on much feedback, we realized that there are many instructors who simply needed a shorter, less technical book than *Principles of Information Systems* for their introductory courses. The *Fundamentals* text is designed to support a first course in information systems and to meet this important need.

Like *Principles of Information Systems*, this new text stands proudly at the beginning of the IS curriculum, offering the basic IS concepts that every business student must learn to be successful. This text has been written specifically for the first course in the IS curriculum, and it treats the appropriate computer and IS concepts in a business context with a managerial emphasis.

Our overall goal in creating *Fundamentals of Information Systems* was to develop an outstanding text that follows the pedagogy and approach of *Principles of Information Systems*, and to retain the breadth of topics presented, but with less detail and fewer subtopics. This text, which has been entirely rewritten with this goal in mind, contains new material and features. New "Information Systems in Action" boxes, for example, were written by guest authors and are included in every chapter. Near the beginning of each chapter, we have also added a few sentences or a short paragraph that describes how learning the material in the chapter can directly benefit any student majoring in accounting, finance, management, marketing, personnel, information systems, or any other business area.

We have always advocated that education in information systems is critical for employment in almost any field. Today, information systems are used for business processes from communications to order processing to number crunching and in business functions ranging from marketing to human resources to accounting and finance. Chances are, regardless of your future occupation, you need to understand what information systems can and cannot do and be able to use them to help you accomplish your work. You will be expected to suggest new uses of information systems and help design solutions to business problems using information systems. You will be challenged to identify and evaluate information systems options. To be successful, you must be able to view information systems from the perspective of business and organizational needs. For your solutions to be accepted, you must identify and address their impact on fellow workers. For these reasons, a course in information systems is essential for students in today's high-tech world.

Fundamentals of Information Systems continues the tradition and approach of the five editions of *Principles of Information Systems*. Our primary objective is to develop the best information systems text and accompanying materials for the first information technology course required of all business students. Using surveys, questionnaires, focus groups, and feedback that we have received from adopters and others who teach in the field, we have been able to develop the highest-quality set of teaching materials available.

GOALS OF THIS TEXT

Fundamentals of Information Systems has four main goals:

1. To present a core of IS principles with which every business student should be familiar.
2. To offer a survey of the IS discipline that will enable all business students to understand the relationship of advanced courses to the curriculum as a whole.
3. To present the changing role of the IS professional.
4. To show the value of the discipline as an attractive field of specialization.

Because *Fundamentals of Information Systems* is written for all business majors, we believe it is important not only to present a realistic perspective on IS in business but also to provide students with the skills they can use to be effective leaders in their companies.

AUTHOR TEAM

Ralph Stair and George Reynolds have teamed up again for this new text. Together, they have more than 50 years of academic and industrial experience. Ralph Stair brings years of writing, teaching, and academic experience. He has written more than 20 books and a large number of articles while at Florida State University. George Reynolds brings a wealth of computer and industrial experience to the project, with more than 30 years' experience working in government, institutional, and commercial IS organizations. He has also authored nine texts and is an adjunct professor at the University of Cincinnati, teaching the introductory IS course. The Stair and Reynolds team brings a solid conceptual foundation, along with practical IS experience, to thousands of readers every year.

RESOURCES FOR INSTRUCTORS

The teaching tools that accompany this text offer many options for enhancing your course. In this edition, we emphasize the importance of distance learning. And, as always, we are committed to providing one of the best teaching resource packages available in this market. Here are your options.

Instructor's Manual with Solutions

The *Instructor's Manual* is available in both electronic and printed formats. This all-new *Instructor's Manual* provides valuable chapter overviews; highlights key principles and critical concepts; offers sample syllabi, learning objectives, and discussion topics; and features possible essay topics, further readings or cases, and solutions to all of the end-of-chapter questions and problems, as well as suggestions for conducting team activities. Additional end-of-chapter questions are also included, as well as the rationale, methodology, and solutions for each.

ExamView®

This textbook is accompanied by ExamView, a powerful testing software package that allows instructors to create and administer printed, computer (LAN-based), and Internet exams. ExamView includes hundreds of questions that correspond to the topics covered in this text, enabling students to generate detailed study guides that include page references for further review. The computer-based and Internet testing components allow students to take exams at their computers and also save the instructor time by grading each exam automatically.

PowerPoint Presentations

Our CD-ROM-based presentation tool developed in Microsoft PowerPoint offers a wealth of resources for use in the classroom. Instead of using traditional overhead transparencies, these PowerPoint Presentations allow you to create impressive computer-generated screen shows including graphics and videos. All of the graphics from the book (except photos) have been included.

Distance Learning

Course Technology is proud to present on-line courses in WebCT and Blackboard, as well as at MyCourse.com, Course Technology's own course enhancement tool, to provide the most complete and dynamic learning experience possible. When you add on-line content to one of your courses, you're adding a lot: self-tests, links, glossaries, and, most of all, a gateway to the twenty-first century's most important information resource. We hope you will make the most of your course, both on-line and off-line. For more information on how to bring distance learning to your course, contact your local Course Technology sales representative.

ACKNOWLEDGMENTS

A book of this size and undertaking is always a team effort. We would like to thank every one of our fellow teammates at Course Technology for their dedication and hard work. Many thanks to our Managing Editor, Jennifer Locke, and our Senior Product Manager, Jennifer Muroff. A number of people behind the scenes made this book a reality. Many thanks to the following reviewers: Kathleen Hartzel, Duquesne University; Terry Fox, Emporia State University; Brian Kovar, Kansas State University; Noushin Ashrafi, University of Massachusetts, Boston; James LaBarre, University of Wisconsin–Eau Claire; Rebecca Lawson, Lansing Community College; Donna Auston, Louisiana State University, Shreveport; Edward Mollette, Baruch College, City University of New York; Vikram Sethi, Southwest Missouri State University; and Cara Gallagher and Thomas Kim, Duquesne University. We would like to acknowledge and thank the team at Elm Street Publishing Services for their hard work on the manuscript. Karen Hill, development editor, helped with all stages of this project. Martha Beyerlein, Emily Friel, Barb Lange, Jonathan Lyzun, Jack Semens, and Abby Westapher helped with production and the final stages of the book.

Many thanks to the sales force at Course Technology. You make this all possible. You helped to get important feedback from current and future adopters. As Course Technology product users, we know how important you are.

Ralph Stair would like to thank the Department of Information and Management Sciences, College of Business Administration, at Florida State University for their support and encouragement. He would also like to thank his family—Lila and Leslie—for their support. George Reynolds thanks his family—Ginnie, Tammy, Kim, Kelly, and Kristy—for their patience and support in this major project.

TO THE GUEST AUTHORS

We sincerely appreciate your assistance as contributors to this text of an "Information Systems in Action" box:

Anne Nelson, *High Point University*
James LaBarre, *University of Wisconsin–Eau Claire*
Brian Kovar, *Kansas State University*
Ken Baldauf, *Florida State University*
Kathleen S. Hartzel, *Duquesne University*
Demetrios D. Mahramas, *Deloitte Consulting*
Gerry Santoro, *Pennsylvania State University*
John Vargo, *University of Canterbury*
Rebecca Lawson, *Lansing Community College*
Roger McHaney, *Kansas State University*

OUR COMMITMENT

We are sincerely committed to serving the needs of our adopters and readers. We have listened to their comments and thank them for their time. As always, we welcome input and feedback. If you have any questions or comments regarding *Fundamentals of Information Systems*, please contact us through Course Technology or your local representative, via e-mail at mis@course.com, or via the Internet at www.course.com; or address your comments, criticisms, suggestions, and ideas to:

Ralph Stair
George Reynolds
Course Technology
25 Thomson Place
Boston, MA 02210

BRIEF CONTENTS

PART I

INFORMATION SYSTEMS IN PERSPECTIVE 1

CHAPTER 1 An Introduction to Information Systems in Organizations 2

PART II

TECHNOLOGY 37

CHAPTER 2 Hardware and Software 38

CHAPTER 3 Organizing Data and Information 90

CHAPTER 4 Telecommunications, the Internet, Intranets, and Extranets 130

PART III

BUSINESS INFORMATION SYSTEMS 171

CHAPTER 5 Electronic Commerce and Transaction Processing Systems 172

CHAPTER 6 Information and Decision Support Systems 210

CHAPTER 7 Specialized Business Information Systems: Artificial Intelligence, Expert Systems, and Virtual Reality 248

PART IV

SYSTEMS DEVELOPMENT AND SOCIAL ISSUES 281

CHAPTER 8 Systems Development 282

CHAPTER 9 Security, Privacy, and Ethical Issues in Information Systems and the Internet 318

GLOSSARY 345
INDEX 355

CONTENTS

PART I

INFORMATION SYSTEMS IN PERSPECTIVE 1

CHAPTER 1

An Introduction to Information Systems in Organizations 2

FARMERS INSURANCE GROUP: FINDING HIGHER REVENUE AND LOWER CLAIMS IN INFORMATION 3

Information Concepts 4

Data versus Information 4

ETHICAL AND SOCIETAL ISSUES
What Is Ethics? 5

The Characteristics of Valuable Information 6
The Value of Information 7

What Is an Information System? 7

Input, Processing, Output, and Feedback 7
Manual and Computerized Information Systems 8
Computer-Based Information Systems 9

Business Information Systems 11

Electronic Commerce 11
Transaction Processing Systems 11
Information and Decision Support Systems 12
Special-Purpose Business Information Systems: Artificial Intelligence, Expert Systems, and Virtual Reality 13

Systems Development 14

Organizations and Information Systems 16

Technology Diffusion, Infusion, and Acceptance 18

Competitive Advantage 18

Factors That Lead Firms to Seek Competitive Advantage 18
Strategic Planning for Competitive Advantage 19

Performance-Based Information Systems 21

Productivity 22
Quality 23
Return on Investment and the Value of Information Systems 23

E-COMMERCE
Finding Quality through E-Commerce 24

Information Systems Personnel 25

Roles and Functions of the Information Systems Department 25
Typical IS Titles and Functions 27

INFORMATION SYSTEMS IN ACTION
Not Just Fun and Games: Building Information Systems for the Olympics 29

Other IS Careers 29

CASE 1 Coors Ceramics Revamps Information Systems 34

CASE 2 Total Cost of Ownership 35

CASE 3 UPS Turns to Technology for a Strategic Advantage 35

PART II

TECHNOLOGY 37

CHAPTER 2

Hardware and Software 38

HOME DEPOT: BUILDING SOFTWARE TO MEET CUSTOMERS' NEEDS 39

Overview of Hardware 40
Hardware Components 40
Hardware Components in Action 40
Processing and Memory Devices: Power, Speed, and Capacity 42
Processing Characteristics and Functions 42
Memory Characteristics and Functions 44
Multiprocessing 45
Secondary Storage and Input and Output Devices 45
Secondary Storage Access Methods 46
Secondary Storage Devices 46
Input Devices 49
Output Devices 52
Computer System Types 54
Personal Computers 54
Midrange Computers 56
Mainframe Computers 56
Supercomputers 57
Overview of Software 58

ETHICAL AND SOCIETAL ISSUES
Employee Monitoring 59

Supporting Individual, Group, and Organizational Goals 59
Systems Software 60
Operating Systems 61
Personal Computer Operating Systems 64
Workgroup Operating Systems 65
Enterprise Operating Systems 67
Consumer Appliance Operating Systems 67
Application Software 68
Types and Functions of Application Software 68
Personal Application Software 69

E-COMMERCE
Samsung Employs ASP to Accelerate E-Commerce Initiative 72

Workgroup Application Software 74
Enterprise Application Software 76

Programming Languages 77
The Evolution of Programming Languages 78
Language Translators 80
Software Bugs 81

INFORMATION SYSTEMS IN ACTION
The Microprocessor—A Brief History 82

CASE 1 Electronic Ink 87
CASE 2 CD-ROM Titles 87
CASE 3 Flash Chips 88

CHAPTER 3

Organizing Data and Information 90

CATALINA MARKETING CORPORATION: PROVIDING DATA SERVICES 91

Data Management 92
The Hierarchy of Data 92
Data Entities, Attributes, and Keys 93
The Traditional Approach versus the Database Approach 94
Data Modeling and Database Models 97
Data Modeling 98
Database Models 99
Database Management Systems (DBMSs) 102
Providing a User View 103
Creating and Modifying the Database 104
Storing and Retrieving Data 105
Manipulating Data and Generating Reports 106
Popular Database Management Systems 108
Selecting a Database Management System 108
Database Developments 110
Data Warehouses, Data Marts, and Data Mining 110

E-COMMERCE
Safeway Customers Involved in E-Commerce Experiment 114

On-Line Analytical Processing (OLAP) 115
Open Database Connectivity (ODBC) 117
Object-Relational Database Management Systems 118
Business Intelligence 120

● **ETHICAL AND SOCIETAL ISSUES**
United Technologies Gathers Competitive
Intelligence 121

● **INFORMATION SYSTEMS IN ACTION**
Grandma and Limp Bizkit, Diapers and Beer 122

CASE 1 Lockheed Martin Implements OLAP
         System 126

CASE 2 Allina Health System Implements a Data
         Warehouse 127

CASE 3 Fifth Third Bank Investing in Internet
         Systems 128

CHAPTER 4

Telecommunications, the Internet,
Intranets, and Extranets 130

   FedEx: SPARRING WITH UPS ON THE
   INTERNET 131

**An Overview of Telecommunications and
Networks 132**
   Telecommunications 133
   Networks and Distributed Processing 136
Use and Functioning of the Internet 141
   How the Internet Works 143
   Accessing the Internet 144
**Internet and Telecommunication
Services 146**
   Voice Mail, Electronic Mail, and Instant
      Messaging 146
   Telecommuting, Videoconferencing, and
      Internet Phone Service 147
   Electronic Data Interchange (EDI) 149
   Public Network and Specialized
      Services 150
   Distance Learning 151
   On-Line Music, Radio, and Video 151
   Telnet, FTP, and Content Streaming 152
   Chat Rooms 152
The World Wide Web 153
   Web Browsers 154
   Search Engines 154
   Java 155
   Push Technology 155
   Business Uses of the Web 156

● **E-COMMERCE**
Home Builders Build Sites on the Internet 157

Intranets and Extranets 157

Net Issues 159
   Management Issues 160
   Service Bottlenecks 160
   Privacy and Security 161

● **ETHICAL AND SOCIETAL ISSUES**
Outrage on the Web 162

● **INFORMATION SYSTEMS IN ACTION**
The Post Office: Remaining Competitive in the Internet
Age 163

CASE 1 Mirage on the Net 167

CASE 2 Kaiser Permanente Looks for a Cure on
         the Web 167

CASE 3 *The Washington Post* Tries the
         Web 168

PART III

**BUSINESS INFORMATION
SYSTEMS 171**

CHAPTER 5

Electronic Commerce and Transaction
Processing Systems 172

   WEIRTON STEEL: E-COMMERCE
   PIONEER 172

**An Introduction to Electronic
Commerce 174**
   Value Chains in E-Commerce   174
   Business to Business (B2B) 175
   Business to Consumer (B2C) 176
E-Commerce Applications 176
   Retail and Wholesale 176
   Manufacturing 177
   Marketing 178
   Investment and Finance 179
   Auctions 181
E-Commerce Technology Components 182
   Hardware 182
   Web Server Software 183
   E-Commerce Software 183
   Network and Packet Switching 184
   Electronic Payment Systems 185

● **E-COMMERCE**
Get Ready for Global E-Commerce 187

Strategies for Successful E-Commerce 188
　Developing an Effective Web Presence 188
　Putting Up a Web Site 189
　Building Traffic to Your Web Site 189
An Overview of Transaction Processing Systems 190
　Traditional Transaction Processing Methods and Objectives 191
　Transaction Processing Activities 192

●ETHICAL AND SOCIETAL ISSUES
Credit Card Fraud and Transaction Processing 193

　Order Processing Systems 197
Enterprise Resource Planning 199
　An Overview of Enterprise Resource Planning 199
　Advantages and Disadvantages of ERP 199

INFORMATION SYSTEMS IN ACTION
The Future of Transaction Processing in Financial Services 202

CASE 1 Starting a Procurement Business 207
CASE 2 ERP in Mergers and Acquisitions 207
CASE 3 Reapplication of General Electric Web Site 208

CHAPTER 6
Information and Decision Support Systems 210

REUTERS GROUP: PROVIDING INFORMATION AND DECISION SUPPORT ON THE INTERNET 211

Decision Making and Problem Solving 212
　Decision Making as a Component of Problem Solving 212
　Programmed versus Nonprogrammed Decisions 213
　Optimization, Satisficing, and Heuristic Approaches 214
An Overview of Management Information Systems 215
　Management Information Systems in Perspective 215
　Inputs to a Management Information System 216
　Outputs of a Management Information System 216

Functional Aspects of the MIS 219
　A Financial Management Information System 219
　A Manufacturing Management Information System 221

E-COMMERCE
Information and Inventory Control 223

　A Marketing Management Information System 223
　A Human Resource Management Information System 224
　Other Management Information Systems 225
An Overview of Decision Support Systems 226
　Capabilities of a Decision Support System 227
　A Comparison of DSSs and MISs 228
Components of a Decision Support System 228
　The Model Base 228

●ETHICAL AND SOCIETAL ISSUES
The Dark Side of E-Trading 231

　The Dialogue Manager 231
The Group Decision Support System 231
　Characteristics of a GDSS 232
　GDSS Software 234
　GDSS Alternatives 234
The Executive Support System 236
　Executive Support Systems in Perspective 236
　Capabilities of an Executive Support System 237

INFORMATION SYSTEMS IN ACTION
Instant Messaging for the Network-Supported Course 239

CASE 1 Collaborative Work Gives a Competitive Edge 244
CASE 2 Marketing Research on the Internet 245
CASE 3 Investment Information from Financial Web Sites 245

CHAPTER 7

Specialized Business Information Systems: Artificial Intelligence, Expert Systems, and Virtual Reality 248

ASK JEEVES: SEARCH ENGINE HELPS "HUMANIZE" ON-LINE EXPERIENCE 249

An Overview of Artificial Intelligence 250
Artificial Intelligence in Perspective 250
The Nature of Intelligence 251
The Difference between Natural and Artificial Intelligence 253
The Major Branches of Artificial Intelligence 253

ETHICAL AND SOCIETAL ISSUES
Neural Networks Provide Hope 257

An Overview of Expert Systems 257
Characteristics of an Expert System 258
When to Use Expert Systems 259
Components of Expert Systems 260
Expert Systems Development 265
Applications of Expert Systems and Artificial Intelligence 267

E-COMMERCE
Bots Automate Corporate Purchasing 269

Virtual Reality 269
Interface Devices 270
Immersive Virtual Reality 271
Other Forms of Virtual Reality 271
Useful Applications 271

INFORMATION SYSTEMS IN ACTION
Artificial Neural Network Satisfies Customers 273

CASE 1 Fuzzy Logic System Designed to Help Travelers 277

CASE 2 Expert Systems Help Develop Business Plans 278

CASE 3 AI Software Used to Monitor and Control Networks 279

PART IV

SYSTEMS DEVELOPMENT AND SOCIAL ISSUES 281

CHAPTER 8

Systems Development 282

AT&T: RENTING SOFTWARE TO BUSINESS CLIENTS 283

An Overview of Systems Development 284
Participants in Systems Development 284
Information Systems Planning 284
Systems Development Life Cycles 284
The Traditional Systems Development Life Cycle 286

E-COMMERCE
Developing an Internet Site for Suppliers and Dealers 288

Prototyping 289
Rapid Application Development and Joint Application Development 290
The End-User Systems Development Life Cycle 291
Use of Computer-Aided Software Engineering (CASE) Tools 291
Systems Investigation 292
Initiating Systems Investigation 292
Feasibility Analysis 292
The Systems Investigation Report 293
Systems Analysis 293
Data Collection 294
Data Analysis 294
Requirements Analysis 296
The Systems Analysis Report 297
Systems Design 298
Generating Systems Design Alternatives 300
Financial Options 300
Evaluating and Selecting a System Design 301
Freezing Design Specifications 301
The Design Report 302

Systems Implementation 303
  Acquiring Hardware from an Information
  Systems Vendor 303
  Acquiring Software: Make or Buy? 304
  Acquiring Database and Telecommunications
  Systems 305
  User Preparation 305
  IS Personnel: Hiring and Training 305
  Site Preparation 305
  Data Preparation 306
  Installation 306
  Testing 306
  Start-Up 306
  User Acceptance 307
Systems Maintenance and Review 308
  Systems Maintenance 308

ETHICAL AND SOCIETAL ISSUES
Monitoring Debit Card Usage 309

  Systems Review 309

INFORMATION SYSTEMS IN ACTION
Developing Distance Learning Courses for
Flexibility 311

CASE 1 IT Projects at Coca-Cola 315

CASE 2 Mergers Drive Systems
         Development 316

CASE 3 Developing a Wireless Net to Improve
         Customer Service 317

CHAPTER 9
Security, Privacy, and Ethical Issues in
Information Systems and the Internet 318

  LOS ALAMOS NATIONAL LABORATORY:
  COMPROMISE OF NUCLEAR WEAPON
  SECRETS 319

Computer Waste and Mistakes 320
  Computer Waste 320
  Computer-Related Mistakes 320
  Preventing Computer-Related Waste and
  Mistakes 320
Computer Crime 322
  The Computer as a Tool to Commit
  Crime 324
  The Computer as the Object of Crime 324
  Preventing Computer-Related Crime 328

E-COMMERCE
Business Entrepreneurs Carve New Path 332

Privacy 332
  Privacy Issues 332
  Individual Efforts to Protect Privacy 333

ETHICAL AND SOCIETAL ISSUES
Monitoring E-Mail—An Invasion of Privacy? 335

Health Concerns 337
  Avoiding Health and Environmental
  Problems 337

INFORMATION SYSTEMS IN ACTION
TRUSTe: Building a Reputation of Trust On-Line 338

CASE 1 Predatory Hiring Practices 341

CASE 2 The Children's Online Privacy Protection
         Act of 1998 341

CASE 3 Taxing Internet Sales 342

Glossary 345
Index 355

PART I

Information Systems in Perspective

An Introduction to Information Systems in Organizations

Companies are more aggressive than ever in adopting and deploying leading-edge technology. The reason: fear of being left behind.

— Rick Whiting and Beth Davis, summarizing results of a 1999 *Information Week* survey of 300 IT managers.

Principles	Learning Objectives
The value of information is directly linked to how it helps decision makers achieve the organization's goals.	• *Distinguish data from information and describe the characteristics used to evaluate the quality of data.*
Knowing the potential impact of information systems and having the ability to put this knowledge to work can result in a successful personal career, an organization that reaches its goals, and a society with a higher quality of life.	• *Identify the basic types of business information systems and discuss who uses them, how they are used, and what kinds of benefits they deliver.*
System users, business managers, and information systems professionals must work together to build a successful information system.	• *Identify the major steps of the systems development process and state the goal of each.*
Information systems add value to an organization and its processes.	• *Identify the seven value-added processes in the supply chain and describe the role of information systems within them.*
Because information systems are so important, businesses need to be sure that improved or completely new systems help lower costs, increase profits, improve service, or achieve a competitive advantage.	• *Identify some of the strategies employed to lower costs or improve service.* • *Define the term* competitive *advantage and discuss how organizations are using information systems to gain such an advantage.*
Information systems personnel are the key to unlocking the potential of any new or modified system.	• *Define the types of roles, functions, and careers available in information systems.*

Farmers Insurance Group

Finding Higher Revenue and Lower Claims in Information

Large companies process millions of business transactions every year and store huge amounts of data on them. Often, the data is spread across a variety of computer systems in different areas of the country or even the world. This raw data, although needed for record keeping, has little value to managers and decision makers unless it can be filtered and processed into meaningful information. But doing so is the real challenge: Finding strategic information in a mountain of data can be like finding a needle in a haystack, but the effort is usually worth it. The results can be a staggering increase in revenues and profits. With today's fast computers and a knowledgeable information systems staff, the possibility of turning raw data into useful and profitable information can become a reality. Farmers Insurance Group increased profits and lowered claims by using information well.

Like other companies, Farmers Insurance Group was sitting on a huge amount of raw data. The data, however, was scattered across different computer systems in different locations. As in all insurance companies, underwriting determines what insurance policies a company can offer and at what premiums. Farmers's underwriting business was responsible for assessing insurance risk, which can make the difference between profits and losses. The people who are responsible for determining insurance risk are called actuaries. According to Tom Boardman, an assistant actuary at Farmers, "As competition has gotten more intense in the insurance industry, the traditional ways of segmenting risk aren't good enough at providing you competitive advantage." Boardman was referring to how most insurance companies categorize risk. For example, high-powered sports cars are more likely to be involved in expensive accidents than ordinary sedans. So, insurance companies can put sports cars in a different risk category than sedans and charge customers who own them a higher premium. In assessing risk, an actuary would traditionally have a hunch—for example, that sports cars are more prone to accidents than sedans. Then the actuary would test that hunch using a computer. According to Boardman, this process was like using the computer "to dig up data to prove or unprove those hunches." One disadvantage of this old approach is that small but profitable market niches may be ignored or not priced correctly. As a result, Farmers decided to look into a computer system to help it find profitable market niches.

The company found the help it needed through IBM, which developed a customized software product for Farmers called DecisionEdge, an advanced decision support system that combined raw data from seven different databases on a staggering 35 million records. Consolidating the raw data into useful information took about twice as long as expected, but the additional wait was worth it. Farmers was able to locate market niches that it hadn't seen before. For example, DecisionEdge helped Farmers determine that not all sports car owners are alike—those who were older and had at least one other car were less likely to be in an expensive accident. Once this market niche was identified, Farmers could offer that segment of the sports car market lower premiums. Using DecisionEdge to find the market niche resulted in millions of dollars of increased revenues for Farmers.

The approach used by Farmers is sometimes called "data scrubbing." It allows a company to consolidate important information and squeeze additional revenues and profits from it. After helping Farmers and seeing a market opportunity, IBM also decided to offer its DecisionEdge software to other insurance companies.

As you read this chapter, consider the following:

- How was Farmers able to transform its raw data into meaningful information and additional revenues?

- Describe how this approach could be used in other industries.

Information systems are everywhere. An advanced information system used by movie theaters provides patrons quick retrieval for advance ticket sales ordered through the telephone, the Internet, or automated kiosks.
(Source: Courtesy of Radiant Systems, Inc.)

information system (IS)

a set of interrelated components that collect, manipulate, and disseminate data and information and provide a feedback mechanism to meet an objective

An **information system (IS)** is a set of interrelated components that collect, manipulate, and disseminate data and information and provide a feedback mechanism to meet an objective. We all interact daily with information systems, both personally and professionally. We use automatic teller machines at banks; checkout clerks scan our purchases using bar codes and scanners; we access information over the Internet; and we get information from kiosks with touchscreens. Major Fortune 500 companies are spending more than $1 billion per year on information technology. In the future, we will depend on information systems even more. Knowing the potential of information systems and being able to put this knowledge to work can result in a successful personal career, organizations that reach their goals, and a society with a higher quality of life.

Regardless of your career choice, information systems will play an important role in your work life. You will find some of the highest salaries and advancement opportunities available if you work in the IS field. Most students, however, will not work in the IS field. Instead, they choose an area such as sales, marketing, accounting, finance, production, or research and development. As you will see throughout this text, information systems play a critical role in all these areas. Throughout this text, we show you how information systems can be used to advance your career and help your organization achieve its goals in an ever-changing environment.

Computers and information systems are constantly changing the way organizations conduct business. Today we live in an information economy. Information itself has value, and commerce often involves the exchange of information, rather than of tangible goods. Systems based on computers are increasingly being used to create, store, and transfer information. Investors are using information systems to make multimillion-dollar decisions, financial institutions are employing them to transfer billions of dollars around the world electronically, and manufacturers are using them to order supplies and distribute goods faster than ever before. Computers and information systems will continue to change our society, our businesses, and our lives (see the "Ethical and Societal Issues" box). In this chapter, we present a framework for understanding computers and information systems and discuss why it is important to study information systems. This understanding will help you unlock the potential of properly applied information systems concepts.

INFORMATION CONCEPTS

Information is a central concept throughout this book. The term is used in the title of the book, in this section, and in almost every chapter. To be an effective manager in any area of business, you need to understand that information is one of an organization's most valuable and important resources. This term, however, is often confused with the term *data*.

data

raw facts, such as an employee's name and number of hours worked in a week, inventory part numbers, or sales orders

information

a collection of facts organized in such a way that they have additional value beyond the value of the facts themselves

Data versus Information

Data consists of raw facts, such as an employee's name and number of hours worked in a week, inventory part numbers, or sales orders. As shown in Table 1.1, several types of data can be used to represent these facts. When the facts are organized or arranged in a meaningful manner, they become information. **Information** is a collection of facts organized in such a way that they have

ETHICAL AND SOCIETAL ISSUES
What Is Ethics?

Every society has a set of moral rules that establish the boundaries of acceptable behavior. Often the rules about such behavior are expressed in statements about what you should or should not do. These rules fit together, more or less consistently, to form the moral code by which a society lives.

Unfortunately, moral codes are seldom completely consistent. Our everyday life raises moral questions that we cannot easily answer. Sometimes that is because there are contradictions among our different values and we are uncertain about which value should be given priority. For example, if we witness a friend breaking a law, we are caught in a conflict between loyalty to our friend and the value of telling the truth. At other times, our traditional values do not cover new situations, and we have to work out how to extend them. For example, we might believe in the value of personal privacy, but in a time in which information technology has had such a profound impact on society that some people call this the Information Age, what rules should we establish to govern access to information in computer databases about private individuals?

Also, no two moral codes are identical. They often vary by cultural group, gender, ethnic background, and religion. In some countries, certain behaviors are frowned upon, but in other cultures the opposite may be true. For example, attitudes toward illegal copying of software (software piracy) vary from strong opposition to strong support.

When we step back and reflect on our moral beliefs, we are engaging in ethical reflection. Ethics, then, is the conscious reflection on our moral beliefs with the aim of improving, extending, or refining those beliefs in some way.

Discussion Questions

1. Recall a situation in which you had to deal with a conflict in values. What process did you use to resolve this issue?
2. Identify three areas in which the changes that have been brought on by the expanded use of information systems raise a conflict with our moral beliefs.

Critical Thinking Questions

3. Provide two other examples in which attitudes toward a particular behavior are viewed differently by people of different cultures, gender, ethnic backgrounds, or religions.
4. How are ethics learned?

Sources: Lawrence M. Hinman, *Ethics: A Pluralistic Approach to Moral Theory,* 2nd ed. (Fort Worth, Tex.: Harcourt, Brace, 1997); and the "ISWorld Net Professional Ethics" Web site at http://www.cityu.edu.hk/is/ethics/ethics.htm, accessed February 16, 2000.

additional value beyond the value of the facts themselves. For example, a particular manager might find the knowledge of total monthly sales to be more suited to his purpose (i.e., more valuable) than the number of sales for individual sales representatives.

Turning data into information is a *process,* or a set of logically related tasks performed to achieve a defined outcome. The process of defining relationships among data to create useful information requires knowledge. *Knowledge* is an awareness and understanding of a set of information and of how that information can be made useful to support a specific task or reach a decision. Part of the knowledge needed for building a railroad layout, for instance, is understanding how large an area is available for the layout and how many trains will run on the track. The process of converting data into information also requires knowledge to select or reject facts based on their relevancy to particular tasks. So, information can be considered data made more useful through the application of knowledge.

Data	Represented by
Alphanumeric data	Numbers, letters, and other characters
Image data	Graphic images or pictures
Audio data	Sound, noise, or tones
Video data	Moving images or pictures

TABLE 1.1

Types of Data

FIGURE 1.1

The Process of Transforming
Data into Information

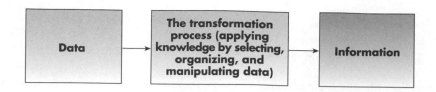

In some cases, data is organized or processed mentally or manually. In others, a computer is used. In the earlier example of the sales manager, the manager could have manually calculated the sum of the sales of each representative, or a computer could calculate this sum. What is important is not so much where the data comes from or how it is processed but whether the results are useful and valuable. This transformation process is shown in Figure 1.1.

The Characteristics of Valuable Information

To be valuable to managers and decision makers, information should have the characteristics described in Table 1.2. These characteristics also make the information more valuable to an organization. If information is not accurate or complete, poor decisions can be made, costing the organization thousands or even millions of dollars. For example, if an inaccurate forecast indicates that sales will be very high when the opposite is true, an organization can invest millions of dollars in a new plant that is not needed. And if information is not pertinent to the situation, not delivered to decision makers in a timely fashion, or too complex to understand, it may be of little value to an organization.

TABLE 1.2

The Characteristics of Valuable
Information

Characteristics	Definitions
Accurate	Accurate information is error free. In some cases, inaccurate information is generated because inaccurate data is fed into the transformation process (this is commonly called garbage in, garbage out [GIGO]).
Complete	Complete information contains all the important facts. For example, an investment report that does not include all important costs is not complete.
Economical	Information should also be relatively economical to produce. Decision makers must always balance the value of information with the cost of producing it.
Flexible	Flexible information can be used for a variety of purposes. For example, information on how much inventory is on hand for a particular part can be used by a sales representative in closing a sale, by a production manager to determine whether more inventory is needed, and by a financial executive to determine the total value the company has invested in inventory.
Reliable	Reliable information can be depended on. In many cases, the reliability of the information depends on the reliability of the data collection method. In other instances, reliability depends on the source of the information. A rumor from an unknown source that oil prices might go up may not be reliable.
Relevant	Relevant information is important to the decision maker. Information that lumber prices might drop may not be relevant to a computer chip manufacturer.
Simple	Information should also be simple, not overly complex. Sophisticated and detailed information may not be needed. In fact, too much information can cause information overload, whereby a decision maker has too much information and is unable to determine what is really important.
Timely	Timely information is delivered when it is needed. Knowing last week's weather conditions will not help when trying to decide what coat to wear today.
Verifiable	Information should be verifiable. This means that you can check it to make sure it is correct, perhaps by checking many sources for the same information.
Accessible	Information should be easily accessible by authorized users and provided in the right format and at the right time to meet their needs.
Secure	Information should be secure from access by unauthorized users.

Useful information can vary widely in the value of each of these quality attributes. With market-intelligence data, some inaccuracy and incompleteness is acceptable, but timeliness is essential. Market intelligence may alert us that our competitors are about to make a major price cut. The exact details and timing of the price cut may not be as important as being warned far enough in advance to plan how to react. On the other hand, accuracy, verifiability, and completeness are critical for data used in accounting for company assets such as cash, inventory, and equipment.

The Value of Information

The value of information is directly linked to how it helps decision makers achieve their organizations' goals. For example, the value of information might be measured in the time required to make a decision or in increased profits to the company. Consider a market forecast that predicts a high demand for a new product. If market forecast information is used to develop the new product and the company can make an additional profit of $10,000, the value of this information to the company is $10,000 minus the cost of the information. Valuable information can also help managers decide whether to invest in additional information systems and technology. A new computerized ordering system may cost $30,000, but it may generate an additional $50,000 in sales. The value added by the new system is the additional revenue from the increased sales of $20,000.

WHAT IS AN INFORMATION SYSTEM?

As mentioned previously, an information system can be defined in a number of different ways. But an information system (IS) is typically considered to be a set of interrelated elements or components that collect (input), manipulate (process), and disseminate (output) data and information and provide a feedback mechanism to meet an objective. (See Figure 1.2.)

Input, Processing, Output, Feedback

Input

input

the activity of gathering and capturing raw data

In information systems, **input** is the activity of gathering and capturing raw data. In producing paychecks, for example, the number of hours every employee worked must be collected before paychecks can be calculated or printed. In a university grading system, student grades must be obtained from instructors before a summary of grades for the semester or quarter can be compiled and sent to students.

Input can take many forms. In an information system designed to produce paychecks, employee time cards might be the initial input. In a 911 emergency telephone system, an incoming call would be considered an input. Input to a marketing system might include customer survey responses. Notice that regardless of the system involved, the type of input is determined by the desired output of the system.

FIGURE 1.2

The Components of an Information System

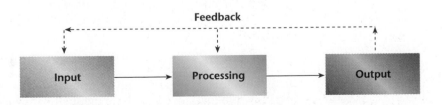

Input can be a manual process, or it may be automated. A scanner at a grocery store that reads bar codes and enters the grocery item and price into a computerized cash register is a type of automated input process. Regardless of the input method, accurate input is critical to achieve the desired output.

Processing

processing

converting or transforming data into useful outputs

In information systems, **processing** involves converting or transforming data into useful outputs. Processing can involve making calculations, making comparisons and taking alternative actions, and storing data for future use.

Processing can be done manually or with the assistance of computers. In payroll, the number of hours each employee worked must be converted into net pay. The required processing can first involve multiplying the number of hours worked by the employee's hourly pay rate to get gross pay. If weekly hours worked are greater than 40 hours, overtime pay may also be determined. Then deductions are subtracted from gross pay to get net pay. For instance, federal and state taxes can be withheld or subtracted from gross pay; many employees have health and life insurance, savings plans, and other deductions that must also be subtracted from gross pay to arrive at net pay.

Output

output

production of useful information, usually in the form of documents and reports

In information systems, **output** involves producing useful information, usually in the form of documents and reports. Outputs can include paychecks for employees, reports for managers, and information supplied to stockholders, banks, government agencies, and other groups. In some cases, output from one system can become input for another. For example, output from a system that processes sales orders can be used as input to a customer billing system. Often output from one system can be used as input to control other systems or devices. For instance, office furniture manufacturing is complicated and contains many variables. A salesperson, customer, and furniture designer often go through several design iterations to meet the customer's needs. Special computer software and hardware is used to create the original design and rapidly revise it. Once the last design mock-up is approved, the design software creates a bill of materials that goes to manufacturing to produce the order.

Output can be produced in a variety of ways. For a computer, printers and display screens are common output devices. Output can also be a manual process involving handwritten reports and documents.

Feedback

feedback

output that is used to make changes to input or processing activities

In information systems, **feedback** is output that is used to make changes to input or processing activities. For example, errors or problems might make it necessary to correct input data or change a process. Consider a payroll example. Perhaps the number of hours an employee worked was entered into a computer as 400 hours instead of 40 hours. Most information systems check to make sure that data falls within certain predetermined ranges. For number of hours worked, the range might be from 0 to 100 hours. It is unlikely that an employee would work more than 100 hours for any given week. In this case, the information system would determine that 400 hours is out of range and provide feedback, such as an error report. The feedback is used to check and correct the input on the number of hours worked to 40. If undetected, this error would result in a very high net pay printed on the paycheck!

Manual and Computerized Information Systems

As discussed earlier, an information system can be manual or computerized. Some investment analysts manually draw charts and trend lines to assist them in making investment decisions. Tracking data on stock prices (input) over the last few months or years, these analysts develop patterns on graph paper (processing) that help them determine what stock prices are likely to do in the next few days or weeks (output). Some investors have made millions of dollars using manual stock analysis information systems. Of course, there are many excellent computerized information systems as well. Many computer systems have been developed to follow stock indexes and markets and to suggest when large blocks of stocks should be purchased or sold (program trading) to take advantage of market discrepancies.

Often information systems begin as manual systems and become computerized. Consider the way the U.S. Postal Service sorts mail. At one time most letters were visually scanned by postal employees to determine the ZIP code and then manually placed in an appropriate bin. Today the bar-coded addresses on letters passing through the postal system are "read" electronically and automatically routed to a bin via conveyors. The computerized sorting system results in speedier processing time and provides information that helps with transportation planning. It is important to stress, however, that simply computerizing a manual information system does not guarantee improved system performance. If the underlying information system is flawed, the act of computerizing it might only magnify the impact of these flaws.

Computer-Based Information Systems

computer-based information system (CBIS)

a single set of hardware, software, databases, telecommunications, people, and procedures that are configured to collect, manipulate, store, and process data into information

technology infrastructure

all the hardware, software, databases, telecommunications, people, and procedures that are configured to collect, manipulate, store, and process data into information

A **computer-based information system (CBIS)** is a single set of hardware, software, databases, telecommunications, people, and procedures that are configured to collect, manipulate, store, and process data into information. For example, a company's payroll system, order entry system, and inventory control system are examples of CBISs. The components of a CBIS are illustrated in Figure 1.3. A business's **technology infrastructure** includes all the hardware, software, databases, telecommunications, people, and procedures that are configured to collect, manipulate, store, and process data into information. The technology infrastructure is a set of shared IS resources that form the foundation of each computer-based information system.

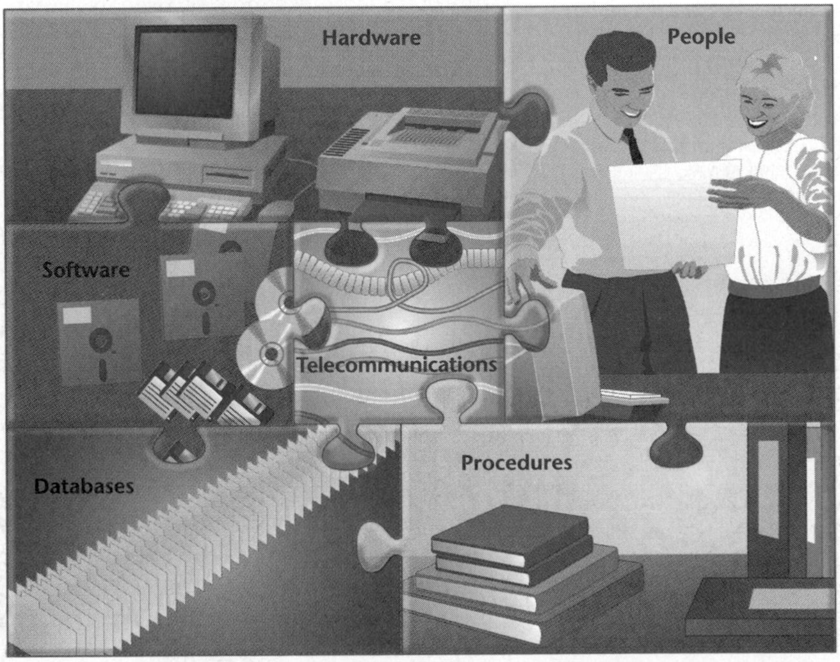

FIGURE 1.3

The Components of a Computer-Based Information System

hardware

computer equipment used to perform input, processing, and output activities

Hardware

Hardware consists of computer equipment used to perform input, processing, and output activities. Input devices include keyboards, automatic scanning devices, equipment that can read magnetic ink characters, and many other devices. Processing devices include the central processing unit and main memory. There are many output devices, including secondary storage devices, printers, and computer screens.

software

the computer programs that govern the operation of the computer

Software

Software is the computer programs that govern the operation of the computer. These programs allow the computer to, for example, process payroll, send bills to customers, and provide managers with information to increase profits, reduce costs, and provide better customer service. There are two basic types of software: system software (which controls basic computer operations such as start-up and printing) and applications software (which allows specific tasks to be accomplished, such as word processing and tabulating numbers). A program that allows users to create a spreadsheet (such as Excel or Lotus) is an example of applications software.

database

an organized collection of facts and information

Databases

A **database** is an organized collection of facts and information. An organization's database can contain facts and information on customers, employees, inventory, competitors' sales information, and much more. Most managers and executives believe a database is one of the most valuable and important parts of a computer-based information system.

telecommunications

the electronic transmission of signals for communications; enables organizations to carry out their processes and tasks through effective computer networks

networks

connected computers and computer equipment in a building, around the country, or around the world to enable electronic communications

Internet

the world's largest computer network, actually consisting of thousands of interconnected networks, all freely exchanging information

Telecommunications, Networks, and the Internet

Telecommunications is the electronic transmission of signals for communications and enables organizations to carry out their processes and tasks through effective computer networks. **Networks** consist of connected computers and computer equipment in a building, around the country, or around the world to enable electronic communications. The **Internet** is the world's largest computer network, actually consisting of thousands of interconnected networks, all freely exchanging information. It is estimated that about half of all U.S. households own personal computers, and perhaps a third of them are connected to the Internet.[1] The technology used to create the Internet is now being applied within companies and organizations to create *intranets*, which allow people within an organization to exchange information and work on projects. For example, KPMG, a management consulting firm, moved its entire body of knowledge to an intranet called KWorld. KWorld provides KPMG's consultants a central online repository of information and fosters more efficient collaboration among its consultants around the world.[2] An *extranet* is a network based on Internet technologies that allows selected outsiders, such as business partners and customers, to access authorized resources of the intranet of a company. Many people use extranets everyday without realizing it—to track packaged goods, order products from their suppliers, or access customer assistance from other companies. When you log on to the FedEx site to check the status of a package, for example, you are using an extranet.

People

People are the most important element in most computer-based information systems. Information systems personnel include all the people who manage, run, program, and maintain the system. Users are any people who use information systems to get results. Users include financial executives, marketing representatives, manufacturing operators, and many others. Certain computer users are also IS personnel.

FIGURE 1.4

Growth of Business-to-Business
E-commerce
(Source: Gartner Group as
published in Clinton Wilder,
"Business Booms for Specialized
Web Marketplaces," *Information
Week*, February 7, 2000, p. 43.
Used with permission.)

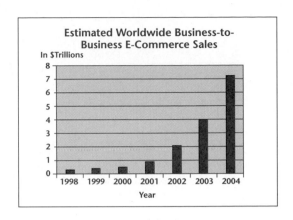

procedures

the strategies, policies, methods,
and rules for using a CBIS

e-commerce

any business transaction executed
electronically between parties such
as companies (business-to-business),
companies and consumers
(business-to-consumer), business
and the public sector, and
consumers and the public sector

Procedures

Procedures are the strategies, policies, methods, and rules for using a CBIS. For example, some procedures describe when each program is to be run. Others describe who can have access to facts in the database. Still other procedures describe what is to be done in case a disaster, such as a fire, an earthquake, or a hurricane, renders the CBIS unusable.

Now that we have looked at computer-based information systems in general, we briefly examine the most common types used in business today. These IS types are covered in more detail later in the book.

BUSINESS INFORMATION SYSTEMS

Workers at all levels, in all kinds of firms, and in all industries are using information systems to improve their effectiveness. These information systems include electronic commerce, transaction processing, information and decision support, and specialized business information systems.

E-commerce is widely used for
business-to-business (or B2B)
transactions. Doctors can use a
transaction processing system
by Claimsnet.com to obtain
payment from patients'
insurance companies.

Electronic Commerce

E-commerce involves any business transaction executed electronically between parties such as companies (business-to-business), companies and consumers (business-to-consumer), business and the public sector, and consumers and the public sector. People may assume that e-commerce is reserved mainly for consumers visiting Web sites for on-line shopping, but Web shopping is only a small part of the e-commerce picture. The major volume of e-commerce—and its fastest-growing segment—is business-to-business transactions that make purchasing easier for corporations (Figure 1.4).[3]

Transaction Processing Systems

Since the 1950s, computers have been used to perform common business applications. The objective of many of these early systems was to reduce costs, which was

FIGURE 1.5

A Payroll Transaction Processing System

The inputs (numbers of employee hours worked and pay rates) go through a transformation process to produce outputs (paychecks).

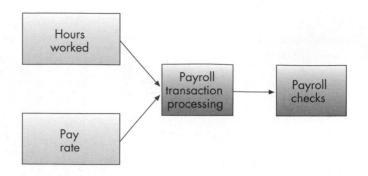

transaction

any business-related exchange such as payments to employees, sales to customers, and payments to suppliers

transaction processing system (TPS)

an organized collection of people, procedures, software, databases, and devices used to record completed business transactions

achieved by automating many routine, labor-intensive business systems. A **transaction** is any business-related exchange such as payments to employees, sales to customers, and payments to suppliers. A **transaction processing system (TPS)** is an organized collection of people, procedures, software, databases, and devices used to record completed business transactions. To understand a transaction processing system is to understand basic business operations and functions. For example, one of the first business systems to be computerized was the payroll system (Figure 1.5). The primary inputs for a payroll TPS are the numbers of employee hours worked during the week and pay rates. The primary output consists of paychecks. Early payroll systems produced employee paychecks along with important employee-related reports required by state and federal agencies, such as the Internal Revenue Service. Other TPSs include systems to process customer orders (order processing), order new supplies and needed parts (purchasing), and keep accurate records (accounting).

Information and Decision Support Systems

Although early accounting and financial transaction processing systems were valuable, it has become clear that the data stored in these systems can be used to help managers make better decisions in their respective business areas, whether human resources, marketing, or administration. Satisfying the needs of managers and decision makers continues to be a major factor in developing information systems. Management information systems and decision support systems can be used to satisfy these needs.

Management Information Systems

management information system (MIS)

an organized collection of people, procedures, software, databases, and devices used to provide routine information to managers and decision makers

A **management information system (MIS)** is an organized collection of people, procedures, software, databases, and devices used to provide routine information to managers and decision makers. The focus of an MIS is primarily on operational efficiency. Marketing, production, finance, and other functional areas are supported by management information systems and linked through a common database. Management information systems typically provide standard reports generated with data and information from the transaction processing system (Figure 1.6).

Decision Support Systems

By the 1980s, dramatic improvements in technology resulted in information systems that were less expensive but more powerful than earlier systems. People at all levels of organizations began using personal computers to do a variety of tasks; they were no longer solely dependent on the information systems department for all their information needs. During this time, people recognized

FIGURE 1.6

Functional management
information systems draw
data from the organization's
transaction processing system.

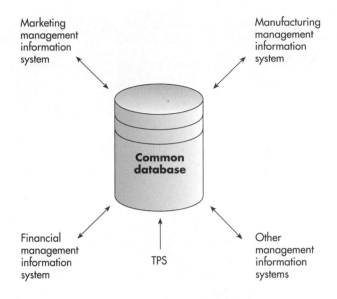

**decision support
system (DSS)**

an organized collection of people,
procedures, software, databases,
and devices used to support
problem-specific decision making

that computer systems could support additional decision-making activities. A **decision support system (DSS)** is an organized collection of people, procedures, software, databases, and devices used to support problem-specific decision making. The focus of a DSS is on decision-making effectiveness. Whereas an MIS helps an organization "do things right," a DSS helps a manager "do the right thing."

The essential elements of a DSS include a collection of models used to support a decision maker or user (model base), a collection of facts and information to assist in decision making (database), and systems and procedures (user interface) that help decision makers and other users interact with the DSS (Figure 1.7 on page 14).

Special-Purpose Business Information Systems: Artificial Intelligence, Expert Systems, and Virtual Reality

artificial intelligence (AI)

a field in which the computer system
takes on the characteristics of
human intelligence

expert system (ES)

a system that gives a computer the
ability to make suggestions and act
as an expert in a particular field

In addition to TPSs, MISs, and DSSs, organizations often use special systems. One of these systems is based on the notion of **artificial intelligence (AI)**, in which the computer system takes on the characteristics of human intelligence. The field of artificial intelligence includes several subfields (see Figure 1.8 on page 15). **Expert systems (ES)** give the computer the ability to make suggestions and act as an expert in a particular field. The unique value of expert systems is that they allow organizations to capture and use the wisdom of experts and specialists. So, years of experience and specific skills are not completely lost when a human expert dies, retires, or leaves for another job. Expert systems can be applied to almost any field or discipline. Expert systems have been used to monitor complex systems such as nuclear reactors, perform medical diagnoses, locate possible repair problems, design and configure information system components, perform credit evaluations, and develop marketing plans for a new product or new investment strategies.

virtual reality

immersive virtual reality, which
means the user becomes fully
immersed in an artificial,
three-dimensional world that is
completely generated by a computer

Virtual reality is another example of a special-purpose business information system. Originally, the term **virtual reality** referred to immersive virtual reality, in which the user becomes fully immersed in an artificial, three-dimensional world that is completely generated by a computer. It may represent any three-dimensional setting, real or abstract, such as a building, an archaeological excavation site, the human anatomy, a sculpture, or a

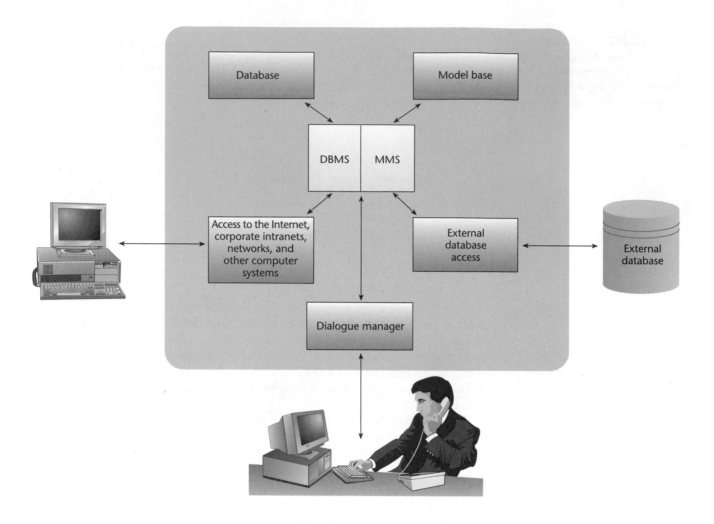

FIGURE 1.7

Essential DSS Elements

crime scene reconstruction. Virtual worlds can be animated, interactive, and shared. Input devices such as head-mounted displays (Figure 1.9), data gloves (Figure 1.10 on page 16), joysticks, and handheld wands allow the user to navigate through a virtual environment and to interact with virtual objects. Virtual reality is used in today's organizations for many purposes, including designing work spaces and buildings and training employees to use specialized equipment such as airplanes and fighter jets. It is also being applied to e-commerce to allow shoppers to try on selected clothing and use data gloves to experience products firsthand.

Now that we have reviewed the basics types of business informations systems, we turn to a brief discussion of how systems are developed in organizations. Systems development is a critical function in today's organizations—and one that can often take many resources to accomplish well.

SYSTEMS DEVELOPMENT

systems development

the activity of creating or modifying existing business systems

Systems development is the activity of creating or modifying existing business systems. Developing information systems to meet business needs is highly complex and difficult; it is common for information systems projects to over-run budgets and exceed scheduled completion dates. Business managers would like the development process to be more manageable, with predictable costs and timing. Systems development includes investigation, analysis, design, implementation, and maintenance and review.

FIGURE 1.8

The Major Elements of Artificial Intelligence

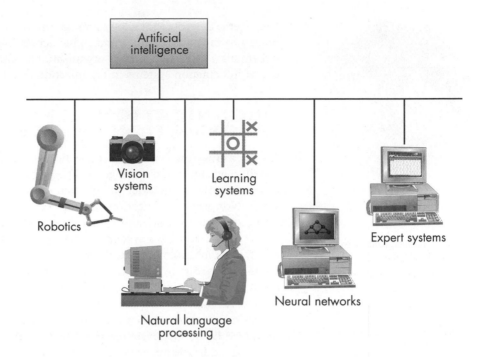

The goal of *systems investigation* is to gain a clear understanding of the problem to be solved or opportunity to be addressed. Once this is understood, the next question is, "Is the problem worth solving?" Since organizations have limited resources—people and money—this question deserves careful consideration. If the decision is to continue with the solution, the next step, *systems analysis*, defines the problems and opportunities of the existing system. *Systems design* determines how the new system will work to meet the business needs defined during systems analysis. *Systems implementation* involves creating or acquiring the various system components (hardware, software, databases, and so on) defined in the design step, assembling them, and putting the new system into operation. The purpose of *systems maintenance and review* is to maintain and modify the system to meet changing business needs.

FIGURE 1.9

A Head-Mounted Display

The head-mounted display (HMD) was the first device providing its wearer with an immersive experience. A typical HMD houses two miniature display screens and an optical system that channels the images from the screens to the eyes, thereby presenting a stereo view of a virtual world. A motion tracker continuously measures the position and orientation of the user's head and allows the image-generating computer to adjust the scene representation to the current view. As a result, the viewer can look around and walk through the surrounding virtual environment.
(Source: Courtesy of Virtual Research Systems, Inc.)

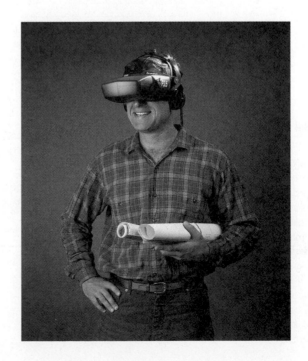

FIGURE 1.10

A Data Glove

Realistic interactions with virtual objects via such devices as a data glove that senses hand position allow for manipulation, operation, and control of virtual worlds. (Source: Courtesy of Virtual Technologies, Inc.)

organization

a formal collection of people and other resources established to accomplish a set of goals

value chain

a series (chain) of activities that includes inbound logistics, warehouse and storage, production, finished product storage, outbound logistics, marketing and sales, and customer service

Systems development allows an organization to effectively use computer technology to increase revenues and reduce costs. Well-designed, well-implemented information systems can help organizations achieve their goals. We discuss the use of information systems in organizations next.

ORGANIZATIONS AND INFORMATION SYSTEMS

An **organization** is a formal collection of people and other resources established to accomplish a set of goals. The primary goal of a for-profit organization is to maximize shareholder value, often measured by the price of the company's stock. Nonprofit organizations include social groups, religious groups, universities, and other organizations that do not have profit as the primary goal.

An organization is a system. Money, people, materials, machines and equipment, data, information, and decisions are constantly in use in any organization. As shown in Figure 1.11, resources such as materials, people, and money are input to the organizational system from the environment, go through a transformation mechanism, and are output to the environment. The outputs from the transformation mechanism are usually goods or services. The goods or services produced by the organization are of higher relative value than the inputs alone. Through adding value or worth, organizations attempt to achieve their goals.

All business organizations contain a number of value-added processes. Providing value to a stakeholder—customer, supplier, manager, or employee—is the primary goal of any organization. The value chain, first described by Michael Porter in a 1985 *Harvard Business Review* article, is a concept that reveals how organizations can add value to their products and services. The **value chain** is a series (chain) of activities that includes inbound logistics, warehouse and storage, production, finished product storage, outbound logistics, marketing and sales, and customer service (Figure 1.12). Each of these activities is investigated to determine what can be done to increase the value perceived by a customer. Depending on the customer, value may mean lower price, better service, higher quality, or uniqueness of product. The value comes from the skill, knowledge, time, and energy invested by the company.

FIGURE 1.11

A General Model of an Organization

Information systems support and work within all parts of an organizational process. Although not shown in this simple model, input to the process subsystem can come from internal and external sources. Just prior to entering the subsystem, data is external. Once it enters the subsystem, it becomes internal. Likewise, goods and services can be output to either internal or external systems.

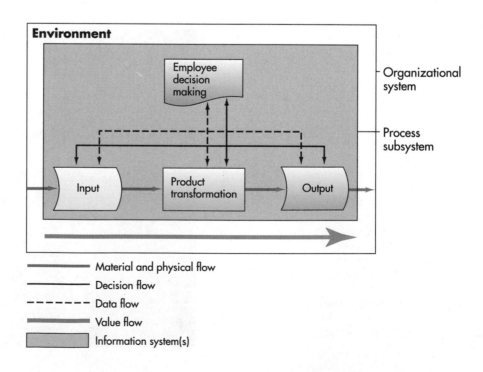

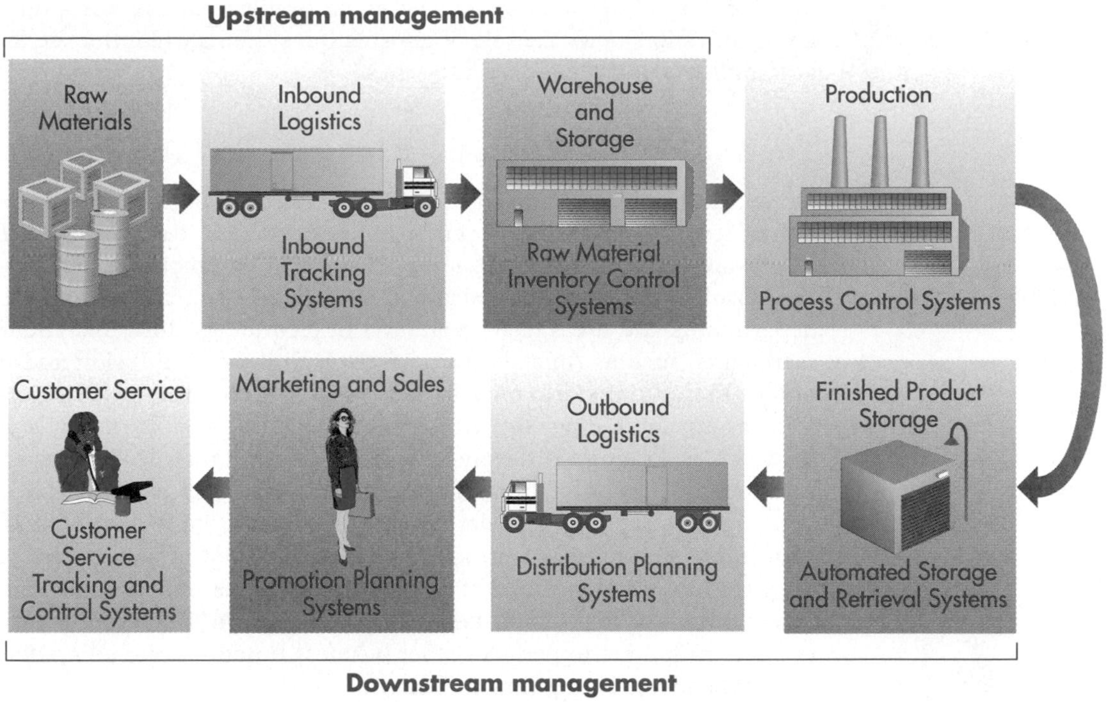

Upstream management

Downstream management

FIGURE 1.12

The Value Chain of a
Manufacturing Company

By adding a significant amount of value to their products and services, companies ensure further organizational success.

What role does an information system play in these value-added processes? A traditional view of information systems holds that they are used by organizations to control and monitor value-added processes to ensure effectiveness and efficiency. An information system can turn feedback from the value-added process subsystems into more meaningful information for employees' use within an organization. This information might summarize the performance of the systems and be used as the basis for changing the way the system operates. Such changes could involve using different raw materials (inputs), designing new assembly-line procedures (product transformation), or developing new products and services (outputs). In this view, the information system is external to the process and serves to monitor or control it.

A more contemporary view, however, holds that information systems are often so intimately intertwined with the underlying value-added process that they are best considered part of the process itself.[4] From this perspective, the information system is internal to and plays an integral role in the process, whether by providing input, aiding product transformation, or producing output. Consider a phone directory business that creates phone books for international corporations. A corporate customer requests a phone directory listing all steel suppliers in Western Europe. Using its information system, the directory business can sort files to find the suppliers' names and phone numbers and organize them into an alphabetical list. The information system itself is an integral part of this process. It does not just monitor the process externally but works as part of the process to transform a product. In this example, the information system turns raw data input (names and phone numbers) into a salable output (a phone directory). The same system might also provide the input (data files) and output (printed pages for the directory).

The latter view brings with it a new perspective on how and why information systems can be used in business. Rather than searching to understand the value-added process independently of information systems, we consider the potential role of information systems within the process itself, often leading to the discovery of new and better ways to accomplish the process. Thus, the way an organization views the role of information systems will influence the ways it accomplishes its value-added processes.

Technology Diffusion, Infusion, and Acceptance

technology diffusion

a measure of how widely technology is spread throughout an organization

The extent to which technology is used throughout an organization can be a function of technology diffusion, infusion, and acceptance. **Technology diffusion** is a measure of how widely technology is spread throughout an organization. An organization in which computers and information systems are located in most departments and areas has a high level of technology diffusion.[5] Some online merchants, such as Amazon.com, have a high level of diffusion and use computer systems to perform most of their business functions, including marketing, purchasing, and billing. **Technology infusion**, on the other hand, is the extent to which technology is deeply integrated into an area or department. Some architectural firms, for example, use computers in all aspects of designing a building or structure. The design area thus has a high level of infusion. Of course, it is possible for a firm to have a high level of infusion in one aspect of its operations and a low level of diffusion overall. The architectural firm may use computers in all aspects of design (high infusion in the design area) but may not use computers to perform other business functions, including billing, purchasing, and marketing (low diffusion).

technology infusion

the extent to which technology is deeply integrated into an area or department

Just because an organization has a high level of diffusion and infusion, with computers throughout the organization, it does not necessarily mean that information systems are being used to their full potential. In fact, the assimilation and use of expensive computer technology throughout organizations varies greatly.[6] One reason is a low degree of acceptance and use of the technology among some managers and employees. Research has attempted to explain the important factors that enhance or hinder the acceptance and use of information systems.[7] A number of possible explanations of technology acceptance and usage have been studied. The **technology acceptance model (TAM)** describes the factors that can lead to higher acceptance and use of technology in an organization, including perceived usefulness of the technology, ease of its use, quality of the information system, and the degree to which the organization supports the use of the information system.[8] Companies hope that a high level of diffusion, infusion, and acceptance will lead to greater performance and profitability.[9] Some experts believe that the booming stock market of the late 1990s was in part a result of the successful implementation and use of technology to increase productivity and profitability.

technology acceptance model (TAM)

a description of the factors that can lead to higher acceptance and use of technology in an organization, including perceived usefulness of the technology, ease of its use, quality of the information system, and the degree to which the organization supports the use of the information system

COMPETITIVE ADVANTAGE

competitive advantage

a significant and (ideally) long-term benefit to a company over its competition

A **competitive advantage** is a significant and (ideally) long-term benefit to a company over its competition. Establishing and maintaining a competitive advantage is complex, but a company's survival and prosperity depend on its success in doing so.

Factors That Lead Firms to Seek Competitive Advantage

five-force model

a widely accepted model that identifies five key factors that can lead to attainment of competitive advantage: rivalry among existing competitors, the threat of new market entrants, the threat of substitute products and services, the bargaining power of buyers, and the bargaining power of suppliers

A number of factors can lead to the attainment of competitive advantage. Michael Porter, a prominent management theorist, suggested a now widely accepted **five-force model**. The five forces are rivalry among existing competitors, the threat of new market entrants, the threat of substitute products and services, the bargaining power of buyers, and the bargaining power of suppliers. The more these forces combine in any instance, the more likely firms will seek competitive advantage and the more dramatic the results of such an advantage will be.

Rivalry among Existing Competitors

Rivalry among existing competitors is an important factor leading firms to seek competitive advantage. Typically, highly competitive industries are characterized by high fixed costs of entering or leaving the industry, low degrees of product differentiation, and many competitors. Although all firms are rivals with their competitors, industries with stronger rivalries tend to have more firms seeking competitive advantage.

Threat of New Entrants

The threat of new entrants is another important force leading an organization to seek competitive advantage. A threat exists when entry and exit costs to the industry are low and the technology needed to start and maintain the business is commonly available. For example, consider a small restaurant. The owner does not require millions of dollars to start the business, food costs do not go down substantially for large volumes, and food processing and preparation equipment is commonly available. When the threat of new market entrants is high, the desire to seek and maintain competitive advantage to dissuade new market entrants is usually high.

Threat of Substitute Products and Services

The more consumers are able to obtain similar products and services that satisfy their needs, the more likely firms are to try to establish competitive advantage. Such an advantage often creates a "new playing field" in which consumers no longer consider "substitute" products acceptable replacements. Consider the personal computer industry and the introduction of low-cost computers. A number of consultants and computer manufacturers made much of the high cost of ownership associated with personal computers in the mid-1990s. They introduced low-cost network computers with minimal hard disk space, slower CPUs, and less main memory than some consumers desired, but at half the cost of a standard workstation. There was considerable interest in these new machines for a while, but traditional personal computer manufacturers fought back. They developed a class of powerful workstations and implemented new pricing strategies to make them available at under $1000. This eliminated the primary advantage of the stripped-down network computers and regained lost customers.

Bargaining Power of Customers and Suppliers

Large buyers exert significant influence on a firm. But their power can be diminished if they cannot make purchases elsewhere. Suppliers can help an organization obtain a competitive advantage. In some cases, suppliers have entered into *strategic alliances* with firms, in which the suppliers act as a part of the company. Suppliers and companies can use telecommunications to link their computers and personnel to obtain fast reaction times and the ability to get parts or supplies when they are needed to satisfy customers.

Strategic Planning for Competitive Advantage

To be competitive, a company must be fast, nimble, flexible, innovative, productive, economical, and customer oriented. It must also align the information system strategy with general business strategies and objectives.[10] Given the five market forces just mentioned, Porter proposed three general strategies to attain competitive advantage: altering the industry structure, creating new products and services, and improving existing product lines and services. Research into the use of information systems to help organizations achieve competitive advantage has confirmed and extended Porter's original work to include additional strategies—such as forming alliances with other companies, developing a niche market, maintaining competitive cost, and creating product differentiation.[11]

The impending strategic merger between AOL and Time Warner will create the world's first fully integrated media and communications company for the "Internet Century."
(Source: AP/World Wide Photos.)

Altering the Industry Structure

Altering the industry structure is the process of changing the industry to become more favorable to the company or organization. This goal can be accomplished by gaining more power over suppliers and customers. A strategic alliance, introduced earlier, is one approach that can be used. Also called a strategic partnership, a strategic alliance is an agreement between two or more companies to jointly produce and distribute goods and services. An example of a successful strategic partnership is the merger of CMGI and Raging Bull.[12] David Wetherell, who runs CMGI as a successful mutual fund company, saw Raging Bull, one of the fastest-growing on-line investor chat rooms, while he was surfing the Web. He quickly purchased 50 percent of the company for $2 million. Raging Bull, which at the time was being run by three college students in a basement, quickly moved to CMGI's headquarters in Andover, Massachusetts, where the advantage of the strategic partnership started to be realized. The four-month-old Raging Bull site, which provided free investment chat rooms and had been financed by one of the founder's credit cards, started generating revenues immediately by using Internet ads. This strategic partnership, like others, was a winning strategy for everyone involved.

Creating New Products and Services

Creating new products and services is always an approach that can help a firm gain a competitive advantage. This is especially true of the computer industry and other high-tech businesses. If a company does not introduce new products and services every few months, it can quickly stagnate, lose market share, and decline. Companies that stay on top are constantly developing new products and services. Vattenfall, Sweden's largest public utility company, for example, has developed a new wireless service that allows people to control machines on the factory floor, to monitor security devices and appliances at home, and to gain access to the Internet through a cell phone.[13] Some believe that this technology could be worth billions of dollars when it is introduced in the United States. New products are particularly important in the dynamic technology industry.

Improving Existing Product Lines and Services

Improving existing product lines and services is another approach to staying competitive. The improvements can be either real or perceived. Manufacturers of household products are always advertising new and improved products. In some cases, the improvements are more perceived than real; usually, only minor changes are made to the existing product. Many food and beverage companies are introducing "Healthy" and "Light" product lines. A popular beverage company introduced "born on" dating for beer.

Using Information Systems for Strategic Purposes

Combining improved understanding of the potential of information systems with the growth of new technology and applications has led organizations to use IS to gain a competitive advantage. In simplest terms, competitive advantage is usually embodied either in a product or service that has the most added value to consumers and that is unavailable from the competition, or in an internal system that delivers benefits to a firm that are not enjoyed by its competition. A classic example is SABRE, a sophisticated computerized reservation system installed by American Airlines and one of the first CBISs recognized for providing competitive advantage. Travel agents used this system for rapid access to flight information, offering travelers reservations, seat assignments, and ticketing. The travel agents also achieved an efficiency benefit from the SABRE system. Because SABRE displayed American Airline flights whenever possible, it also gave the airline a long-term, significant competitive advantage.

The extent to which companies are using computers and information technology for competitive advantage continues to grow.[14] Investments in information systems that result in happy customers and efficient suppliers can do as much to achieve a competitive advantage as internal systems, such as payroll and billing. Table 1.3 lists several examples of how companies have attempted to gain a competitive advantage.

PERFORMANCE-BASED INFORMATION SYSTEMS

There have been at least three major stages in the business use of IS. The first stage started in the 1960s and was oriented toward cost reduction and productivity. This stage generally ignored the revenue side, not looking for opportunities to increase sales via the use of IS. The second stage started in the 1980s and was defined by Porter and others. It was oriented toward gaining a competitive advantage. In many cases, companies spent large amounts on IS

TABLE 1.3

Competitive Advantage Factors and Strategies

Factors That Lead to Attainment of a Competitive Advantage	Strategies		
	Alter Industry Structure	**Create New Products and Services**	**Improve Existing Product Lines and Services**
Rivalry among existing competitors	Blockbuster changes the industry structure with its chain of video and music stores.	Dell, Gateway, and other PC makers develop computers that excel at downloading Internet music and playing the music on high-quality speakers.	Food and beverage companies offer "healthy" and "light" product lines.
Threat of new entrants	AOL and Time Warner plan to merge to form a large Internet and media company.	Apple Computer introduces an easy-to-use iMac computer that can be used to create and edit home movies.	Starbucks offers new coffee flavors at premium prices.
Threat of substitute products and services	Ameritrade and other discount stockbrokers offer low costs and research on the Internet.	Wal-Mart uses technology to monitor inventory and product sales to determine the best mix of products and services to offer at various stores.	Cosmetic companies add sunscreen to their product lines.
Bargaining power of buyers	Ford, GM, and others require that suppliers locate near their manufacturing facilities.	Investors and traders of the Chicago Board of Trade (CBOT) put pressure on the institution to implement electronic trading.	Retail clothing stores require manufacturing companies to reduce order lead times and improve materials used in the clothing.
Bargaining power of suppliers	American Airlines develops SABRE, a comprehensive travel program used to book airline, car rental, and other reservations.	Intel develops SpeedStep, a chip for laptop computers, that operates at faster speeds when connected to an electrical outlet.	Hayworth, a supplier of office furniture, has a computerized-design tool that helps it design new office systems and products.

FIGURE1.13

Three Stages in the Business
Use of IS

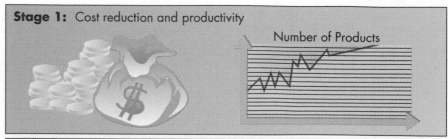

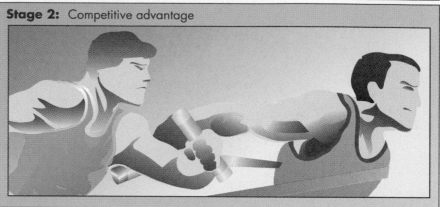

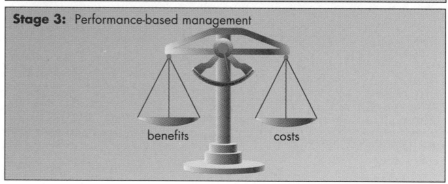

and ignored the costs. Today we are seeing a shift from strategic management to performance-based management in many organizations. This third stage carefully considers both strategic advantage and costs, using productivity, quality, return on investment (ROI), and other measures of performance.[15] Figure 1.13 illustrates these stages.

Productivity

productivity

a measure of the output achieved
divided by the input required

Developing information systems that measure and control productivity is a key element for most organizations. **Productivity** is a measure of the output achieved divided by the input required. A higher level of output for a given level of input means greater productivity; a lower level of output for a given level of input means lower productivity. Consider a tax preparation firm, for example, where productivity can be measured by the tax returns prepared divided by the total hours the employee worked. In a 40-hour week, an employee may have prepared 30 tax returns. The productivity is thus equal to 30/40, or 75 percent. With administrative and other duties, a productivity level of 75 may be excellent. The numbers assigned to productivity levels are not always based on labor hours—productivity may be based on factors such as the amount of raw materials used, resulting quality, or time to produce the goods or service. In any case, what is important is not the value of the productivity number but how it compares with other time periods, settings, and organizations.

$$\text{Productivity} = (\text{Output}/\text{Input}) \times 100\%$$

Measuring productivity is important because improving productivity boosts a nation's standard of living. In an era of intense international competition, the need to improve productivity is critical to the well being of any enterprise or country. If a company does not take advantage of technological and management innovation to improve productivity, its competitors will. The ability to apply information technology to improve productivity will separate successful enterprises from failures.

Quality

quality

the ability of a product (including services) to meet or exceed customer expectations

The definition of the term *quality* has evolved over the years. In the early years of quality control, firms were concerned with meeting design specifications—that is, conformance to standards. If a product performed as designed, it was considered a high-quality product. A product can perform its intended function, however, and still not satisfy customer needs. Today, **quality** means the ability of a product (including services) to meet or exceed customer expectations. This view of quality is completely customer oriented. A high-quality product satisfies customers by functioning correctly and reliably, meeting needs and expectations, and being delivered on time with courtesy and respect. Companies such as Boeing combine computer technology and continuous-improvement programs to both enhance quality and cut costs.[16] Boeing uses portable computing devices to monitor the quality of its airplane manufacturing process. After the quality data is entered into the portable computer, it is transferred to larger computers for analysis and report writing. Read the "E-Commerce" box to see how Internet companies, such as Drugstore.com, are implementing quality programs.

Return on Investment and the Value of Information Systems

return on investment (ROI)

one measure of IS value that investigates the additional profits or benefits that are generated as a percentage of the investment in information systems technology

One measure of IS value is **return on investment (ROI)**. This measure investigates the additional profits or benefits that are generated as a percentage of the investment in information systems technology. A small business that generates an additional profit of $20,000 for the year as a result of an investment of $100,000 for additional computer equipment and software would have a return on investment of 20 percent ($20,000/$100,000).

Earnings Growth

Another measure of IS value is the increase in profit, or earnings growth, it brings. For instance, suppose a mail-order company, after installing an order processing system, had a total earnings growth of 15 percent compared with the previous year. Sales growth before the new ordering system was only about 8 percent annually. Assuming that nothing else affected sales, the earnings growth brought by the system was 7 percent.

Market Share

Market share is the percentage of sales that one company's products or services have in relation to the total market. If installing a new on-line Internet catalog increases sales, it might help a company increase its market share by 20 percent.

E-COMMERCE

Finding Quality through E-Commerce

Companies have long recognized that superior customer quality is a key to attracting and retaining customers. Slogans such as "Quality is Job One" or "Our Customers Always Come First" are typical. Companies often spend thousands of dollars attracting new customers with special offers and one-time discounts, but then ignore the importance of keeping current customers satisfied. One dissatisfied customer can easily tell five to ten friends and family members of a bad experience. The result can be a massive loss, and the company may not even know what hit it. All the company sees is a drastic reduction in sales and profits. The lesson is that quality pays off in higher sales and profits, but have the new e-commerce companies learned this important lesson?

Internet-based companies have entered most markets that were once dominated by traditional companies. Travel, stock trading, investment advice, auctions, and shopping are just a few examples of traditional industries that are now thriving on the Internet. Yet quality and customer satisfaction seem to be lacking in many of these industries. In a survey conducted by the NPD Group, fewer than one-third of respondents "strongly agreed" that they got accurate information from Web shopping. Fewer than one-fourth thought that the customer service areas of on-line companies responded quickly or really understood their needs. Without question, quality is a major concern for on-line customers in general. Some Internet companies, however, are starting to get the message that quality does indeed count.

Many on-line companies, such as Drugstore.com, are starting to realize the importance of high quality. Drugstore.com has instituted a quality program with two different types of quality specialists. One type works on developing a better customer interface for the system and ensures that customers are satisfied with their on-line experience with Drugstore.com. The other quality specialists are the technical people who make sure that all the systems are running correctly with minimal downtime and interruptions. To achieve high quality at both the customer interface level and the technical level requires computer systems that work well at both the front end and the back end of the business.

The front-end systems for an Internet company are those the customer sees. They include the Internet Web site and systems that allow customers to get information and place orders. Once an order is placed, the back-end systems are used to fill the orders and ship products to customers. The back-end systems are more traditional uses of technology. So, for an existing company, the back-end computer sys-

tems may already be in place. The problem, though, is to make sure that the two systems are communicating with each other and working smoothly. Some Internet companies have great front-end systems that provide information and take orders, and terrible back-end systems that fill orders and ship products. According to Steven Nevill, CIO of Gerald Stevens, which sells flowers and gifts on-line, "You want your customers—whether shopping on-line, calling an 800 number, or walking into a store—to have the same experience. You want whoever is interacting with them to know everything about them, where they have been before, and be able to service them in the same way." This desire for a consistently high-quality experience is why some companies have waited to go on-line—they want to be sure that their sites meet expectations.

Many companies, such as Wal-Mart, are taking their time to get their Web sites the best they can be. General Electric, which uses a rigorous quality program, has the same approach to quality for its e-business. Talking about the program, General Electric's e-business leader, Camille Farhat, said, "We cannot relax that standard." But this "go slow and get it right" approach may not bring in the highest revenues or profits. While companies are taking their time to get it right, other companies are launching on-line sites without careful quality planning. Companies such as E*trade and eBay started quickly and had technical problems and crashes. But according to Thomas Eisenmann, who teaches a course on e-commerce at the Harvard Business School, "E*trade and eBay have both suffered many hiccups in terms of reliability, but it doesn't seem fundamentally to have hurt the trust in their brands."

Discussion Questions

1. What are the features of the quality program developed by Drugstore.com?
2. What is the difference between a front-end and a back-end system?

Critical Thinking Questions

3. Why is quality important for on-line companies? How would you increase quality for an on-line business?
4. Is it always best for a company to take its time to develop a high-quality Web site?

Sources: Adapted from Gary Anthes, "The Quest for E-Quality," *Computerworld*, December 13, 1999, p. 46; and Janet Rae-Dupree, "The Neupert Treatment," *Business Week*, February 14, 2000, p. 83.

Customer Awareness and Satisfaction

Although customer satisfaction can be difficult to quantify, about half of today's best global companies measure the performance of their information systems based on feedback from internal and external users. Some companies use surveys and questionnaires to determine whether the investment in information systems has increased customer awareness and satisfaction.

Total Cost of Ownership

total cost of ownership (TCO)

a measure of the total cost of owning computer equipment, including desktop computers, networks, and large computers

In addition to the preceding measures, some companies also track total costs. One measure, developed by the Gartner Group and explored in more detail in a case at the end of the chapter, is the **total cost of ownership (TCO)**.[17] This approach breaks total costs into such areas as the cost to acquire the technology, technical support, administrative costs, and end-user operations. Other costs in TCO include retooling and training costs. TCO can be used to get a more accurate estimate of the total costs for systems that range from small PCs to large mainframe systems.

Most organizations today realize that they must look at both sides of the equation—benefits as well as costs—in evaluating potential information system investments. Also, determining return on investment can help the IS staff prove its contribution to the organization and ensure that its efforts are aligned with the company's overall business objectives.

INFORMATION SYSTEMS PERSONNEL

With huge corporate investments in information systems, it is likely that you will be working with information systems personnel during your career. So, it is important to know the roles and functions of these individuals and their departments. A few of you will become information systems personnel. For many schools and departments, information systems majors attain some of the highest starting annual salaries—sometimes exceeding $50,000. In this section, we will explore the roles and functions of information systems personnel and information systems departments.

Roles and Functions of the Information Systems Department

IS personnel typically work in an information systems department that employs system operators, systems analysts, computer programmers, and other information systems personnel. They may also work in other functional departments or areas in a support capacity. In addition to technical skills, IS personnel need written and verbal communication skills, an understanding of organizations and the way they operate, and the ability to work with people (system users). In general, information systems personnel must maintain the broadest perspective on organizational goals. For most medium- to large-sized organizations, information resources are typically managed through an IS department. In smaller businesses, one or more people may manage information resources, with support from outside services (outsourcing). As shown in Figure 1.14, the information systems organization has three primary responsibilities: operations, systems development, and support.

Operations

The operations component of a typical IS department focuses on the use of information systems in corporate or business-unit computer facilities. It tends to focus more on the efficiency of information system functions rather than their effectiveness.

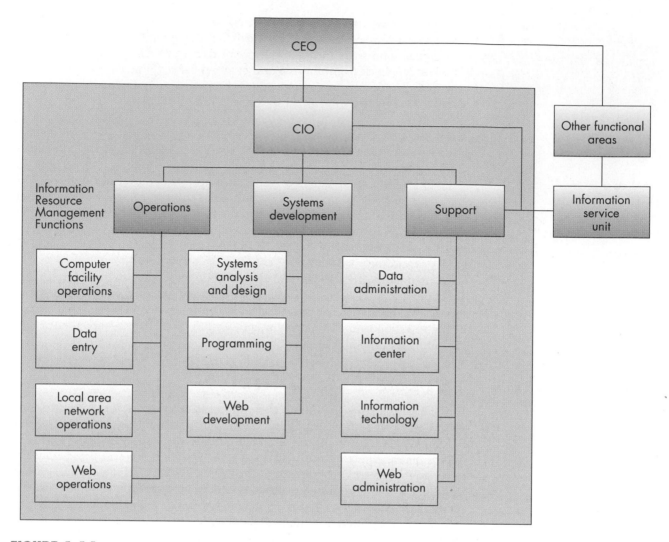

FIGURE 1.14

The Three Primary
Responsibilities of Information
Systems

Each of these elements—operations, systems development,
and support—contains sub-
elements critical to the efficient
and effective operation of the
organization.

The primary function of a system operator is to run and maintain IS equipment, such as mainframe systems, networks, tape drives, disk devices, printers, and so on. System operators are typically trained at technical schools or through on-the-job experience. Other operations include logging, scheduling, hardware maintenance, and preparation of input and output. Data-entry operators convert data into a form the computer system can use. They may use terminals or other devices to enter business transactions, such as sales orders and payroll data. Increasingly, data entry is being automated—captured at the source of the transaction rather than being entered later. In addition, companies may have local area network and Web or Internet operators who are responsible for running the local network and any Internet sites the company may have.

Systems Development

The systems development component of a typical IS department focuses on specific development projects and ongoing maintenance and review. Systems analysts and programmers focus on these concerns.

The role of a systems analyst is multifaceted. Systems analysts help users determine what outputs they need from the system and construct the plans to develop the programs that produce these outputs. Systems analysts then work with programmers to ensure that the appropriate programs are purchased, modified from existing programs, or developed. The major responsibility of a computer programmer is to use the plans developed by the systems analyst to develop or adapt one or more computer programs that produce the desired outputs. The main focus

System operators focus on the efficiency of information system functions rather than their effectiveness. Their primary function is to run and maintain IS equipment.
(Source: Image provided by PhotoDisc © 2000.)

information center

a support function that provides users with assistance, training, application development, documentation, equipment selection and setup, standards, technical assistance, and troubleshooting

of systems analysts and programmers is to achieve and maintain information system effectiveness.

Support

The support component of a typical IS department focuses on providing user assistance in hardware and software acquisition and use, data administration, and user training and assistance. Because information systems hardware and software is costly, especially if mistakes are made, the acquisition of computer hardware and systems software is often managed by a specialized group within the support component. This group sets guidelines and standards for the rest of the organization to follow in making purchases. Gaining and maintaining an understanding of available technology is an important part of the acquisition of information systems. Also, developing good relationships with vendors is important.

The support component typically operates the information center. An **information center** provides users with assistance, training, application development, documentation, equipment selection and setup, standards, technical assistance, and troubleshooting. Although many firms have attempted to phase out information centers, others have changed the focus of this function from technical training to helping users find ways to maximize the benefits of the information resource.

Typical IS Titles and Functions

The organizational chart shown in Figure 1.14 is a simplified model of an IS department in a typical medium- or large-sized organization. Many organizations have even larger departments, with increasingly specialized positions such as librarian, quality assurance manager, and the like. Smaller firms often combine the roles depicted in Figure 1.14 into fewer formal positions.

Chief Information Officer

The overall role of the chief information officer (CIO) is to use the IS department's equipment and personnel to help the organization attain its goals. The CIO, usually a manager at the vice president level, is concerned with the overall needs of the organization. He or she is responsible for corporate wide policy, planning, management, and acquisition of information systems. Some of the CIO's top concerns include integrating information systems operations with corporate strategies, keeping up with the rapid pace of technology, and defining and assessing the value of systems development projects in terms of performance, cost, control, and complexity. The high level of the CIO position is consistent with the idea that information is one of the organization's most important resources. This individual works with other high-level officers of the organization, including the chief financial officer (CFO) and the chief executive officer (CEO), in managing and controlling total corporate resources.

Depending on the size of the information systems department, there may be several people at senior IS managerial levels. Some of the job titles associated with information systems management are CIO, vice president of information systems, and manager of information systems. A central role of all these individuals is to communicate with other areas of the organization to determine changing needs. Often these individuals are part of an advisory or steering committee that helps the CIO and other IS managers with their decisions about the use of information systems to support corporate goals. CIOs must work closely with advisory committees, stressing effectiveness and teamwork and viewing information systems as an integral part of the organization's business processes—not an adjunct to the organization.

LAN Administrators

Local area network (LAN) administrators set up and manage network hardware, software, and security processes. They manage the addition of new users, software, and devices to the network. They isolate and fix operations problems. LAN administrators are currently in high demand.

Internet Careers

Explosive growth in the use of the Internet to conduct business has caused a need for skilled personnel to develop and coordinate Internet use. In 2000, the amount of business-to-business commerce exceeded $40 billion, and this number is expected to exceed $1 trillion in a few years.[18] This surge in Internet use has resulted in career opportunities with traditional and on-line companies. These careers, which include top-level administrative positions, are in the areas of Web operations, Web development, and Web administration.

Internet jobs within a traditional company include Internet strategists and administrators, Internet systems developers, Internet programmers, and Web site operators. The Internet has become so important to some companies that they have created a new position, chief Internet officer, with responsibilities and salary similar to the CIO's.

certification

a process for testing skills and knowledge that results in an endorsement by the certifying authority that an individual is capable of performing a particular job

In addition to traditional companies, there are many exciting career opportunities in companies that offer products and services over the Internet.[19] These companies include Amazon.com, Yahoo!, eBay, and many others. Systest, for example, specializes in finding and eliminating digital bugs that could halt the operation of a computer system.[20] According to Christopher Hardesty, chief financial officer, "The Internet has come in and really revolutionized the kind of performance testing we do." Typically, these companies start as private operations and then are offered to the public through initial public offerings (IPOs). To achieve their initial funding, many startup Internet companies seek funding from venture capital (VC) groups. These groups often provide seed money in return for stock and stock options. Unlike traditional companies, many start-up Internet companies offer their employees low salaries but generous stock options that can be worth millions if the company becomes successful. Some top-level managers, politicians, and national newscasters have abandoned their traditional jobs to work for an Internet start-up company, hoping to become rich from stock options when the company goes public.

Internet job sites such as Monster.com allow job hunters to browse job opportunities and post their résumés.

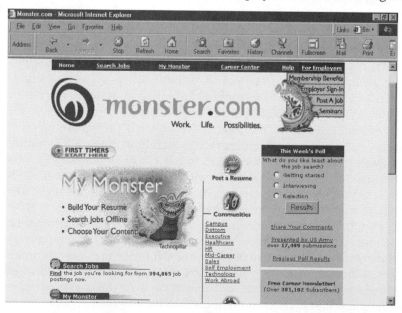

A number of Internet sites post job opportunities for Internet careers and more traditional careers, such as Monster.com. These sites allow prospective job hunters to browse job opportunities, job locations, salaries, benefits, and other factors. In addition, some of these sites allow job hunters to post their résumés.

Quite often, the people filling IS roles have completed some form of certification. **Certification** is a process for testing skills and knowledge resulting in an endorsement by the certifying authority that an individual

INFORMATION SYSTEMS IN ACTION
Not Just Fun and Games: Building Information Systems for the Olympics

Dr. Anne Nelson, High Point University,
High Point, North Carolina

How do you measure the value of an information system? Every day, organizations cash in on the dynamic use of information systems to help solve organizational problems. Another way to determine the value of these systems is when they are used to save people's lives.

E Team Inc. in Canoga Park, California, is focusing on technology to build and manage a Web-based information system for crisis control at the 2002 Winter Olympics in Salt Lake City, Utah. The Olympic games will test the value of E Team's information system not only to help solve organizational problems but to prevent them from happening in the first place.

Building and managing the emergency response system for the Olympic games may be the biggest and most challenging contract for E Team yet, a two-year-old company whose experience includes building military networks designed to battle everything from traffic jams to terrorist attacks. It also built information systems to handle the 2000 Democratic National Convention and allowed Los Angeles city officials to monitor problems related to the Y2K bug. Other E Team customers include the cities of San Francisco and Philadelphia, the state of Louisiana, the U.S. Department of Transportation, and Japanese automaker Toyota.

E Team is going for the gold at the 2002 Winter Olympics with its focus on technology. The information system will utilize a wireless Internet-based tactical network similar to that used by the U.S. military. The system will allow participants to connect to mapping software from geographic information systems incorporating document-sharing and messaging system software. The Internet will provide the backbone to bring together business managers, information systems professionals, police and fire departments, government agencies, hospitals, utilities, and other approved users. In an atmosphere that fosters cooperation, the system's stakeholders will be linked to vital information needed to watch, track, and monitor the Olympic events.

Chris Kramer, spokesperson for the Utah Olympic Public Safety Command, says that his organization uses similar information systems to manage day-to-day "normal stuff," such as a car accident at the root of a traffic jam on a major highway, to more extreme cases where there might be a major hazardous spill or even a bomb. E Team's system for the 2002 Winter Olympics demonstrates the potential impact of information systems. When measuring the value of an information system in terms of people's lives, the value is significant. For E Team, its gold medal will be measured in how successfully the information system helps the decision makers achieve their critical safety goals.

Sources: Karen Kaplan, "Olympics Will Test E Team's Mettle," and "Internet: Canoga Park Firm Will Manage a Web-based Information System for Crisis Control at 2002 Winter Games," *The Los Angeles Times*, December 11, 2000.

is capable of performing a particular job. Certification frequently involves specific, vendor-provided or vendor-endorsed coursework. There are several popular certification programs, including Novell Certified Network Engineer, Microsoft Certified Professional Systems Engineer, and Certified Project Manager.

One of the greatest fears of every IS manager is spending several thousand dollars to help an employee get certified and then to lose that person to a higher-paying position with a new firm. As a consequence, some organizations request a written commitment from individuals to stay a certain time after obtaining their certification. Needless to say, this requirement can create some ill will with the employee. To provide newly certified employees with incentives to remain, other organizations provide salary increases based on additional credentials they acquire.

Other IS Careers

In addition to working for an IS department in an organization, information systems personnel can work for one of the large IS consulting firms, such as Andersen Consulting, EDS, and others. These jobs often entail a large amount of travel, because consultants are assigned to work on various projects wherever the client is. Such roles require excellent people and project management skills in addition to IS technical skills.

Another IS career opportunity is to be employed by a hardware or software vendor developing or selling products. Such a role enables the individual to work on the cutting edge of technology and can be extremely challenging and exciting!

● SUMMARY

PRINCIPLE • The value of information is directly linked to how it helps decision makers achieve the organization's goals.

Data consists of raw facts; information is data organized into a meaningful form. The process of defining relationships between data requires knowledge. Knowledge is an awareness and understanding of a set of information and of how that information can be made useful to support a specific task. To be valuable, information must have several characteristics: it should be accurate, complete, economical, flexible, reliable, relevant, simple, timely, verifiable, accessible, and secure. The value of information is directly linked to how it helps people achieve their organization's goals.

Information systems are sets of interrelated elements that collect (input), manipulate (process), and disseminate (output) data and information. Input is the activity of capturing and gathering new data; processing involves converting or transforming data into useful outputs; and output involves producing useful information. Feedback is output that is used to make changes to input or processing activities. The components of a computer-based information system include hardware, software, databases, telecommunications and networks, people, and procedures.

PRINCIPLE • Knowing the potential impact of information systems and having the ability to put this knowledge to work can result in a successful personal career, an organization that reaches its goals, and a society with a higher quality of life.

Information systems play an important role in today's businesses and society. The key to understanding the existing variety of systems begins with learning their fundamentals. The types of systems used within organizations can be classified into four basic groups: e-commerce, TPSs, MISs and DSSs, and special-purpose business information systems.

E-commerce involves any business transaction executed electronically between parties such as companies (business-to-business), companies and consumers (business-to-consumer), business and the public sector, and consumers and the public sector. The major volume of e-commerce—and its fastest-growing segment—is business-to-business transactions that make purchasing easier for big corporations.

The most fundamental system is the transaction processing system (TPS). A transaction is any business-related exchange. TPSs include order processing, purchasing, accounting, and related systems.

The management information system (MIS) uses the information from a TPS to generate information useful for management decision making. The focus of an MIS is primarily on operational efficiency. A decision support system (DSS) is an organized collection of people, procedures, software, databases, and devices used to support problem-specific decision making.

Special-purpose business information systems include artificial intelligence systems, expert systems, and virtual reality systems. Artificial intelligence (AI) includes a wide range of systems in which the computer system takes on the characteristics of human intelligence. An expert system (ES) acts as an expert consultant to a user seeking advice about a specific situation. Originally, the term *virtual reality* referred to immersive virtual reality, in which the user becomes fully immersed in an artificial, three-dimensional world that is completely generated by a computer. Virtual reality is used in today's organizations for many purposes, such as designing buildings and training employees to use specialized equipment.

PRINCIPLE • System users, business managers, and information systems professionals must work together to build a successful information system.

Systems development involves creating or modifying existing business systems. The major steps of this process and their goals include systems investigation (gain a clear understanding of the problem); systems analysis (define the problems and opportunities of the existing system); systems design (determine how the system will work to meet business needs); systems implementation (create or acquire the various system components defined in the design step); and systems maintenance and review (maintain and modify the system to meet changing business needs).

PRINCIPLE • Information systems add value to an organization and its processes.

An organization is a formal collection of people and other resources established to accomplish a set of goals. The primary goal of a for-profit organization is to maximize shareholder value. Nonprofit organizations include social groups, religious groups, universities, and other organizations that do not have profit as the primary goal.

Organizations are systems with inputs, transformation mechanisms, and outputs. Value-added processes increase the relative worth of the combined inputs on their way to becoming final outputs of the organization. The value chain is a series (chain) of activities that includes inbound logistics, warehouse and storage, production, finished product storage, outbound logistics, marketing and sales, and customer service.

Organizations use information systems to support organizational goals. Before deciding on an information system for an organization, managers should identify the firm's critical success factors, which must be supported by the system. Because information systems typically are designed to improve productivity, methods for measuring the system's impact on productivity should be devised.

PRINCIPLE • Because information systems are so important, businesses need to be sure that improved or completely new systems help lower costs, increase profits, improve service, or achieve a competitive advantage.

The extent to which technology is used throughout an organization can be a function of technology diffusion, infusion, and acceptance. Technology diffusion is a measure of how widely technology is spread throughout an organization. Technology infusion is the extent to which technology is deeply integrated into an area or department. The technology acceptance model (TAM) investigates factors, such as perceived usefulness of the technology, ease of its use, quality of the information system, and the degree to which the organization supports the use of the information system, to predict information system use and performance.

•   •   •

A competitive advantage is a significant and (ideally) long-term benefit to a company over its competition. A five-force model covers factors that lead firms to seek competitive advantage: rivalry among existing competitors, the threat of new market entrants, the threat of substitute products and services, the bargaining power of buyers, and the bargaining power of suppliers. Three strategies to attain competitive advantage are altering the industry structure, creating new products and services, and improving existing product lines and services.

The ability of an information system to provide or maintain competitive advantage should also be determined. Competitive advantage is usually embodied either in a product or service that has the most added value to consumers and that is unavailable from the competition, or in an internal system that delivers benefits to a firm that are not enjoyed by its competition.

Developing information systems that measure and control productivity is a key element for most organizations. One measure of IS value is return on investment (ROI). This measure investigates the additional profits or benefits that are generated as a percentage of the investment in information systems technology. Earnings growth, market share, customer satisfaction, and total cost of ownership (TCO) can also be useful measures.

PRINCIPLE • Information systems personnel are the key to unlocking the potential of any new or modified system.

IS personnel typically work in an information systems department that employs system operators, systems analysts, computer programmers, and other information systems personnel. They may also work in other functional departments or areas in a support capacity. In addition to technical skills, IS personnel also need written and verbal communication skills, an understanding of organizations and the way they operate, and the ability to work with people (system users). In general, information systems personnel must maintain the broadest perspective on organizational goals.

System operators run and maintain IS equipment, such as mainframe systems, networks, tape drives, disk devices, printers, and so on. Systems analysts help users determine what outputs they need from the system and construct the plans to develop the programs that produce these outputs. Systems analysts then work with programmers to ensure that the appropriate programs are purchased, modified from existing programs, or developed. The major responsibility of a computer programmer is to use the plans developed by the systems analyst to develop or adapt one or more computer programs that produce the desired outputs.

The overall role of the chief information officer (CIO) is to use the IS department's equipment and personnel to help the organization attain its goals. LAN administrators set up and manage network hardware, software, and security processes. There is also an increasing need for trained personnel to set up and manage a company's Internet presence, including Internet strategists and administrators, Internet systems developers, Internet programmers, and Web site operators.

In addition to working for an IS department in an organization, information systems personnel can work for one of the large IS consulting firms, such as Andersen Consulting, EDS, and others. Another IS career opportunity is to be employed by a hardware or software vendor developing or selling products.

● REVIEW QUESTIONS

1. What is an information system? What are some of the ways information systems are changing our lives?
2. How would you distinguish data and information? Information and knowledge?
3. Identify at least six characteristics of valuable information.
4. What is a computer-based information system? What are its components?
5. What is a business's technology infrastructure?
6. What are the most common types of computer-based information systems used in business organizations today? Give an example of each.
7. What are some of the benefits organizations seek to achieve through using information systems?
8. What is a value-added process? Give several examples.
9. What are some general strategies organizations employ to achieve competitive advantage?
10. Define *productivity*. Why is it difficult to measure the impact that investments in information systems have on productivity?
11. Briefly define *technology diffusion*, *technology infusion*, and *technology acceptance model*.
12. What is the total cost of ownership?
13. What is an information systems unit?

● DISCUSSION QUESTIONS

1. Describe the "ideal" automated car license plate renewal system for your state. Describe the input, processing, output, and feedback associated with this system.
2. How can useful information vary widely from the quality attributes of valuable information?
3. Discuss the potential use of virtual reality to enhance the learning experience for new automobile drivers. How might such a system operate? What are the benefits and potential drawbacks of such a system?
4. Discuss how information systems are linked to the business objectives of an organization.
5. You have been hired to work in the IS area of a manufacturing company that is starting to use the Internet to order parts from its suppliers and offer sales and support to its customers. What types of Internet positions would you expect to see at the company?
6. You have been asked to participate in the preparation of your company's strategic plan. Specifically, your task is to analyze the competitive marketplace using Porter's five-force model. Prepare your analysis, using your knowledge of a business you have worked for or have an interest in working for.
7. Based on the analysis you performed in discussion question 6, what possible strategies could your organization adopt to address these challenges? What role could information systems play in these strategies? Use Porter's strategies as a guide.
8. Imagine that you are the CIO for a large multi-national company. Outline a few of your key responsibilities.
9. What sort of information systems position would be most appealing to you—working as a member of an IS organization, working as a consultant, or working for an information systems hardware or software vendor? Why?

● PROBLEM-SOLVING EXERCISES

1. Prepare a data disk and a backup disk for the problem-solving exercises and other computer-based assignments you will complete in this class. Create one directory for each chapter in the textbook (you should have nine directories). As you work through the problem-solving exercises and complete other work using the computer, save your assignments for each chapter in the appropriate directory. On the label of each disk be sure to include your name, course, and section. On one disk write "Working Copy"; on the other write "Backup."

2. Search through several business magazines (*Business Week, Computerworld, PC Week,* and so on) for a recent article that discusses the use of information technology to deliver significant business benefits to an organization. Now use other resources to find additional information about the same organization (*Reader's Guide to Periodical Literature,* on-line search capabilities available at your school's library, the company's public relations department, Web pages on the Internet, and so on). Use word processing software to prepare a one-page summary of the different resources you tried and their ease of use and effectiveness.

3. Create a simple spreadsheet to help manage your "to do" list of tasks.
   a. For each item on your "to do" list, define the date the task must be completed, briefly describe the task, and indicate whether the task is Urgent (important and it must be done by the due date), Pressing (important, but the due date can slide a day or two), or Trivial (not important and the due date is not critical). Your spreadsheet might look like the following:

   Date Due   Task   Importance

   b. Now enter the items for your "to do" list. Use the features of the spreadsheet software to sort all tasks by "Importance"; under "Importance" sort by "Date Due."
   c. As you complete tasks, delete them from the spreadsheet. As you identify new tasks to be done, add them to the spreadsheet. Always sort the list by importance and due date.

 4. A new IS project has been proposed that will produce not only cost savings but also an increase in revenue. The initial costs to establish the system are estimated at $500,000. The rest of the cash flow data is presented in the following table.

	Year 1	Year 2	Year 3	Year 4	Year 5
Increased Revenue	$0	$100	$150	$200	$250
Cost Savings	0	50	50	50	50
Depreciation	0	75	75	75	75
Initial Expense	500				

(All amounts in thousands)

   a. Using your spreadsheet program, calculate the return on investment for this project. Assume that the cost of capital is 7 percent.
   b. How would the rate of return change if the project were able to deliver $50,000 in additional revenue and generate cost savings of $25,000 in the first year?

● TEAM ACTIVITIES

1. Before you can do a team activity, you need a team! The class members may self-select their teams, or the instructor may assign members to groups. Once your group has been formed, meet and introduce yourselves to each other. You will need to find out the first name, hometown, major, and e-mail address and phone number of each member. Find out one interesting fact about each member of your team, as well. Come up with a name for your team.

   With the other members of your group, use word processing software to write a one-page summary of what your team hopes to gain from this course and what you are willing to do to accomplish these goals. Send the report to your instructor via e-mail.

2. Have your team interview a company, university, or governmental agency that is quality conscious. Write a brief report that describes the quality efforts taken and the impact of these quality initiatives on how successful the organization has been in achieving its goals.

● WEB EXERCISES

1. Throughout this book, you will see how the Internet provides a vast amount of information to individuals and organizations. We will stress the World Wide Web, or simply the Web, which is an important part of the Internet. Most large universities and organizations have an address on the Internet, called a Web site or home page. The address of the Web site for this publisher is http://www.course.com. You can gain access to the Internet through a browser, such as Internet Explorer or Netscape. Using an Internet browser, go to the Web site for this publisher. What did you find? Try to obtain information on this book. You may be asked to develop a report or send an e-mail message to your instructor about what you found.

2. This book emphasizes the importance of information. You can get information from the Internet by going to a specific address, such as http://www.ibm.com, http://www.whitehouse.gov, or http://www.fsu.edu. These addresses will give you access to the home pages of IBM Corporation, the White House, and Florida State University, respectively. Note that "com" is used for businesses or commercial operations, "gov" is used for governmental offices, and "edu" is used for educational institutions. Another approach is to use a search engine. A search engine is a Web site that allows you to enter key words or phrases to find information. There are also lists or menus that can be used. The search engine returns other

Web sites (hits) that correspond to a search request. Yahoo!, developed by two Tulane University students, was one of the first search engines on the Internet. Using Yahoo! at http://www.yahoo.com,

search for information about a company or topic discussed in Chapter 1. You may be asked to develop a report or send an e-mail message to your instructor about what you found.

● CASES

 Coors Ceramics Revamps Information Systems

Coors Ceramics was spun off from the Adolph Coors Company in December 1992. Today the company is one of the leading suppliers of ceramic materials and components to the semiconductor and laser industries and has developed a worldwide reputation for quality and precision.

Coors's old information systems took as long as two days to process new orders. Because of delays and inaccuracies in processing, there was no way a salesperson could track the exact status of a particular customer's order. With 1,500 orders coming in monthly, that was a huge problem. To compensate for the processing delays, Coors would produce more components than it had received orders for so that it could build up inventory to meet customers' desired delivery dates. Although it did help meet customer demand, this approach raised inventory levels, production costs, and overhead costs. Customer delivery was also a problem. The old system could track shipments only on a weekly basis. If a customer wanted an order on Monday, and Coors shipped it by the following Saturday, the system logged that order as being on time. When customers called to complain, the salesperson would get no valid data from the system other than an incorrect "shipped on time" report.

It was clear that improvements were needed; however, before investing in new information systems, Coors defined three key business goals that the new systems must achieve: First, Coors had to increase customer satisfaction. Salespeople were under tremendous pressure to get information for customers—which in turn prevented them from developing new orders and selling product. Second, Coors wanted to reduce lead times. If work-in-progress, inventory, and delivery schedules could be reduced, then Coors could fulfill more customer orders. Third, Coors needed to reduce operating costs.

Coors's approach to meeting these goals required streamlining many fundamental business work processes. The project team focused first on how to meet the needs of the customers before they thought about how to update their outdated information

systems. This rethinking often required challenging fundamental assumptions about how the business should operate. Once the work processes were redesigned, the project team implemented an integrated set of information systems. These new systems automated the work processes associated with acquiring raw materials, transforming raw materials into top-quality products, and delivering them to customers within the shortest possible time.

The project proved to be highly successful. Since the systems were installed, Coors's product cycle has been cut from an average of 12 weeks to 8 weeks, and on-time shipments have improved to over 95 percent. Coors salespeople can now be confident that "shipped on time" means the order was delivered on time—not just that it shipped within a seven-day period.

The new information systems have also improved business decision making. Each morning, the general manager of sales and marketing meets with key people from manufacturing, engineering, and sales. They review the previous day's sales and requests for new products. They discuss how things are going and can check on the current status because everything that happened as of that morning is already in Coors's information systems, ready for decision making.

Discussion Questions

1. How has implementation of an integrated set of information systems enabled Coors to meet customer needs more effectively?

2. Did this system meet all three key business goals for new systems at Coors? Why or why not?

Critical Thinking Questions

3. Identify three key decisions that must be made at the business review meeting each morning. Identify six questions that are likely to be asked by the general manager at the morning business review meeting.

4. What additional features or benefits might you want this basic system to deliver?

Sources: Adapted from the "Investor Relations" and "Products and Services" portions of the Coors Tek Web site at http://www.coorstek.com, accessed February 9, 1999; and the "About QAD's Applications" portion of the QAD Web site at http://www.qad.com/product, accessed February 9, 2000.

2 Total Cost of Ownership

For decades, companies have been investing millions of dollars in information systems in an attempt to increase revenues, reduce costs, or both. Finally, with the booming stock market and increased productivity, it seems that this investment in information systems has paid off. But individual companies can find it very difficult to measure the impact of their information systems investment.

Using traditional financial measures, many companies have attempted and failed to measure the impact of information systems investments on profitability. Doctoral students doing research on the impact on information systems investments have also failed to show a significant relationship between IS investment and results. Having to deal with so many other factors, such as general economic conditions, the competition's activities, and corporate wide programs in such areas as marketing, has made it almost impossible to determine whether an IS investment or some other factor has contributed to increased profits. Even with these difficulties, most firms agree that measuring the impact of any investment is important. As a result, a number of consulting organizations are proposing other measures, including Gartner Group's total cost of ownership (TCO) model.

The TCO model investigates the total cost of owning computer equipment or systems. The model attempts to include all costs, including direct and indirect costs, of owning a computer device. In addition, TCO can include what is called a "futz factor" to take into account other computing activities, such as nonbusiness uses of a computer system. The model was first applied to desktop computers. Since then, it has been applied to software, networks, telecommunications, larger mainframe computers, handheld organizers, and other computer devices. According to Bill Kirwin, vice president of the Gartner Group, "We use TCO to look at the overall impact of the implementation. Cost is the numerator. The denominator might be service, customer satisfaction, quality levels or productivity, for example. . . . TCO can be very revealing. It might show that certain departments are too autonomous or that the IT department is being poorly managed."

A number of companies have successfully used the TCO approach. A pharmaceuticals company, for example, was able to use the TCO approach to save about 30 percent of what it paid for basic computer services. In addition, some companies offer products and services that include the TCO approach. IBM, for example, offers TCO Baseline Analysis, TCO Advanced Analysis, and Universal Management Tools Validation Pilot. These services can cost from about $3,000 to more than $7,500.

Although TCO has been widely used and accepted, some people believe that the approach is not worth the effort. According to a spokesman for the (U.K.) National Computing Centre (NCC), "We can't take TCO as a reliable guide to realistic expenditures. We've found from our members that it's a meaningless figure."

Discussion Questions

1. What is the TCO model?
2. What are the disadvantages of the TCO approach?

Critical Thinking Questions

3. Describe how TCO can be used to help a company determine its total computer-related expenditures.
4. If you had to measure the effectiveness of a complete computer system, what measures would you use?

Sources: Adapted from Jacqueline Emich, "Total Cost of Ownership," *Computerworld*, December 20, 1999, p. 52; and Simon Goodley, "Forget TCO," *Computing*, February 3, 2000, p. 1.

3 UPS Turns to Technology for a Strategic Advantage

People often claim that the saying "You can't teach an old dog new tricks" applies to old, traditional companies. It is often said that it takes a new, upstart company to take full advantage of changing times and the Internet age. Although this may be true for some old, traditional companies, it is not true for one of the oldest and most respected companies in America—UPS.

UPS began in the early 1900s by moving a limited number of packages in the Seattle area. The first vehicles it used were Model T Fords. With its coffee-brown uniforms and vehicles, UPS has not only survived for almost a century, it has thrived. Company income for 1999 exceeded $2 billion on revenues of $27 billion. In recent years, the company has seen annual growth rates that exceed 20 percent. Today, the company moves 13 million packages daily. The company's ability to change and adapt is a key reason for its continued success for almost 100 years. According to Chief Executive James Kelly, "We have to be more adaptable. We have to know when to add and when to subtract." Clearly, Kelly, who started with UPS as a part-time driver, knows how to compute the way to success for UPS. And that way is through technology. The success has reached to all levels of the company. After a recent initial public offering, a number of long-term UPS truck drivers and other employees became instant millionaires as a result of their stock options.

This long-term success story is a result of the staggering investment UPS has made in computer technology. Over the last ten years, UPS has invested about $11 billion in computer systems and related equipment. In the past, UPS could be categorized as a trucking company that used technology. Today, UPS thinks of itself as a technology company that uses trucks. All aspects of its business have been automated, with the Internet playing a central part in its long-term business strategy. Each driver, for example, uses an electronic tracking device, called a Delivery Information Acquisition Device (DIAD). Using this device, a company can track its shipment even before the UPS truck leaves its driveway. But UPS does much more than deliver packages. For example, UPS delivers Gateway computers to customers with a cash-on-delivery system, in which UPS collects payments from customers receiving Gateway computers and deposits the payments directly into Gateway bank accounts.

However, UPS hasn't always had an easy or successful time. A few years ago, the Teamsters walkout cost UPS about $200 million in lost sales. For many inside UPS, this was a wake-up call to be even more aggressive in using technology to propel the company into the next century. According to one observer of the impact of the Teamsters walkout on UPS, "You never want to wound a tiger. You want to kill it, because if you wound it, it only becomes more ferocious." From all accounts, UPS is becoming more ferocious in its use of technology to increase profits and give it a long-term competitive advantage.

Discussion Questions

1. Describe the history and success of UPS.
2. How was UPS able to use technology to its competitive advantage?

Critical Thinking Questions

3. How could the lessons of UPS be used in other industries?
4. If you were the CEO of another shipping company, such as FedEx, what would you do to keep your company competitive over UPS?

Sources: Adapted from Kelly Barron, "UPS Company of the Year," *Forbes*, January 10, 2000, p. 79; and Beth Bacheldor, "Ford-UPS Alliance," *Information Week*, February 7, 2000.

● NOTES

Source for the opening vignette on page 3: Adapted from Thomas Hoffman, "Finding a Rich Niche," *Computerworld*, February 8, 1999, p. 44; and "Farmers Trim Paper Trail," *Future Banker*, September 6, 1999, p. 24.

1. Stephen H. Wildstrom, "Age of the E-pliance," *Business Week*, January 17, 2000, p. 20.
2. John Madden, "KPMG Sharing Knowledge," *PC Week*, August 9, 1999, p. 51.
3. Clinton Wilder, "Business Booms for Specialized Web Marketplaces," *Information Week*, February 7, 2000, p. 43.
4. Peter Keen, "IT's Value in the Chain," *Computerworld*, February 14, 2000, p. 48.
5. Christoph Loch and Berndao Huberman, "A Punctuated-Equilibrium Model of Technology Diffusion," *Management Science*, February 1999, p. 160.
6. Curtis Armstrong and V. Sambamurthy, "Information Technology Assimilation in Firms," *Information Systems Research*, December 1999, p. 304.
7. Ritu Agarwal and Jayesh Prasad, "Are Individual Differences Germane to the Acceptance of New Information Technology," *Decision Sciences*, Spring 1999, p. 361.
8. Kwon et al., "A Test of the Technology Acceptance Model," *Proceedings of the Hawaii International Conference on System Sciences*, January 4–7, 2000.
9. Henry Lucas and V. K. Spitler, "Technology Use and Performance," *Decision Sciences*, Spring 1999, p. 291.
10. Hellen Pukszta, "Don't Split IT Strategy from Business Strategy," *Computerworld*, January 11, 1999, p. 35.
11. M. Porter and V. Millar, "How Information Systems Give You Competitive Advantage," *Journal of Business Strategy*, Winter 1985. See also M. Porter, *Competitive Advantage* (New York: Free Press, 1985).
12. Paul Judge, "One Big Happy Family—But for How Long," *Business Week*, October 25, 1999, p. 148.
13. Eva Sohlman, "The New Remote Control," *ABCNews Online*, April 18, 2000.
14. "The State of IT: 2010," *Computerworld*, January 17, 2000, p. 50.
15. Eileen Birge, "How to Measure IT Performance," *Beyond Computing*, January 1999, p. 32.
16. Matt Hamblen, "Handhelds Help Boeing Boost Quality Inspections," *Computerworld*, November 8, 1999, p. 38.
17. Jacqueline Emigh, "Total Cost of Ownership," *Computerworld*, December 20, 1999. p. 52.
18. Peter Sappal, "E-Commerce Recruitment," *The Wall Street Journal*, January 25, 2000, p. B20.
19. Vicky Uhland, "Tech Jobs Abound This Year: Top Positions Involve E-Commerce," *Denver Rocky Mountain News*, February 27, 2000, p. J1.
20. Sandy Graham, "Systest Goes after Digital Bugs," *Denver Rocky Mountain News*, February 27, 2000, p. 3G.

PART II

Technology

CHAPTER 2

Hardware and Software

Software rates among the most poorly constructed, unreliable, and least maintainable technological artifacts invented by man. To be fair, software also shares credit for the most spellbinding advances of the 20th century. In today's world, banks, hospitals, and space missions would be inconceivable without it. The challenge of the next century will be to bring software quality to the same level as we expect from cars, televisions, and other relatively dependable hunks of hardware.

— Paul Strassman, former CIO for Xerox Corporation and the CIO for the U.S. Defense Department who now heads a private consulting firm

Principles

Information system users must work closely with information system professionals to define business needs, evaluate options, and select the hardware and software that provide a cost-effective solution to those needs.

Organizations do not develop proprietary application software unless doing so will meet a compelling business need that can provide a competitive advantage.

End users and IS professionals use a programming language whose functional characteristics are appropriate to the task at hand.

Learning Objectives

- *Identify and discuss the role of the essential hardware components of a computer system.*
- *List and describe popular classes of computer systems and discuss the role of each.*
- *Outline the role of the operating system and discuss how operating systems have evolved over time.*
- *Identify and briefly describe the functions of the two basic kinds of software.*

- *Discuss how application software can support personal, workgroup, and enterprise business objectives.*
- *Identify three basic approaches to developing application software.*

- *Outline the overall evolution of programming languages and highlight the differences between programming languages used by end users and IS professionals.*

Home Depot

*Putting Hardware and Software
Together to Meet Customers' Needs*

Home Depot is the world's largest home improvement retailer, with over 1,900 stores planned for 2004. The company's stores cater to do-it-yourselfers, as well as home improvement, construction, and building maintenance professionals. Home Depot's Expo Design Centers are stand-alone stores that include interior design showrooms staffed with employees who can plan and coordinate home improvement projects for customers. A one-stop full-service interior design center, each Expo store is filled with showroom after showroom of exciting ideas, hundreds of full-size displays, and lots of friendly, knowledgeable professionals who can help customers with any project, big or small.

Expo Design Center customers have their own project designer to work with them from the very beginning, assisting in the selection of products, styles, and colors that best suit their lifestyles. Once the details are settled, a project superintendent can take over and work with the contractor and installers to complete the job. Until recently, customer planning and ordering in the Expo Design Centers was done manually—a complicated nightmare for both customers and Home Depot staff, with workers flipping through paper catalogs and then entering special purchase orders into Home Depot's existing ordering system. Meanwhile, interior designers and project coordinators who supervise installations had to use file folders and pieces of paper to track the many details.

Since the key to success of the Expo Design Centers is personal one-on-one attention and follow-up with customers, the current system clearly wasn't acceptable. Home Depot recognized the need for improved computer hardware and software to provide this high level of customer support. The company bought powerful computers, linked them in a network, and developed software to enable employees to view products, create purchase orders for customers, and track the progress of projects. Now products can be viewed easily and ordered directly with the new system. Room measurements and installation schedules are also entered into the tracking system, so all the details needed to track and follow up on home improvement projects are at the fingertips of the Home Depot project superintendents. The new software uses "storyboards" that lead users through a task step by step, thus minimizing training requirements for employees, most of whom have construction and trade backgrounds.

As you read this chapter, consider the following:

- What are the various kinds of hardware and software and how are they used?

- Where can software be obtained, and what are the pros and cons of each approach?

Appropriate use of information systems can reap huge benefits in business, as Home Depot did to provide a high level of customer service and convenience. Employing information systems and providing additional processing capabilities can increase employee productivity, expand business opportunities, and allow for more flexibility. Information system users must work closely with information system professionals to define business needs, evaluate options, and select the hardware and software that provide a cost-effective solution to those needs. To meet this responsibility, you must understand the basic concepts of hardware and software presented in this chapter.

OVERVIEW OF HARDWARE

In building a car, manufacturers try to match the intended use of the vehicle to its components. Racing cars require special types of engines, transmissions, and tires. The selection of a transmission for a racing car, then, requires not only consideration of how much of the engine's power can be delivered to the wheels (efficiency and effectiveness) but also how expensive the transmission is (cost), how reliable it is (control), and how many gears it has (complexity). Similarly, organizations assemble the hardware components of a computer system so that they are effective, efficient, and well suited to the tasks that need to be performed. Users and IS professionals often need to make these decisions together, combining their knowledge of systems and business functions as well as forecasting their future needs.

Consider the role of an end user in selecting the hardware used by eBay, one of the most popular on-line auction Web sites.[1] Meg Whitman was hired as CEO in part to help get a handle on the hardware and technological needs of the fast-growing company even though her technical savvy was that of a casual user. She literally lived in the operations center through most of the summer after a catastrophic 22-hour outage convinced her that she was battling for e-Bay's life. Ultimately, eBay came out of the experience with a technically astute CEO, a beefed-up IS department, a strong new IS executive, and a commitment to build a powerful and reliable e-commerce system using state-of-the-art hardware components.

Like other companies facing explosive growth, eBay is constantly striving to keep its hardware working and in line with its business. The goal is often to upgrade hardware to keep pace with ever-increasing business. In eBay's case, the solution was to get a person who knew how to match hardware with the growing demands of the business.

To stay successful, eBay must ensure that its hardware system can support 7.7 million registered users. The site has more than 1.7 million visitors each day.

Hardware Components

Computer system hardware components include devices that perform the functions of input, processing, data storage, and output (Figure 2.1). To understand how these hardware devices work together, consider an analogy from a paper-based office environment. Imagine a one-room office occupied by a single individual. The human (the central processing unit) organizes and manipulates data. The person's mind (primary storage) and the desk occupied by the human (secondary storage) are places to temporarily store data. Filing cabinets

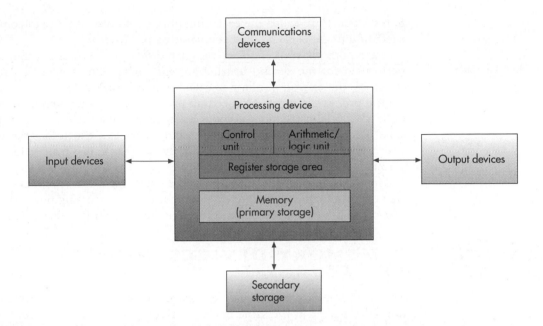

FIGURE 2.1

Computer System Components

These components include input devices, output devices, communications devices, primary and secondary storage devices, and the central processing unit (CPU). The control unit and the arithmetic/logic unit (ALU) constitute the CPU.

central processing unit (CPU)

the part of the computer that consists of two primary elements: the arithmetic/logic unit and the control unit

arithmetic/logic unit (ALU)

portion of the CPU that performs mathematical calculations and makes logical comparisons

control unit

part of the CPU that sequentially accesses program instructions, decodes them, and coordinates the flow of data in and out of the ALU, primary storage, and even secondary storage and various output devices

instruction time (I-time)

the time it takes to perform the fetch-instruction and decode-instruction steps of the instruction phase

provide additional data storage (secondary storage). In this analogy, the incoming and outgoing mail trays can be understood as sources of new data (input) or as a place to put the processed paperwork (output).

The ability to process (organize and manipulate) data is a critical aspect of a computer system, in which processing is accomplished by an interplay between one or more of the central processing units and primary storage. Each **central processing unit (CPU)** consists of two primary elements: the arithmetic/logic unit and the control unit. The **arithmetic/logic unit (ALU)** performs mathematical calculations and makes logical comparisons. The **control unit** sequentially accesses program instructions, decodes them, and coordinates the flow of data in and out of the ALU, primary storage, and even secondary storage and various output devices.

Primary memory, which holds program instructions and data, is closely associated with the CPU. To understand the function of processing and the interplay between the CPU and memory, let's examine the way a typical computer executes a program instruction.

Hardware Components in Action

The execution of any machine-level instruction involves two phases: the instruction phase and the execution phase. During the instruction phase, the following takes place:

- *Step 1: Fetch instruction.* The instruction to be executed is accessed from memory by the control unit.
- *Step 2: Decode instruction.* The instruction is decoded so the central processor can understand what is to be done, relevant data is moved from memory, and the location of the next instruction is identified.

Steps 1 and 2 are called the instruction phase, and the time it takes to perform this phase is called the **instruction time (I-time)**.

The second phase is the execution phase. During the execution phase, the following steps are performed:

- *Step 3: Execute the instruction.* The ALU does what it is instructed to do. This could involve making either an arithmetic computation or a logical comparison.
- *Step 4: Store results.* The results are stored in memory.

execution time (E-time)

the time it takes to execute an instruction and store the results

machine cycle

the instruction phase followed by the execution phase

Steps 3 and 4 are called the execution phase. The time it takes to complete the execution phase is called the **execution time (E-time)**.

After both phases have been completed for one instruction, they are again performed for the second instruction, and so on. The instruction phase followed by the execution phase is called a **machine cycle** (Figure 2.2). Some central processing units can speed up processing by using *pipelining*, where the CPU gets one instruction, decodes another, and executes a third at the same time. The Pentium processor, for example, uses two execution unit pipelines, which gives the processing unit the ability to execute two instructions in a single machine cycle.

Now that you have learned about the basic hardware components and the way they function, we turn to an examination of processing power, speed, and capacity. These three attributes determine the capabilities of a hardware device.

PROCESSING AND MEMORY DEVICES: POWER, SPEED, AND CAPACITY

The components responsible for processing—the CPU and memory—are housed together in the same box or cabinet, called the system unit. All other computer system devices, such as the monitor and keyboard, are linked either directly or indirectly into the system unit housing. As discussed previously, achieving information system objectives and organizational goals should be the primary consideration in selecting processing and memory devices. In this section, we investigate the characteristics of these important devices.

Processing Characteristics and Functions

Because efficient processing and timely output is important, organizations use a variety of measures to gauge processing speed. These measures include the time it takes to complete a machine cycle, clock speed, and others.

Machine Cycle Time

The time it takes to execute the instruction phase and the execution phase is the machine cycle time. Machine cycle time is one measure of processing speed.

clock speed

a series of electronic pulses, produced at a predetermined rate, that affect machine cycle time

Clock Speed

Each CPU produces a series of electronic pulses at a predetermined rate, called the **clock speed**, which affects machine cycle time. The control unit executes an instruction in accordance with the electronic cycle, or pulses of the CPU "clock." Each instruction takes at least the same amount of time as the interval

FIGURE 2.2

Execution of an Instruction

In the instruction phase, the computer's control unit fetches the instruction to be executed from memory (1). Then the instruction is decoded so the central processor can understand what is to be done (2). In the execution phase, the ALU does what it is instructed to do, making either an arithmetic computation or a logical comparison (3). Then the results are stored in memory (4). The instruction and execution phases together make up one machine cycle.

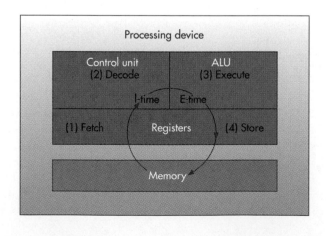

between pulses. The shorter the interval between pulses, the faster each instruction can be executed.

Clock speed is often measured in megahertz (MHz), or millions of cycles per second. The clock speed for personal computers can range from 200 MHz to 700 MHz or more.[2] An AMD Athlon chip can reach speeds of 1,000 MHz.[3]

Physical Characteristics of the CPU

CPU speed is also limited by physical constraints. Most CPUs are collections of digital circuits imprinted on silicon wafers, or chips, each no bigger than the tip of a pencil eraser. To turn a digital circuit within the CPU on or off, electrical current must flow through a medium (usually silicon) from point A to point B. The speed at which it travels between points can be increased by either reducing the distance between the points or reducing the resistance of the medium to the electrical current.

Reducing the distance between points has resulted in ever-smaller chips, with the circuits packed closer together. In the 1960s, shortly after patenting the integrated circuit, Gordon Moore, former chairman of the board of Intel (the largest microprocessor chip maker), formulated what is now known as **Moore's Law**. This hypothesis states that transistor (the microscopic on/off switches, or the microprocessor's brain cells) densities on a single chip will double every 18 months. Moore's Law has held up amazingly well over the years. For Moore's law to continue to hold up over the long run, however, improved chip-fabrication techniques must be developed, and research into optical and laser chips must be successful.[4] For the short term, Moore's Law remains in effect. In addition to increased speeds, Moore's law has had an impact on costs and overall system performance. As seen in Figure 2.3, the number of megabytes per dollar and the number of transistors on a chip also continue to climb.

Another substitute material for silicon chips is superconductive metal. **Superconductivity** is a property of certain metals that allows current to flow with minimal electrical resistance. Traditional silicon chips create some electrical resistance that slows processing. Chips built from less resistant superconductive metals offer increases in processing speed.

Moore's Law

a hypothesis that states that transistor densities on a single chip will double every 18 months

superconductivity

a property of certain metals that allows current to flow with minimal electrical resistance

FIGURE 2.3

Moore's Law Affects More than Processor Speeds
(Source: Reprinted with permission from "Moore's Law Will Continue to Drive Computing," *PC Magazine*, June 22, 1999, p. 146. Copyright © 1999, Ziff Davis Publishing Holdings, Inc. All rights reserved.)
*Data from International Data Corp. for 2002; Disk/Trend Inc. for all other years.
†Data from *PC Magazine*, August 1988 (Compaq Deskpro 386S).
‡Projected data.

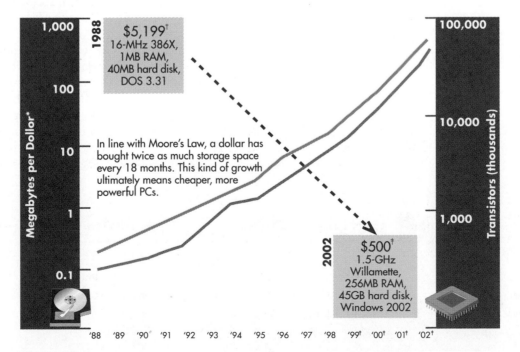

Memory Characteristics and Functions

Located physically close to the CPU (to decrease access time), memory provides the CPU with a working storage area for program instructions and data. The chief feature of memory is that it rapidly provides the data and instructions to the CPU.

Storage Capacity

Like the CPU, memory devices contain thousands of circuits imprinted on a silicon chip. Each circuit is either conducting electrical current (on) or not (off). By representing data as a combination of on or off circuit states, the data is stored in memory. Usually eight bits are used to represent a character, such as the letter *A*. Eight bits together form a **byte**. Table 2.1 below summarizes commonly used measurements. Storage capacity is measured in bytes, abbreviated with the letter *B*, with one byte usually equal to one character.

Types of Memory

There are several forms of memory, as shown in Figure 2.4. Instructions or data can be temporarily stored in **random access memory (RAM)**. RAM is temporary and volatile—RAM chips lose their contents if the current is turned off or disrupted (as in a power surge, brownout, or electrical noise generated by lightning or nearby machines). RAM chips are mounted directly on the computer's main circuit board or in chips mounted on peripheral cards that plug into the computer's main circuit board. These RAM chips consist of millions of switches that are sensitive to changes in electric current.

RAM comes in many different varieties. The mainstream type of RAM is *extended data out,* or *EDO, RAM*. Another kind of RAM memory is called *SDRAM,* or *synchronous DRAM,* which has the advantage of a faster transfer speed between the microprocessor and the memory. *Dynamic RAM (DRAM)* chips need high or low voltages applied at regular intervals—every two milliseconds (two one-thousandths of a second) or so—if they are not to lose their information.

Another type of memory, **ROM**, an acronym for **read-only memory**, is usually nonvolatile. In ROM, the combination of circuit states is fixed, and therefore its contents are not lost if the power is removed. ROM provides permanent storage for data and instructions that do not change, such as programs and data from the computer manufacturer.

There are other types of nonvolatile memory as well. Programmable read-only memory (PROM) is a type in which the desired data and instructions—and hence the desired circuit state combination—must first be programmed into the memory chip. After that, PROM behaves like ROM. PROM chips are used where the CPU's data and instructions do not change, but the application is so specialized or unique that custom manufacturing of a true ROM chip would be too costly. A common use of PROM chips is for storing the instructions for popular video games, such as those from Nintendo and Sega. Game instructions are

byte

eight bits together that represent a single character of data

random access memory (RAM)

a form of memory in which instructions or data can be temporarily stored

read-only memory (ROM)

a nonvolatile form of memory

TABLE 2.1

Name	Abbreviation	Exact Number of Bytes	Approximate Number of Bytes
Byte	B	8 bits	One
Kilobyte	KB	1,024 bytes	One thousand
Megabyte	MB	1,024 × 1,024 bytes	One million
Gigabyte	GB	1,024 × 1,024 × 1,024 bytes	One billion
Terabyte	TB	1,024 × 1,024 × 1,024 × 1,024 bytes	One trillion

FIGURE 2.4

Basic Types of Memory Chips

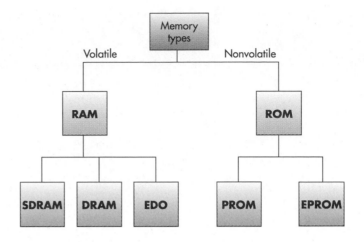

FIGURE 2.4

Basic Types of Memory Chips

programmed onto the PROM chips by the game manufacturer. Instructions and data can be programmed into a PROM chip only once. Erasable programmable read-only memory (EPROM) is similar to PROM except, as the name implies, the memory chip can be erased and reprogrammed.

Multiprocessing

multiprocessing

simultaneous execution of two or more instructions

coprocessor

part of the computer that speeds processing by executing specific types of instructions while the CPU works on another processing activity

There are a number of forms of **multiprocessing**, which involves the simultaneous execution of two or more instructions. One form of multiprocessing involves **coprocessors**. A coprocessor speeds processing by executing specific types of instructions while the CPU works on another processing activity. Coprocessors can be internal or external to the CPU and may have different clock speeds than the CPU. Each type of coprocessor best performs a specific function. For example, a math coprocessor chip can be used to speed mathematical calculations, and a graphics coprocessor chip decreases the time it takes to manipulate graphics.

Parallel Processing

parallel processing

a form of multiprocessing that speeds processing by linking several processors to operate at the same time, or in parallel

Another form of multiprocessing, called **parallel processing**, speeds processing by linking several processors to operate at the same time, or in parallel. The most frequent business uses for parallel processing are modeling, simulation, and analysis of large amounts of data. In today's marketplace, consumers demand quick response and customized service, so companies are gathering and reporting more information about their customers. Collecting and organizing the enormous amount of customer data is no easy task, but parallel processing can help companies organize data on existing consumer buying patterns and process them more quickly to build an effective marketing program. As a result, a company can gain a competitive advantage.

SECONDARY STORAGE AND INPUT AND OUTPUT DEVICES

secondary storage

devices that store larger amounts of data, instructions, and information more permanently than allowed with main memory; also called permanent storage

As we have seen, memory is an important factor in determining overall computer system power. However, memory provides only a small amount of storage area for the data and instructions the CPU requires for processing. Computer systems also need to store larger amounts of data, instructions, and information more permanently than main memory allows. **Secondary storage**, also called permanent storage, serves this purpose.

Compared with memory, secondary storage offers the advantages of nonvolatility, greater capacity, and greater economy. Most forms of secondary storage are considerably less expensive than memory (see Figure 2.5). Because of

Device	DAT tape	Hard drive	CD-RW	3.5" diskette	ZIP Plus Drive	RAM
Cost	$49.95	$230.00	$200.00	$.50	$100.00	$269.95
Storage	10,000MB	18GB	740MB	1.4MB	100MB–250MB	64MB
Cost per megabyte	$.005	$.08	$.27	$.35	$1.00	$4.21

FIGURE 2.5

Cost Comparison for Various Forms of Data Storage

All forms of secondary storage cost considerably less per megabyte of capacity than RAM, although they have slower access times. A diskette costs about 35 cents per megabyte, while RAM can cost around $4 per megabyte, 11 times more. (Source: Data from CompUSA Direct Catalog, February 1998.)

sequential access

retrieval method in which data must be accessed in the order in which it is stored

direct access

retrieval method in which data can be retrieved without the need to read and discard other data

sequential access storage device (SASD)

device used to sequentially access secondary storage data

direct access storage device (DASD)

device used for direct access of secondary storage data

magnetic tape

common secondary storage medium; Mylar film coated with iron oxide, with portions of the tape magnetized to represent bits

the electromechanical processes involved in using secondary storage, however, it is considerably slower than memory. The selection of secondary storage media and devices requires an understanding of their primary characteristics—access method, capacity, and portability.

Secondary Storage Access Methods

Data and information access can be either sequential or direct. **Sequential access** means that data must be accessed in the order in which it is stored. For example, inventory data stored sequentially may be stored by part number, such as 100, 101, 102, and so on. If you want to retrieve information on part number 125, you need to read and discard all the data relating to parts 001 through 124.

Direct access means that data can be retrieved directly, without the need to pass by other data in sequence. With direct access, it is possible to go directly to and access the needed data—say, part number 125—without reading through parts 001 through 124. For this reason, direct access is usually faster than sequential access. The devices used to sequentially access secondary storage data are simply called **sequential access storage devices (SASDs)**; those used for direct access are called **direct access storage devices (DASDs)**.

Secondary Storage Devices

The most common forms of secondary storage include magnetic tapes, magnetic disks, and optical disks. Some of these media (magnetic tape) allow only sequential access, while others (magnetic and optical disks) provide direct and sequential access. Figure 2.6 shows some different secondary storage media.

Magnetic Tapes

One common secondary storage medium is **magnetic tape**. Similar to the kind of tape found in audio- and videocassettes, magnetic tape is a Mylar film coated with iron oxide. Portions of the tape are magnetized to represent bits. Magnetic tape is a sequential access storage medium. Although access is slower, magnetic tape is usually less expensive than disk storage. In addition, magnetic tape is often used to back up disk drives and to store data off-site for recovery in case of disaster.

FIGURE 2.6

Types of Secondary Storage

Secondary storage devices such as magnetic tapes and disks, optical disks, and CD-ROMs are used to store data for easy retrieval at a later date.
(Source: Courtesy of Imation.)

magnetic disk

common secondary storage medium; bits are represented by magnetized areas

redundant array of independent/inexpensive disks (RAID)

method of storing data that allows the system to create a "reconstruction map" so that if a hard drive fails, it can rebuild lost data

storage area network (SAN)

technology that uses computer servers, distributed storage devices, and networks to tie the storage system together

FIGURE 2.7

Hard Disk

Hard disks give direct access to stored data. The read/write head can move directly to the location of a desired piece of data, dramatically reducing access times, as compared with magnetic tape.
(Source: Courtesy of Seagate)

Magnetic Disks

Magnetic disks are also coated with iron oxide; they can be thin steel platters (see Figure 2.7) or Mylar film (diskettes). As with magnetic tape, magnetic disks represent bits by small magnetized areas. When reading from or writing onto a disk, the disk's read/write head can go directly to the desired piece of data. Thus, the disk is a direct access storage medium and allows for fast data retrieval. For example, if a manager needs information on the credit history of a customer, the information can be obtained in a matter of seconds if the data is stored on a direct access storage device. Magnetic disk storage varies widely in capacity and portability.

RAID

Companies' data storage needs are expanding rapidly. Today's storage configurations routinely entail many hundreds of gigabytes. However, putting the company's data on-line involves a serious business risk—the loss of critical business data can put a corporation out of operation. The concern is that the most critical mechanical components inside a disk storage device—the disk drives, the fans, and other input/output devices—can break.

Organizations now require that their data storage devices be fault tolerant—the ability to continue with little or no loss of performance in the event of a failure of one or more key components. **Redundant array of independent/inexpensive disks (RAID)** is a method of storing data so that if a hard drive fails, the "lost data" on that drive can be rebuilt. With this approach, data is stored redundantly on different physical disk drives using a technique called stripping to evenly distribute the data.

SAN

Storage area network (SAN) uses computer servers, distributed storage devices, and networks to tie everything together. To increase the speed of storing and retrieving data, fiber-optic channels are often used. Although SAN technology is relatively new, a number of companies are using SAN to successfully and efficiently store critical data. Morgan Stanley Dean Witter Trust uses

SAN technology to eliminate paper mail.[5] When mail comes into Morgan Stanley's mailroom, it is opened and scanned into the computer and routed to the appropriate person for action. The scanned documents, however, require a huge amount of storage. Using SAN technology, Morgan Stanley can combine the resources of its network and main computer system to store 3 terabytes at its headquarters in New Jersey City and 1.5 terabytes of backup data in a center about 15 miles away.

Optical Disks

optical disk

a rigid disk of plastic onto which data is recorded by special lasers that physically burn pits in the disk

Another type of secondary storage medium is the **optical disk**. Similar in concept to a ROM chip, an optical disk is simply a rigid plastic disk onto which data is recorded by special lasers that physically burn pits in the disk. Data is directly accessed from the disk by an optical disk device, which operates much like a stereo's compact disk player. This optical disk device uses a low-power laser that measures the difference in reflected light caused by a pit (or lack thereof) on the disk. Each pit represents the binary digit 1; each unpitted area represents the binary digit 0. Once a master optical disk has been created, duplicates can be manufactured using techniques similar to those used to produce music CDs.

compact disk read-only memory (CD-ROM)

a common form of optical disk on which data, once it has been recorded, cannot be modified

A common form of optical disk is called **compact disk read-only memory (CD-ROM)**. Once data has been recorded on a CD-ROM, it cannot be modified—the disk is "read only." CD-writable (CD-W) disks allow data to be written once to a CD disk.[6] CD-rewritable (CD-RW) technology allows personal computer users to replace their diskettes with high-capacity CDs that can be written upon and edited over. The CD-RW disk can hold roughly 500 times the capacity of a 1.4-MB diskette.

Magneto-Optical Disk

magneto-optical disk

a hybrid between a magnetic disk and an optical disk

A **magneto-optical disk** is a type of disk drive that combines magnetic disk technologies with CD-ROM technologies. Like magnetic disks, MO disks can be read and written to. And like diskettes, they are removable. However, their storage capacity can be much greater than magnetic diskettes. In terms of data access speed, they are faster than diskettes but not as fast as hard disk drives. This type of disk uses a laser beam to change the molecular configuration of a magnetic substrate on the disk, which in turn creates visual spots. In conjunction with a photodetector, another laser beam reflects light off the disk and measures the size of the spots; the presence or absence of a spot indicates a bit. The disk can be erased by demagnetizing the substrate, which in turn removes the spots, allowing the process to begin again. Some MO drives can store more than a gigabyte on a single, removable disk. PowerMO by Olympus and the DynaMo by Fujitsu are two examples of magneto-optical devices.

digital video disk (DVD)

storage format used to store digital video or computer data

Digital Video Disk

The **digital video disk (DVD)** brings together the formerly separate worlds of home computing and home video. A DVD is a five-inch CD-ROM look-alike

FIGURE 2.8

Digital Video Disk and Player

DVDs look like CDs but have a much greater storage capacity and can transfer data at a much faster rate.

(Source: Courtesy of Creative Labs)

(Figure 2.8) with the ability to store about 135 minutes of digital video. When used to store video, the picture quality far surpasses anything seen on tape, cable, or standard broadcast TV—sharp detail, true color, no flicker, no snow. The sound is recorded in digital Dolby, creating clear "surround" effects by completely separating all the audio channels in a home theater. The DVD costs less to duplicate and ship than videocassettes, takes less shelf space, and delivers higher quality.

DVD can double as a computer storage disk and provide up to 17-GB capacity. The physical disks resemble CD-ROMs, only they are thinner, so DVD players can also read current CD-ROMs, but

The PC memory card is like a portable hard disk that fits into any Type II PC card slot and can store up to 1 gigabyte. (Source: Courtesy of Kingston Technology.)

expandable storage devices

storage that uses removable disk cartridges to provide additional storage capacity

FIGURE 2.9

Expandable Storage

Expandable storage drives allow users to add storage capacity by simply inserting a removable disk or cartridge. The disks can be used to back up hard disk data or to transfer large files to colleagues. (Source: Courtesy of Iomega.)

current CD-ROM players cannot read the DVDs. The access speed of a DVD drive is faster than the typical CD-ROM drive. DVD manufacturers include Sony, Philips, and Toshiba. These companies are also actively involved in making and improving standard CD-ROM drives. Newer DVD technology provides write-once disks and rewrite disks, often called DVD RAM.

Memory Cards

A group of computer manufacturers formed the Personal Computer Memory Card International Association (PCMCIA) to create standards for a peripheral device known as a PC memory card. These PC memory cards are credit-card-size devices that can be installed in many personal computers. To the rest of the system, the PC memory card functions as though it were a fixed hard disk drive. Although the cost per megabyte of storage is greater than for a traditional hard disk storage, these cards are less failure prone than hard disks, are portable, and are relatively easy to use. Software manufacturers often store the instructions for their program on a memory card for use with laptop computers.

Expandable Storage

Expandable storage devices use removable disk cartridges (Figure 2.9). When your storage needs increase, you can use more removable disk cartridges. The storage capacity can range from under 100 MB to several gigabytes per cartridge. In recent years, the access speed of expandable storage devices has increased so that some are about as fast as an internal disk drive. Expandable storage devices can be internal or external, and a few personal computers now include internal expandable storage devices as standard equipment. CD-RW drives by Hewlett-Packard, Iomega, and others can also be used for expandable storage. These expandable storage devices are ideal for backups of the critical data on your hard drive. They can hold at least 80 times as much data and operate five times faster than the existing 1.44-MB diskette drives. Although more expensive than fixed hard disks, removable disk cartridges combine hard disk storage capacity and diskette portability.

The overall trend in secondary storage is toward more direct-access methods, higher capacity, and increased portability. The business needs and needs of individual users should be considered when selecting a specific type of storage. In general, the ability to store large amounts of data and information and access it quickly can increase organizational effectiveness and efficiency. Table 2.2 lists the most common secondary storage devices and their capacities for easy reference.

Input Devices

A user's first experience with computers is usually through input and output devices. Through these devices—the gateways to the computer system—people provide data and instructions to the computer and receive results from it. Input and output devices are part of the overall user interface, which includes other hardware devices and software that allow humans to interact with a computer system.

As with other computer system components, the selection of input and output devices depends on the needs of the users and business objectives. For example, many restaurant chains use handheld input devices or computerized terminals that let waiters enter orders quickly and accurately. These systems have also cut costs by making inventory tracking more efficient and marketing to customers more effective.

TABLE 2.2

Comparison of Secondary
Storage Devices

Storage Device	Year First Introduced	Maximum Capacity
3.5-inch diskette	1987	1.44 MB
CD-ROM	1990	650 MB
Zip	1995	100–250 MB
DVD	1996	17 GB

Literally hundreds of devices can be used for data input, ranging from special-purpose devices used to capture specific types of data to more general-purpose input devices. We will now discuss several.

Personal Computer Input Devices

A keyboard and a computer mouse are the most common devices used for entry and input of data such as characters, text, and basic commands. Some companies are developing newer keyboards that are more comfortable, adjustable, and faster to use. These keyboards, such as the split keyboard by Microsoft and others, are designed to avoid wrist and hand injuries caused by hours of keyboarding. Using the same keyboard, you can enter sketches on the touchpad and text using the keys.

A computer mouse is used to point to and click on symbols, icons, menus, and commands on the screen. The computer responds by taking a number of actions, such as copying data into the computer system or opening files.

Voice-Recognition Devices

voice-recognition device

an input device that recognizes
human speech

Another type of input device can recognize human speech. Called **voice-recognition devices**, these tools use microphones and special software to record and convert the sound of the human voice into digital signals. Speech recognition can be used on a factory floor to allow equipment operators to give basic commands to machines while their hands perform other operations. Voice recognition can also be used by security systems to allow only authorized personnel into restricted areas. Voice recognition has also been used in medicine. Doctors at Quincy Medical Center in Massachusetts use voice recognition to create and update medical records for 32,000 patients per year.[7] The primary advantage is saving time. "I can handwrite a chart for a sprained ankle in four to five minutes," Dr. Daiz at Quincy Medical Center says. "I can dictate it in one minute and 10 seconds." In some cases, special dictation devices, such as the Dragon Naturally Speaking Mobile, can be used to capture memos and reports. Once captured, the speech can be easily uploaded through a port into the computer. Software can then convert the stored speech to text.

digital computer camera

input device used with a PC to
record and store images and video
in digital form

Voice recognition has found its way into everyday products, such as autos.[8] The Clarion AutoPC, available on many Ford cars and trucks, is a voice-recognition system that allows a driver to activate radio programs and CDs. It can even tell you the time. Asking "What time is it?" will get a response such as, "Eleven thirty-four." The cost is about $1,000, and the system includes a built-in computer.

Digital Computer Cameras

Some personal computers work with **digital computer cameras**, which record and store images and video in electronic form. When you take pictures, the images are electronically stored in the camera. A cable is then used to connect the camera to the computer, and the images can be downloaded. During the download process, the visual images are converted into digital codes by a computer board. The images can then be modified and included in other applications. For example, a digital photo of a company's office can be captured and pasted into

An ergonomic keyboard
is designed to be more
comfortable to use.
(Source: Courtesy of Adesso)

a word processing document for a company brochure. You can even add sound and handwriting to the photo. Some digital cameras, such as the Sony Mavica, can store images on diskettes, which can then be inserted into a computer to transfer the photos to a hard disk. Also, some personal computers, as shown in Figure 2.10, have a video camera that records full-motion video.

The current state-of-the-art cameras can deliver 2-megapixel resolution, enough resolution to deliver snapshot-sized photos that provide crisp, clear images. A **pixel** is a dot of color on a photo image or a point of light on a display screen. The number one advantage of digital cameras is saving time and money by eliminating the need to process film. When Kodak print film is developed, Kodak offers the option of placing pictures on a CD in addition to traditional prints.[9] Once stored on the CD, the photos can be edited, placed on an Internet site, or sent electronically to business associates or friends around the world.

Terminals

Inexpensive and easy to use, terminals are input devices that perform data input. A terminal is connected to a complete computer system, including a processor, memory, and secondary storage. General commands, text, and other data are entered via a keyboard or mouse, converted into machine-readable form, and transferred to the processing portion of the computer system. Terminals are normally connected directly to the computer system by telephone lines or cables and can be placed in offices, in warehouses, and on the factory floor.

Scanning Devices

Image and character data can be input using a scanning device. A page scanner is like a copy machine. The page to be scanned is typically inserted into the scanner or placed face down on the glass plate of the scanner, covered, and scanned. With a handheld scanner, the scanning device is moved or rolled manually over the image to be scanned. Both page and handheld scanners can convert monochrome or color pictures, forms, text, and other images into machine-readable digits. Many companies use scanning devices to help them manage their documents and cut down on the high cost of using and processing paper.

Point-of-Sale (POS) Devices

Point-of-sale (POS) devices are terminals used in retail operations to enter sales information into the computer system. The POS device then computes the total charges, including tax. Many POS devices also use other types of input and output devices, such as keyboards, bar code readers, printers, and screens. A large portion of the money that businesses spend on computer technology involves POS devices.

Automatic Teller Machine (ATM) Devices

Another type of special-purpose input/output device, the automatic teller machine (ATM), is a terminal most bank customers use to perform withdrawals and other transactions for their bank accounts. The ATM, however, is no longer used only for cash and bank receipts. Companies use various ATM devices to support their business processes. Some ATMs dispense tickets for airlines, concerts, and soccer games. Some colleges use them to output transcripts. For this reason, the input and output capabilities of ATMs are quite varied. Like POS devices, ATMs may combine other types of input and output devices. Unisys, for example, has developed an ATM kiosk that allows bank customers to make cash withdrawals and pay bills, and also receive advice on investments and retirement planning.[10]

pixel

a dot of color on a photo image or a point of light on a display screen

point-of-sale (POS) device

terminal used in retail operations to enter sales information into the computer system

FIGURE 2.10

A PC Equipped with a Computer Camera

Digital video cameras make it possible for people at distant locations to conduct videoconferences, eliminating the need for expensive travel to attend physical meetings.
(Source: Stone/Andreas Pollok [Image #828420-015])

Computer-readable bar codes on products provide retailers with data on the product, such as pricing, size, and color. The data is used to track sales and maintain adequate inventory levels. (Source: Courtesy of PSC Inc.)

Touch-Sensitive Screens

Advances in screen technology allow display screens to function as input as well as output devices. By touching certain parts of a sensitive screen, you can execute a program or cause the computer to take an action. Touch-sensitive screens are popular input devices for some small computers because they avoid bulky keyboard input devices. They are frequently used at gas stations for customers to select grades of gas and request a receipt, at fast-food restaurants for order clerks to enter customer choices, at information centers in hotels to allow guests to request facts about local eating and drinking establishments, and at amusement parks to provide directions to patrons. They also are used in kiosks at airports and department stores.

Bar Code Scanners

A bar code scanner employs a laser scanner to read a bar-coded label. This form of input is used widely in grocery store checkouts and in warehouse inventory control.

Output Devices

Computer systems provide output to decision makers at all levels of an organization to solve a business problem or capitalize on a competitive opportunity. In addition, output from one computer system can be used as input into another computer system within the same information system. The desired form of this output might be visual, audio, and even digital. Whatever the output's content or form, output devices function to provide the right information to the right person in the right format at the right time.

Display Monitors

The display monitor is a TV-screen-like device on which output from the computer is displayed. Because the monitor uses a cathode ray tube to display images, it is sometimes called a CRT. The monitor works in much the same way as a TV screen—one or more electron beams are generated from cathode ray tubes. As the beams strike a phosphorescent compound (phosphor) coated on the inside of the screen, a dot on the screen called a pixel lights up. The electron beam sweeps back and forth across the screen so that as the phosphor starts to fade, it is struck again and lights up again.

An upgrade from a 14-inch monitor to a 19-inch monitor can provide a productivity increase for users working with certain applications. The larger display allows for easier, more precise reading of detailed reports. With today's wide selection of monitors, price and overall quality can vary tremendously.

Progress has been made in enabling one device to display both TV and computer output. PC Theatre from Compaq is a consumer living room entertainment device that merges computing and traditional forms of media and entertainment content. This system combines the best features of a TV with a 36-inch monitor and multimedia capabilities. The consumer may watch TV, use the computer, or do both at the same time. The Gateway Destination Big Screen TV comes with a 31-inch monitor, which can be used with a personal computer or for TV viewing.

The quality of a screen is often measured by the number of pixels used to create it. As mentioned previously, a pixel is a dot of color on a photo image or a point of light on a display screen. It is either on or off. A larger number of pixels per square inch means a higher resolution, or clarity and sharpness of the

image. For example, a screen with a 1,024 × 768 resolution (786,432 pixels) has a higher sharpness than one with a resolution of 640 × 350 (224,000 pixels). The distance between one pixel on the screen and the next nearest pixel is known as dot pitch. The common range of dot pitch is from .25 to .31 mm. The smaller the number, the better the picture. A dot pitch of .28 mm or smaller is considered good. Greater pixel densities and smaller dot pitches yield sharper images of higher resolution.

CRT monitors are large and bulky in comparison with LCD monitors (flat displays).
(Source: Courtesy of ViewSonic)

Liquid Crystal Displays (LCDs)

Because CRT monitors use an electron gun, there must be a distance of one foot between the gun and screen, causing them to be large and bulky. Thus, a different technology, flat-panel display, is used for portable personal computers and laptops. One common technology used for flat screen displays is the same liquid crystal display technology used for pocket calculators and digital watches. LCD monitors are flat displays that use liquid crystals—organic, oil-like material placed between two polarizers—to form characters and graphic images on a backlit screen.

The primary choices in LCD screens are passive-matrix and active-matrix LCD displays. In a passive-matrix display, the CPU sends its signals to transistors around the borders of the screen, which control all the pixels in a given row or column. In an active-matrix display, each pixel is controlled by its own transistor attached in a thin film to the glass behind the pixel. Passive-matrix displays are typically dimmer and slower, but less expensive than active-matrix ones. Active-matrix displays are bright and clear and have wider viewing angles than passive-matrix displays. Active-matrix displays, however, are more expensive and can increase the weight of the screen.

Printers and Plotters

One of the most useful and popular forms of output is called hard copy, which is simply paper output from a device called a printer. Printers with different speeds, features, and capabilities are available. Some can be set up to accommodate different paper forms such as blank check forms, invoice forms, and so forth. Newer printers allow businesses to create customized printed output for each customer from standard paper and data input using full color.

The speed of the printer is typically measured by the number of pages printed per minute (ppm). Like a display screen, the quality, or resolution, of a printer's output depends on the number of dots printed per inch. A 600-dpi (dots-per-inch) printer prints more clearly than a 300-dpi printer. A recurring cost of using a printer is the ink-jet or laser cartridge that must be replaced every few thousand pages of output. Figure 2.11 shows a laser printer and an example of its output.

Plotters are a type of hard-copy output device used for general design work. Businesses typically use these devices to generate paper or acetate blueprints, schematics, and drawings of buildings or new products onto paper or transparencies. Standard plot widths are 24 inches and 36 inches, and the length can be whatever meets the need—from a few inches to several feet.

FIGURE 2.11

Laser Printers

Laser printers, available in
a wide variety of speeds
and price ranges, have
many features, including color
capabilities. They are the most
common solution for outputting
hard copies of information.
(Source: Courtesy of Epson
America, Inc.)

personal computer (PC)

relatively small, inexpensive computer system, sometimes called a
microcomputer

Handheld (palm) computers
include a wide variety of
software and communication
abilities, including the ability
to connect to the Internet.
(Source: 3Com and the 3Com logo
are registered trademarks. Palm IIIc™
and the Palm III™ logo are trademarks
of Palm Computing, Inc., 3Com
Corporation or its subsidiaries.)

COMPUTER SYSTEM TYPES

Computer systems can range from desktop (or smaller)
portable computers to massive supercomputers that require
housing in large rooms. Let's examine the types of computer
systems in more detail. Table 2.3 shows general ranges of
capabilities for various types of computer systems.

Personal Computers

As previously noted, **personal computers (PCs)** are relatively small, inexpensive computer systems. Although personal
computers are designed primarily for single users, they are
often tied into larger computer systems as well. Personal computers can be purchased from retail stores or on-line.

There are several types of personal computers. Desktop computers are the
most common personal computer system configuration. Increasingly powerful
desktop computers can provide sufficient memory and storage for most business computing tasks. Desktop PCs have become standard business tools; more
than 30 million are in use in large corporations.

In addition to traditional PCs that use Intel processors and Microsoft software, there are other options. One of the most popular is the iMac by Apple
Computer.[11] The iMac computer is noted for its excellent video and graphics
capabilities, including the ability to create and edit home movies.

Various smaller personal computers can be used for a variety of purposes. A
laptop computer is a small, lightweight PC about the size of a briefcase. Apple,
for example, has the iBook computer, which is a laptop system compatible with
the iMac. Newer PCs include the even smaller and lighter *notebook* and *subnotebook* computers that provide similar computing power. Some notebook and
subnotebook computers fit into docking stations of desktop computers to provide additional storage and processing capabilities. Small PCs continue to rise
in popularity because of their portability and performance. In the past, desktop
computers typically ran at faster speeds than notebook computers, primarily
because of the limitations of running a notebook computer from a battery.
Today's chips, such as Intel's SpeedStep, can reduce the performance difference between notebook computers and desktop systems.[12] The SpeedStep
allows a laptop to run at speeds that approach those of a desktop when using
power from a wall outlet. When powered by its battery, the SpeedStep runs at
slower speeds more typical of notebook computers.

Handheld (palmtop) computers are PCs that provide increased portability
because of their smaller size—some are as small as a credit card. These systems
often include a wide variety of software and communications
capabilities. Boeing, for example, uses handheld computers to
help improve quality inspections of aircraft and aircraft parts.[13]
Boeing's plant in St. Louis manufactures F-14 and F-15
fighter planes. Using Palm IIIX computing devices, Boeing
inspectors have been able to increase inspection quality while
halving inspection times. Boeing saves even more time when
the data is uploaded into computer systems. The old approach
used clipboards to manually record inspection data, which was
then typed into the computer to produce reports.

Embedded computers are computers placed inside other products to add features and capabilities. In automobiles, embedded
computers can help with navigation, engine performance, braking, and other functions. Household appliances, stereos, and
some phone systems also use embedded computers. Embedded

Characteristic	Network Computer	Personal Computer	Workstation	Midrange Computer	Mainframe Computer	Super-computer
Processor Speed	1–5 MIPs	5–20 MIPs	50–100 MIPs	25–100 MIPs	40–4,550 MIPs	60 billion–3 trillion instructions per second
Amount of RAM	4–16 MB	16–128 MB	32–256 MB	32–512 MB	256–1,024 MB	8,192 MB+
Approximate Cost	$500–$1,500	$1,000–$5,000	$4,000 to over $20,000	$20,000 to over $100,000	$250,000 to over $2 million	$2.5 million to over $10 million
How Used	Supports "heads-down" data entry; connects to the Internet	Improves individual worker's productivity	Engineering; CAD; software development	Meets computing needs for a department or small company	Meets computing needs for a company	Scientific applications; marketing; customer support; product development
Example	Oracle Network computer	Dell Pentium computer	Sun Microsystems computer	Hewlett-Packard HP-9000	IBM ES/9000	Cray C90

TABLE 2.3

Types of Computer Systems
(Source: Photos courtesy of Wyse Technology, IBM Corporation, and Los Alamos National Laboratory)

network computer

a cheaper-to-buy and cheaper-to-run version of the personal computer that is used primarily for accessing networks and the Internet

computers and videos are being used in a "smart road" pilot project in Atlanta to send messages to oncoming vehicles to choose alternate routes if needed. In the Eisenhower Tunnel at the top of the continental divide in Colorado, embedded computers and video systems monitor truck traffic and notify trucks to slow down if they are going too fast on the descent to Denver.

A **network computer** is a cheaper-to-buy and cheaper-to-run version of the personal computer that is used primarily for accessing networks and the Internet. These stripped-down computers do not have the storage capacity or power of typical desktop computers, nor do they need it for the role they play. Unlike personal computers, network computers download software from a network when needed. This can make it much easier and less expensive to manage the support, distribution, and updating of software applications. The initial target user is someone who performs what is called heads-down data entry—customer inquiry, phone order taking, and classic data entry. The network computer is designed to have no moving parts to avoid expensive equipment repairs. IBM, Oracle, and Sun Microsystems were the first companies to develop prototypes of such systems, with a purchase price in the $500 to $1,500 range. Advocates of network computers argue that they not only cost less to purchase compared with a standard desktop PC but also cost less to operate. However, the network computer's flexibility is extremely limited when compared with the personal computer.

In addition, PC companies have responded strongly to network computers with lower prices and more competitive products.

Workstations are computers that fit between high-end personal computers and low-end midrange computers in terms of cost and processing power. They cost from $3,000 to $40,000. Workstations are small enough to fit on an individual's desktop. Workstations are used to support engineering and technical users who perform heavy mathematical computing, computer-aided design (CAD), and other applications requiring a high-end processor. Such users need very powerful CPUs, large amounts of main memory, and extremely high-resolution graphic displays to meet their needs. Engineers use CAD programs to create two- and three-dimensional engineering drawings and product designs.

Many companies, including Microsoft and Intel, are developing inexpensive **Web appliances**.[14] A Web appliance is a device that can connect to the Internet, typically through a phone line. It can be used to check stock prices, check e-mail messages, search the Internet for information, and more. Web appliances come in a variety of configurations. Some have a keyboard, a passive-matrix display, a 200-MHz processor, and Web and e-mail software. These devices can cost under $200 to purchase and about $20 per month for an Internet connection.[15] Some Web appliances have the appearance of a cellular phone, with the capabilities of a standard phone and basic Internet connections. Other Web appliances are being attached to everyday products, such as TVs, stoves, and refrigerators. Once attached, the Web appliance will be able to get movie schedules, alert people when their stove may require maintenance, or advertise grocery specials. A number of companies are now making Web appliances, including Netpliance, Compaq, Dell, and Microsoft. In the future, Web appliances may be built into many of the products we use every day.

Midrange Computers

Midrange computers are systems about the size of a three-drawer file cabinet and can accommodate several users at one time. These systems often have secondary storage devices with more capacity than workstation computers and can support a variety of transaction processing activities, including payroll, inventory control, and invoicing. Midrange computers often have excellent processing and decision support capabilities. Many small to medium-size organizations—such as manufacturers, real estate companies, and retail operations—use midrange computers.

Mainframe Computers

Mainframe computers are large, powerful computers often shared by hundreds of concurrent users connected to the machine via terminals. The mainframe computer must reside in an environment-controlled computer room or data center with special heating, venting, and air-conditioning (HVAC) equipment to control the temperature, humidity, and dust levels around the computer. In addition, most mainframes are kept in a secured data center with limited access to the room through some kind of security system. The construction and maintenance of such a controlled-access room with HVAC can add hundreds of thousands of dollars to the cost of owning and operating a mainframe computer. Mainframe computers also require specially trained individuals (called system engineers and system programmers) to care for them. Mainframe computers start at $200,000 for a fully configured system.

The role of the mainframe is a large information processing and data storage utility for a corporation—running jobs too large for other computers, storing files and databases too large to be stored elsewhere, and storing backups of files and databases created elsewhere (these large stores of data are sometimes called *data warehouses).* The mainframe is capable of handling the millions of daily transactions associated with airline, automobile, and hotel/motel reservation systems. It can process the tens of thousands of daily queries necessary to provide data to decision support systems. Its massive storage and input/output capabilities enable it to play the role of a video computer, providing full-motion video to users. Over time, mainframes have been evolving into smaller, faster, less expensive systems as a result of *complementary metal oxide semiconductor (CMOS),* a computer chip fabrication technology that uses special semiconductor material. Mainframe computers provide support for large packaged software products, Web technologies, and communications protocols, much like their smaller cousins, the midrange computers.

Table 2.4 lists the major mainframe computer manufacturers. IBM remains a dominant force in the mainframe market.

Supercomputers

supercomputers

the most powerful computer systems, with the fastest processing speeds

Supercomputers are the most powerful computer systems, with the fastest processing speeds. Supercomputers are used by government agencies to perform the high-speed number crunching needed in weather forecasting and military applications. They are also used to perform the enormous number of calculations required to draw and animate Disney cartoons. To produce those special effects requires handling a gigabyte (one billion characters) of data for every second of film time—obviously, requiring a very powerful computer. They are also used by universities and large corporations involved with research or high-technology businesses. Some large oil companies, for example, use supercomputers to perform sophisticated analysis of detailed data to help them explore for oil.

Although all of the preceding computer system types can be used for general processing tasks, they can also serve a specific and unique purpose, such as supporting Internet and network applications. A computer server is a computer designed for a specific task, such as network or Internet applications. Servers typically have large memory and storage capacities, along with fast and efficient communications abilities. They can range in size from a PC to a mainframe system, depending on the needs of the organization. A Web server is used to handle Internet traffic and communications. An Internet caching server stores Web sites that are frequently used by a company. A file server, discussed in more detail in Chapter 4, stores and coordinates program and data files. An application server provides services to support Web-based applications that connect users to corporate databases. As with general computers, there are benchmarks to help a company determine the performance of a server, such as ZD ServerBench, WebBench, and NetBench.

Companies are supplying their managers and employees with PCs in record numbers and are increasingly monitoring how their employees are using company-purchased equipment. Often their findings are disappointing—employees using corporate PCs to invest in the stock market, play games, and even visit

TABLE 2.4

Mainframe Computer
Manufacturers

Company	Product	MIPS (Millions of Instructions per Second)
IBM	Systems/390 Parallel	178–5,000
Unisys	A Series	1–250
Amdahl	Millennium	122–350
Hitachi Data Systems	Pilot	40–350

Blue Mountain is a supercomputer located at the U.S. Department of Energy's Los Alamos National Laboratory. At the heart of Blue Mountain are 48 commercially available Silicon Graphics® Cray® Origin2000 servers containing a total of 6,144 processors. Blue Mountain runs at 1.6 trillion operations per second (teraOps) and has one of the most advanced graphics systems in the world. With this visualization system, answers to complex scientific problems that would have taken weeks or more to display can now be displayed in minutes.
(Source: Courtesy of Los Alamos National Laboratory.)

pornographic Internet sites while at work. Read the "Ethical and Societal Issues" box to learn about the monitoring of employee computer use.

We now turn to the other critical component of effective computer systems—software. Like hardware, software has made technological leaps in a relatively short time span.

AN OVERVIEW OF SOFTWARE

In the 1950s, when computer hardware was relatively rare and expensive, software costs were a comparatively small percentage of total information systems costs. Today, that situation has dramatically changed. Software can represent 75 percent or more of the total cost of an information system for three major reasons: advances in hardware technology have dramatically reduced hardware costs, increasingly complex software requires more time to develop and so is more costly, and salaries for software developers have increased because the demand for these workers far exceeds the supply. In the future, software is expected to make up an even greater portion of the cost of the overall information system. The critical functions software serves, however, make it a worthwhile investment.

One of software's most critical functions is to direct the workings of the computer hardware. **Computer programs** are sequences of instructions for the computer. **Documentation** describes the program functions that help the user operate the computer system. The program displays some documentation on screen, while other forms appear in external resources, such as printed manuals. There are two basic types of software: systems software and application software. **Systems software** is the set of programs designed to coordinate the activities and functions of the hardware and various programs throughout the computer system. A particular systems software package is designed for a specific CPU design and class of hardware. The combination of a particular hardware configuration and systems software package is known as a **computer system platform**. **Application software** consists of programs that help users solve particular computing problems.

Both systems and application software can be used to meet the needs of an individual, a group, or an enterprise. Application software can support individuals, groups, and organizations to help them realize business objectives. Application software has the greatest potential to affect the processes that add value to a business because it is designed for specific organizational activities and functions, as we saw in the case of Home Depot. The effective implementation

computer programs

sequences of instructions for the computer

documentation

text that describes the program functions to help the user operate the computer system

systems software

the set of programs designed to coordinate the activities and functions of the hardware and various programs throughout the computer system

computer system platform

the combination of a particular hardware configuration and system software package

application software

programs that help users solve particular computing problems

ETHICAL AND SOCIETAL ISSUES

Employee Monitoring

The information technology manager for an engineering company noticed that employees had a high amount of Internet use, but the second-busiest Internet site they visited was not a business site. The manager investigated further and found the site's greeting: "Surf in Style: The Sex Tracker." Realizing that one or more employees were spending company time at a pornographic site troubled the technology manager. And he disliked being a computer snoop, investigating employees' e-mails and Internet use. But the technology manager often tells employees, "You live in a democracy; you don't work in one." Because of the number of employees visiting pornographic, gambling, and other nonbusiness sites, the technology manager started blocking these sites from employee PCs. He also implemented monitoring software that could generate reports on which employees were the most frequent users of Internet sites. As a result, he found that one month a popular stock trading site had more than 3 percent of the total Internet activity for the entire firm. He then accessed reports to see which employee was involved. A detailed investigation of that employee's stock trading activities soon followed.

With today's technology, it is becoming easier to spy on employees. Specialized software can monitor what employees are doing with their PCs. These programs can see whether an employee is using his or her PC to hunt for a new job, buy stock, gamble on-line, or visit pornographic sites. Some believe that the surveillance power available today is staggering. It is possible to quickly pull up e-mails sent by a company employee, even if the e-mail is stored on the employee's own hard disk. Some employees mistakenly believe that if they delete files on their PCs they are safe. But using monitoring software, companies can review all files that employees have deleted from their PCs. They can also monitor virtually any activity performed on corporate PCs—from one employee making retirement calculations on his PC, to another employee sending jokes via e-mail to friends and family, to another using the Internet to search for other job opportunities.

Yet although most information systems managers agree that monitoring employees' PCs is legal and necessary for corporate security, there can be a fine line between corporate security and invasion of employee privacy.

Discussion Questions

1. Despite the legalities, do you think it is right for companies to monitor their employees?
2. What types of employee PC activities should a company monitor?

Critical Thinking Questions

3. If you were the information technology manager for the engineering firm, what would you monitor and how would you respond to inappropriate use of corporate PCs?
4. Should there be different monitoring systems for employees and senior-level managers?

Sources: Adapted from Michael McCarthy, "Web Surfers Beware," *The Wall Street Journal,* January 10, 2000, p. Q1; and John Morris, "Protect Your PC," *PC Magazine,* September 1, 1999, p. 107.

and use of application software can provide significant internal efficiencies and support corporate goals. Before an individual, a group, or an enterprise decides on the best approach for acquiring application software, goals and needs should be analyzed carefully.

Supporting Individual, Group, and Organizational Goals

Every organization relies on the contributions of individuals, groups, and the entire enterprise to achieve business objectives. To help them achieve these objectives, the organization provides them with specific application software and information systems. One useful way of classifying the many potential uses of information systems is to identify the scope of the problems and opportunities addressed by a particular organization, called the **sphere of influence.** These spheres of influence are personal, workgroup, and enterprise, as shown in Table 2.5.

Information systems that operate within the *personal sphere of influence* serve the needs of an individual user. These information systems enable users to improve their personal effectiveness, increasing the amount of work that can be

sphere of influence

the scope of problems and opportunities addressed by a particular organization

Software	Personal	Workgroup	Enterprise
Systems software	Personal computer and workstation operating systems	Network operating systems	Midrange computer and mainframe operating systems
Application software	Word processing, spreadsheet, database, graphics	Electronic mail, group scheduling, shared work	General ledger, order entry, payroll human resources

TABLE 2.5

Classifying Software by Type and Sphere of Influence

personal productivity software

software that enables users to improve their personal effectiveness, increasing the amount of work they can do and its quality

done and its quality. Such software is often referred to as **personal productivity software**. There are many examples of such applications operating within the personal sphere of influence—a word processing application to enter, check spelling of, edit, copy, print, distribute, and file text material; a spreadsheet application to manipulate numeric data in rows and columns for analysis and decision making; a graphics application to perform data analysis; and a database application to organize data for personal use.

A *workgroup* is two or more people who work together to achieve a common goal. A workgroup may be a large, formal, permanent organizational entity such as a section or department or a temporary group formed to complete a specific project. The human resource department of a large firm is an example of a formal workgroup. It consists of several people, is a formal and permanent organizational entity, and appears on a firm's organization chart. An information system that operates in the *workgroup sphere of influence* supports a workgroup in the attainment of a common goal. Users of such applications are operating in an environment where communication, interaction, and collaboration are critical to the success of the group. Applications include systems that support information sharing, group scheduling, group decision making, and conferencing. These applications enable members of the group to communicate, interact, and collaborate.

Information systems that operate within the *enterprise sphere of influence* support the firm in its interaction with its environment. The surrounding environment includes customers, suppliers, shareholders, competitors, special-interest groups, the financial community, and government agencies. Every enterprise has many applications that operate within the enterprise sphere of influence. The input to these systems is data about or generated by basic business transactions with someone outside the business enterprise. These transactions include customer orders, inventory receipts and withdrawals, purchase orders, freight bills, invoices, and checks. One of the results of processing transaction data is that the records of the company are updated. The order entry, finished product inventory, and billing information systems are examples of applications that operate in the enterprise sphere of influence. These applications support interactions with customers and suppliers.

Regardless of the sphere of influence that software supports, all information systems need software programs to control basic computer functions such as memory management and providing an interface for users. Such software is called *systems software* and is the foundation on which applications are built.

SYSTEMS SOFTWARE

Controlling the operations of computer hardware is one of the most critical functions of systems software. Systems software also supports the applications programs' problem-solving capabilities. Different types of systems software include operating systems and utility programs.

Operating Systems

An **operating system (OS)** is a set of computer programs that controls the computer hardware and acts as an interface with application programs (Figure 2.12). The operating system, which plays a central role in the functioning of the complete computer system, is usually stored on disk. After a computer system is started, or "booted up," portions of the operating system are transferred to memory as they are needed. The group of programs, collectively called the operating system, executes a variety of activities, including the following:

- Performing common computer hardware functions
- Providing a user interface
- Providing a degree of hardware independence
- Managing system memory
- Managing processing tasks
- Providing networking capability
- Controlling access to system resources
- Managing files

Common Hardware Functions

All application programs must perform certain tasks—for example, getting input from the keyboard or some other input device, retrieving data from disks, storing data on disks, and displaying information on a monitor or printer. Each of these basic functions requires a more detailed set of instructions to complete. The operating system converts a simple, basic instruction into the set of detailed instructions the hardware requires. In effect, the operating system acts as intermediary between the application program and the hardware. The typical OS performs hundreds of such functions, each of which is translated into one or more instructions for the hardware. The OS notifies the user if input/output devices need attention, if an error has occurred, or if anything abnormal occurs in the system.

User Interface

One of the most important functions of any operating system is providing a **user interface**. A user interface allows individuals to access and command the computer system. The first user interfaces for mainframe and personal computer systems were command based.

A **command-based user interface** requires text commands to be given to the computer to perform basic activities. For example, the command ERASE 00TAXRTN would cause the computer to erase or delete a file called 00TAXRTN. RENAME and COPY are other examples of commands used to rename files and copy files from one location to another.

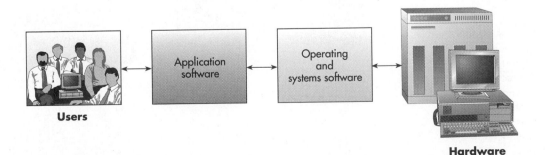

Users   Application software   Operating and systems software   **Hardware**

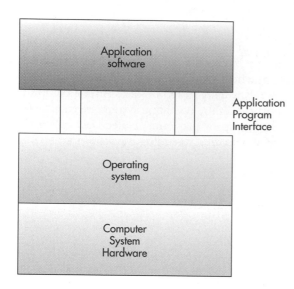

FIGURE 2.13

Application Program Interface Links Application Software to the Operating System

graphical user interface (GUI)

an interface that uses icons and menus displayed on screen to send commands to the computer system

application program interface (API)

interface that allows applications to make use of the operating system

A **graphical user interface (GUI)** uses pictures called *icons* and menus displayed on screen to send commands to the computer system. Many people find that GUIs are easier to use because users intuitively grasp the functions. Today, the most widely used graphical user interface is Windows by Microsoft. As the name suggests, Windows is based on the use of a window, or a portion of the display screen dedicated to a specific application. The screen can display several windows at once. The use of GUIs has contributed greatly to the increased use of computers because users no longer need to know command-line syntax to accomplish a task.

Hardware Independence

The applications use the operating system by making requests for services through a defined **application program interface (API)**, as shown in Figure 2.13. Programmers can use APIs to create application software without understanding the inner workings of the operating system.

Memory Management

The memory management feature of operating systems converts a user's request for data or instructions (called a logical view of the data) to the physical location where the data or instructions are stored. A computer understands only the physical view of data—that is, the specific location of the data in storage or memory and the techniques needed to access it. This concept is described as logical versus physical access. For example, the current price of an item, say, a Texas Instruments BA-35 calculator with an item code of TIBA35, might always be found in the logical location "TIBA35$." If the CPU needed to fetch the price of TIBA35 as part of a program instruction, the memory management feature of the operating system would translate the logical location "TIBA35$" into an actual physical location in memory or secondary storage (Figure 2.14).

Processing Tasks

Task management features of today's operating systems manage all processing activities. Task management allocates computer resources to make the best use of each system's assets. Task management software can permit one user to run several programs or tasks at the same time (multitasking) and allow several users

FIGURE 2.14

An Example of the Operating System Controlling Physical Access to Data

The user prompts the application software for specific data. The operating system translates this prompt into instructions for the hardware, which finds the data the user requested. Having completed this task, the operating system then relays the data back to the user via the application software.

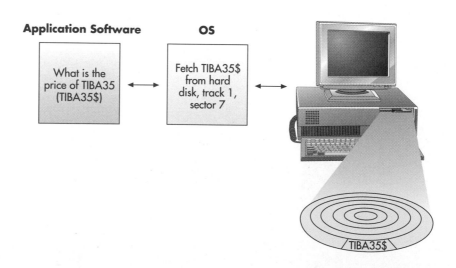

multitasking

capability that allows a user to run more than one application at the same time

to use the same computer at the same time (time-sharing). With **multitasking**, a user can run more than one application at the same time. Without having to exit a program, you can work in one application, easily pop into another, and then jump back to the first program, picking up where you left off. Better still, while you're working in the *foreground* in one program, one or more other applications can be churning away, unseen, in the *background*—sorting a database, printing a document, or performing other lengthy operations that otherwise would monopolize your computer and leave you staring at the screen unable to get other work done. Multitasking can save users a considerable amount of time and effort.

time-sharing

capability that allows more than one person to use a computer system at the same time

Time-sharing allows more than one person to use a computer system at the same time. For example, 15 customer service representatives may be entering sales data into a computer system for a mail-order company simultaneously. In another case, thousands of people may be simultaneously using an on-line computer service to get stock quotes and valuable business news. Time-sharing works by dividing time into small CPU processing time slices, which can be a few milliseconds or less in duration. During a time slice, some tasks for the first user are done. The computer then goes from that user to the next and completes some tasks for that user during that time slice. This process continues for each user and cycles back to the first user. Because the CPU processing time slices are small, it appears that tasks for all users are being completed at the same time. In reality, each user is sharing the time of the computer with other users.

scalability

the ability of the computer to handle an increasing number of concurrent users smoothly

The ability of the computer to handle an increasing number of concurrent users smoothly is called **scalability**. Scalability is a critical feature for systems that must handle a large number of users, such as a mainframe computer or a Web server. Because personal computer operating systems usually are oriented toward single users, management of multiple-user tasks often is not needed.

Networking Capability

The operating system can provide features and capabilities that aid users in connecting to a computer network. For example, Apple computer users have built-in network access through the AppleTalk feature, and the Microsoft Windows operating systems come with the capability to link users to the Internet.

Access to System Resources

Computers often handle sensitive data that can be accessed over networks, so the operating system needs to provide a high level of security against unauthorized access to users' data and programs. Typically, the operating system establishes a log-on procedure that requires users to enter an identification code and a matching password. If the identification code is invalid or if the password does not match the identification code, the user cannot gain access to the computer. The operating system also requires that user passwords be changed frequently—say, every 30 days. If the user is successful in logging on to the system, the operating system records who is using the system and for how long. In some organizations, such records are also used to bill users for time spent using the system. The operating system also reports any attempted breaches of security.

File Management

The operating system performs a file management function to ensure that files in secondary storage are available when needed and that they are protected from unauthorized access. Many computers support multiple users who store files on centrally located disks or tape drives. The operating system keeps track of where each file is stored and who may access it. The operating system must also be able to resolve what to do if more than one user requests access to the same file at the same time.

Personal Computer Operating Systems

Early operating systems for personal computers were very basic. In the last several years, however, more advanced operating systems have been developed, incorporating some features previously available only with mainframe operating systems. Table 2.6 classifies a number of current operating systems by sphere of influence. This section reviews selected popular personal computer operating systems.

Microsoft PC Operating Systems

There has been a continuous and steady evolution of personal computer operating systems since a formerly small company called Microsoft developed PC-DOS and MS-DOS to support the IBM personal computer introduced in the 1970s. Each new version of operating system has improved in terms of ease of use, processing capability, reliability, and ability to support new computer hardware devices. Table 2.6 lists the progression of Microsoft operating systems.

PC-DOS and *MS-DOS* had command-driven interfaces that were difficult to learn and use. *Windows 95* evolved from these early operating systems and from one called *DOS with Windows* and converted to a graphical user interface with a desktop metaphor that showed files as icons within folders (directories). Windows 95 also came with communications software to simplify connection to the Internet and the sending of electronic mail and faxes. It was a multitasking system improving users' ability to complete tasks. Windows 95 also introduced Plug 'n Play capabilities so that end users could add new hardware devices (for example, a printer or a scanner) to their machines with minimal effort. *Windows 98* let an organization's IS department install and configure the operating system and all applications on one machine and then copy the configuration to each end user's machine. System start-up and shutdown were also speeded up.

The *Windows New Technology (NT) Workstation* operating system is designed to take advantage of the newer 32-bit processors, and it features multitasking and advanced networking capabilities. NT can also run programs written for other operating systems. NT supports symmetric multiprocessing, the ability to simultaneously use multiple processors. The many features and capabilities of NT make it very attractive for use on many computers. Microsoft renamed the

TABLE 2.6

Popular Operating Systems

Personal	Workgroup	Enterprise	Consumer
MS-DOS			
DOS with Windows			
Windows 95			
Windows 98			
			Pocket PC
Windows NT Workstation	Windows NT Server	Windows NT Server	
Windows 2000	Windows 2000	Windows 2000 Server	
Unix	Unix	Unix	
Linux	Linux		Linux
MAC OS 9			
MAC OS X	MAC OS X	MAC OS X	
Solaris	Solaris	Solaris	
	Netware		
	OS/390	OS/390	
	MPE/iX	MPE/iX	

- PC-DOS
- MS-DOS
- DOS with Windows
- Windows 95
- Windows 98
- Windows NT Workstation
- Windows 2000 Professional
- Windows ME

TABLE 2.7

Microsoft Personal Computer Operating Systems

next release of the Windows NT line of operating systems *Windows 2000*. This operating system, with 30 million lines of code, took four years to complete and cost Microsoft more than $1 billion to develop.[16] Microsoft designed Windows 2000 Professional as the easiest Windows yet, with high-level security and significant enhancements for laptop users. The operating system is also designed to provide high reliability. *Windows Millennium Edition (ME)* was designed for home use and enables even novice computer users to organize photos, make home movies and records, and play music, as well as the usual computer tasks such as accessing the Internet, playing games, and performing word processing.

Apple Computer PC Operating Systems

While IBM system platforms traditionally use Intel microprocessors and one of the Windows operating systems, Apple computers typically use Motorola microprocessors and a proprietary Apple operating system—the Mac OS. Although IBM and IBM-compatible computers hold the largest share of the business PC market, Apple computers are also quite popular, especially in the fields of publishing, education, and graphic arts. The Apple operating systems have also evolved over a number of years and often provide features not available from Microsoft. For example, *Mac OS 9* provides a Sherlock 2 feature that is your personal search detective and personal shopper. It searches the Internet to locate multiple sources for products you request and compares prices and availability. Mac OS 9 also has a Multiple Users feature that allows you to safely share your Macintosh computer with other people. To keep your computer running smoothly, Auto Updating delivers the latest software updates and system enhancements from Apple.[17] *Mac os x ("Ten")* is a completely new implementation of the Mac operating system and includes an entirely new user interface called "Aqua," which provides a new visual appearance with luminous and semitransparent elements such as buttons, scroll bars, windows, and fluid animation to enhance the user's computing experience.

Linux

Linux is an operating system developed under the GNU General Public License, and its source code is freely available to everyone. This doesn't mean, however, that Linux and its assorted distributions are free—companies and developers may charge money for it as long as the source code remains available.[18] Linux is actually only the *kernel* of an operating system, the part that controls hardware, manages files, separates processes, and so forth. Several combinations of Linux are available with sets of utilities and applications to form a complete operating system. Each of these combinations is called a *distribution* of Linux. Cendant, a $5.3 billion travel services, real estate, and direct marketing company, deployed Linux servers and workstations in about 60 percent of the 4,500 Super 8, Days Inn, and Howard Johnson hotels connected to its new Unix reservation management system. The stability and ease of administration of Linux have been remarkable.[19]

Workgroup Operating Systems

To keep pace with today's high-tech society, the technology of the future must support a world in which Internet use doubles every 100 days. It must support a world in which a single on-line service provider can consume 28 terabytes in 45 days—activity equal to all use on the Internet for one month in 1997. This massive scaling pushes the boundaries of computer science and physics. Powerful and sophisticated operating systems are needed to run the servers that meet these business needs for workgroups. Figure 2.15 summarizes the server market share for 1998, the most current data available.

Windows 2000 Server

Microsoft designed Windows 2000 to do a host of new tasks that are vital for Web sites and corporate Web applications that run on the Internet.[20] Besides being more reliable than Windows NT, it's capable of handling extremely demanding computer tasks such as order processing. It can be tuned to run on machines with up to 32 microprocessors—satisfying the needs of all but the most demanding of Web operators. Four machines can be clustered together to prevent service interruptions that are disastrous for Web sites.[21] With Windows 2000, Microsoft introduced Active Directory, which lets corporations keep track of every employee, computer, software package, and scrap of data in one place.[22]

Unix

Unix is a powerful operating system originally developed by AT&T for minicomputers. Unix can be used on many computer system types and platforms, from personal computers to mainframe systems. Unix also makes it much easier to move programs and data among computers or to connect mainframes and personal computers to share resources. Unix is considered to have a complex user interface with strange and arcane commands, so software developers have provided shells such as Motif from Open Systems Foundation and Open Look from Sun Microsystems. These shells provide a graphical user interface and shield users from the complexity of the underlying operating system. There are many variants of Unix—including HP/UX from Hewlett-Packard, AIX from IBM, UNIX SystemV from UNIX Systems Lab, Solaris from Sun Microsystems, and SCO from Santa Cruz Operations.

Netware

Netware is a network operating system sold by Novell that can support end users on Windows, Macintosh, and Unix platforms. Netware provides directory software to track computers, programs, and people on a network, making it easier for large companies to manage complex networks. Netware users can log on from any computer on the network and still get their own familiar desktop with all their applications, data, and preferences.

Red Hat Linux

Red Hat Linux 6.1 is a network operating system from Red Hat Software that taps into tens of thousands of volunteer programmers who generate a steady stream of improvements for the Linux operating system. The Red Hat Linux network operating system is very efficient at serving up Web pages and can manage a cluster of up to eight servers.

Solaris

Solaris is the Sun Microsystems variation of the Unix operating system that is the current server operating system of choice for large Web sites.[23] Solaris is

FIGURE 2.15

1998 Server Market Share
(Data Source: Steve Hamm, Peter Burrows, and Andy Reinhardt, "Is Windows Ready to Run E-Business?" *Business Week,* January 24, 2000, pp. 154–160.)

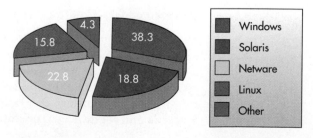

highly reliable and handles the most demanding tasks. It can supervise servers with as many as 64 microprocessors, and eight such computers can be clustered together to work as one. The Solaris operating system can run on Sun's Sparc family of microprocessors as well as computers with Intel microprocessors.[24] An example of the unique features of Solaris is fault detection and analysis that lets IS administrators establish policies for problematic conditions. For example, if a processor gets too hot, the capability may instruct a system to shut down the processor and reboot. Another feature, the reconfiguration coordination manager, lets administrators write policies that automatically redistribute system capacity. Current processing capacity is a staggering one million concurrent processes.[25]

Mac OS X Server

The Mac OS X Server is the first modern server operating system from Apple Computer. It provides Unix-style process management. Protected memory puts each service in its own well-guarded chunk of dynamically allocated memory, preventing a single process from going awry and bringing down the system or other services. Preemptive multitasking ensures that each process gets the right amount of CPU time and system resources it needs for optimal efficiency and responsiveness.

Enterprise Operating Systems

A few years ago, computer industry pundits were predicting the end of large-scale computing systems. The future of enterprise computing, they claimed, was in networks of servers that were less expensive to buy, more flexible, and more adaptable to the changing demands facing businesses in today's marketplace. In fact, Windows NT, Windows 2000, Solaris, and other workgroup operating systems have versions that are designed to operate with extremely powerful network servers to meet enterprise computing needs.

In spite of this vision, there has been a renewed appreciation for the traditional strengths of mainframe computers. The new generation of mainframe computers provides the computing and storage capacity to meet massive data processing requirements and provide a large number of users with high performance and excellent system availability, strong security, and scalability. In addition, a wide range of application software has been developed to run in the mainframe environment, making it possible to purchase software to address almost any business problem. As a result, mainframe computers remain the computing platform of choice for mission-critical business applications for many companies. OS/390 for IBM and MPE/iX for Hewlett Packard are two examples of mainframe operating systems.

Consumer Appliance Operating Systems

New operating systems and other software are changing the way we interact with home entertainment appliances, including TVs, stereos, and other appliances. We look at just two new products here.

Microsoft Pocket PC

Pocket PC is the new name for the Windows CE (for Compact Edition) operating system that comes installed in read-only memory (ROM) on devices such as digital TV set-top devices, PCs in automobiles, and handheld PCs, which are available in computer and consumer electronics stores and from Microsoft Certified Solution Providers. The Pocket PC platform makes possible new categories of non-PC business and consumer devices that can communicate with each other, share information with Windows-based PCs, and connect to the Internet.

FIGURE 2.16

Ericsson Cordless Screen
Phone HS210
(Source: Courtesy of Ericsson)

It provides a graphical user interface, incorporating many elements of the familiar Windows user interface, making it easy to use.[26]

Mobil Linux

Mobil Linux is an operating system that runs a new category of tablet-sized Internet-browsing devices called Web pads, which cost under $1,000. This version of Linux has additional support for certain video and sound cards, application programs for hand-writing recognition, and power management capabilities to minimize battery consumption.[27] The Ericsson Cordless Screen Phone HS210 (see Figure 2.16) allows users to access the Internet, e-mail, and the telephone with a single, contained package. The Screen Phone's interface is a color touch screen, through which the Internet and other features can be quickly and easily accessed. The product is also equipped with a speakerphone to make it more convenient for Web surfers to talk on the phone while they sail through cyberspace. The Screen Phone runs on the Linux operating system.[28]

APPLICATION SOFTWARE

As discussed earlier, the primary function of application software is to apply the power of a computer to give individuals, workgroups, and the entire enterprise the ability to solve problems and perform specific tasks. Application programs perform those specific computer tasks by interacting with systems software to direct the computer hardware. Programs that complete sales orders, control inventory, pay bills, write paychecks to employees, and provide financial and marketing information to managers and executives are examples of application software. Most of the computerized business jobs and activities discussed in this book involve the use of application software.

Types and Functions of Application Software

The key to unlocking the potential of any computer system is application software. A company can either develop a one-of-a-kind program for a specific application (called **proprietary software**) or purchase and use an existing software program (sometimes called **off-the-shelf software**). It is also possible to modify some off-the-shelf programs, giving a blend of off-the-shelf and customized approaches. These different sources of software are shown in Figure 2.17. The relative advantages and disadvantages of proprietary software and off-the-shelf software are summarized in Table 2.8 on page 70.

proprietary software

a one-of-a-kind program for a specific application

off-the-shelf software

existing software program

Proprietary Application Software

Software to solve a unique or specific problem is called *proprietary application software*. This type of software is usually built, but it can also be purchased from an outside company. If an organization has the time and IS talent, it may opt for *in-house development* for all aspects of the application programs. Alternatively, an organization may obtain customized software from external vendors. For example, a third-party software firm, often called a value-added software vendor, may develop or modify a software program to meet the needs of a particular industry or company. A specific software program developed for a particular company is called **contract software**.

contract software

software developed for a particular company

Off-the-Shelf Application Software

Software can also be purchased, leased, or rented from a software company that develops programs and sells them to many computer users and organizations. Software programs developed for a general market are called off-the-shelf software

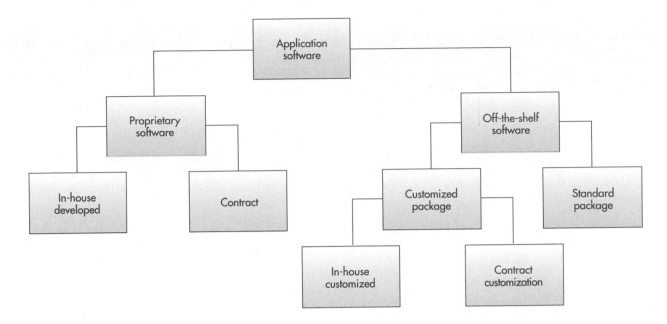

FIGURE 2.17

Sources of Software: Proprietary and Off-the-Shelf

Some off-the-shelf software may be modified to allow some customization.

packages because they can literally be purchased "off the shelf" in a store. Many companies use off-the-shelf software to support business processes.

Customized Package

In some cases, companies use a blend of external and internal software development. That is, off-the-shelf software packages are modified or customized by in-house or external personnel. For example, a software developer may write a collection of programs to be used in an auto body shop that includes features to generate estimates, order parts, and process insurance. Designed properly—and allowing for minor tailoring for each body shop—the same software package can be sold to many businesses. However, since each body shop has slightly different requirements, software vendors would probably provide a wide range of services, including installation of their standard software, modifications for unique customer needs, training of end users, and other consulting services.

Some software companies encourage their customers to make changes to their software and will sometimes make the necessary changes themselves for a fee. Other software companies will not allow their software to be modified by those purchasing or renting it.

Another approach to obtaining a customized software package is to use an **application service provider (ASP)**—a company that provides both end user support and the computers on which to run the software from the user's facilities. They can also simplify a complex corporate software package for users so that it is easier to set up and manage. ASPs also provide contract customization of off-the-shelf software, assist in speeding deployment of new applications, and help IS managers avoid implementation headaches, reducing the need for skilled IS staff members and reducing project start-up expenses. Perhaps the biggest advantage of employing an application service provider is that it frees in-house corporate resources from staffing and managing complex computing projects so that they can focus on more important things. Read more about the use of application service providers in the "E-Commerce" box on page 72.

application service provider (ASP)

a company that provides both end user support and the computers on which to run the software from the user's facilities

Personal Application Software

There are literally hundreds of computer applications that can help individuals at school, home, and work. For example, a graphics program can help a sales manager develop an attractive sales presentation to give at the annual sales meeting. A

Proprietary Software		Off-the-Shelf Software	
Advantages	**Disadvantages**	**Advantages**	**Disadvantages**
You can get exactly what you need in terms of features, reports, and so on.	It can take a long time and significant resources to develop required features.	The intial cost is lower since the software firm is able to spread the development costs over a large number of customers.	An organization might have to pay for features that are not required and never used.
Being involved in the development offers a further level of control over the results.	In-house system development staff may become hard pressed to provide the required level of ongoing support and maintenance because of pressure to get on to other new projects.	There is a lower risk that the software will fail to meet the basic business needs—you can analyze existing features and the performance of the package.	The software may lack important features, thus requiring future modification or customization. This can be very expensive because users must adopt future releases of the software.
There is more flexibility in making modifications that may be required to counteract a new initiative by one of your competitors or to meet new supplier and/or customer requirements. A merger with another firm or an acquisition also will necessitate software changes to meet new business need.	There is more risk concerning the features and performance of the software that has yet to be developed.	Package is likely to be of high quality since many customer firms have tested the software and helped identify many of its bugs.	Software may not match current work processes and data standards.

TABLE 2.8

A Comparison of Proprietary and Off-the-Shelf Software

spreadsheet program allows a financial executive to test possible investment outcomes. The primary personal application programs are word processing, spreadsheet analysis, database, graphics, and on-line services. Advanced software tools—such as project management, financial management, desktop publishing, and creativity software—are finding more and more use in business. The features of personal application software are summarized in Table 2.9. In addition, there are literally thousands of other personal computer applications to perform specialized tasks: to help you do your taxes, get in shape, lose weight, get medical advice, write wills and other legal documents, repair your computer, fix your car, write music, and edit your pictures and videos (see Figures 2.18 and 2.19 on page 73). This type of software, often called user software or personal productivity software, includes general-purpose tools and programs that support individual needs.

Word Processing

If you write reports, letters, or term papers, word processing applications can be indispensable. Word processing applications can be used to create, edit, and print documents. Most come with a vast array of features, including those for checking spelling, creating tables, inserting formulas, creating graphics, and much more. This book (and most like it) was entered into a word processing application using a personal computer (see Figure 2.20 on page 74).

Spreadsheet Analysis

People use spreadsheets to prepare budgets, forecast profits, analyze insurance programs, summarize income tax data, and analyze investments. Whenever numbers and calculations are involved, spreadsheets should be considered.

Type of Software	Explanation	Example	Vendor
Word processing	Create, edit, and print text documents	Word WordPerfect	Microsoft Corel
Spreadsheet	Provide a wide range of built-in functions for statistical, financial, logical, database, graphics, and data and time calculations	Excel Lotus 1-2-3 Quattro Pro	Microsoft Lotus/IBM Originally developed by Borland
Database	Store, manipulate, and retrieve data	Access Approach FoxPro dBASE	Microsoft Lotus/IBM Microsoft Borland
On-line information services	Obtain a broad range of information from commercial services	America Online CompuServe Prodigy	America Online CompuServe Prodigy
Graphics	Develop graphs, illustrations, and drawings	Illustrator FreeHand	Adobe Macromedia
Project management	Plan, schedule, allocate, and control people and resources (money, time, and technology) needed to complete a project according to schedule	Project for Windows On Target Project Schedule Time Line	Microsoft Symantec Scitor Symantec
Financial management	Provide income and expense tracking and reporting to monitor and plan budgets (some programs have investment portfolio management features)	Managing Your Money Quicken	Meca Software Intuit
Desktop publishing (DTP)	Works with personal computers and high-resolution printers to create high-quality printed output, including text and graphics; various styles of pages can be laid out; art and text files from other programs can also be integrated into "published" pages	QuarkXPress Publisher PageMaker Ventura Publisher	Quark Microsoft Adobe Corel
Creativity	Helps generate innovative and creative ideas and problem solutions. The software does not propose solutions, but provides a framework conducive to creative thought. The software takes users through a routine, first naming a problem, then organizing ideas and "wishes," and offering new information to suggest different ideas or solutions	Organizer Notes	Macromedia Lotus

TABLE 2.9

Examples of Personal
Productivity Software

Features of spreadsheets include graphics, limited database capabilities, statistical analysis, built-in business functions, and much more (see Figure 2.21 on page 74).

Database Applications

Database applications are ideal for storing, manipulating, and retrieving data. These applications are particularly useful when you need to access a large amount of data and produce reports and documents. The uses of a database application are varied. You can keep track of a CD collection, the items in your apartment, and tax records using a database application. In business, a database application can help process sales orders, control inventory, order new supplies, send letters to customers, and pay employees. A database can also be a front end to another application. For example, a database application can be used to enter and store income tax information. The stored results can then be exported to other applications, such as a spreadsheet or tax preparation application (see Figure 2.22 on page 75).

E-COMMERCE

Samsung Employs ASP to Accelerate E-Commerce Initiative

Application service providers (ASPs) are transforming software from expensive, complex products into more affordable and easy-to-use services. The goal of ASPs is to make sophisticated software readily available. Rather than taking many months, or even years, to install large, complex programs, ASPs can do it in a few weeks or months. As a result, they are making it much easier and cheaper for companies to get into e-commerce.

Samsung Electronics, with 1999 sales revenue of $23 billion, is a world leader in the electronics industry. The Korea-based concern has operations in about 50 countries, with 54,000 employees worldwide. The company consists of three main business units: Digital Media Systems, Semiconductors, and Information & Communications. As the world's largest manufacturer of computer displays based on TFT-LCDs (thin film transistor-liquid crystal displays), Samsung has displays in one out of every four notebook PCs and one in every five LCD monitors.

Samsung announced in mid-February 2000 that it had developed the first-ever 64-MB SDRAM with 266-MHz speed and expected to lead the formation of a graphical digital environment. The new SDRAM is 30 percent faster than competing devices and will support three-dimensional moving pictures. Samples were provided to companies specializing in graphics controller chip sets starting in June 2000, and Samsung expected to earn $300 million from sales of this device alone in 2000.

To increase sales, Samsung planned to build a Web site for selling this chip and other memory chips—a potentially huge advantage in a market dominated by Asian competitors that are reluctant to sell on the Web. But the cost to buy the computers and software to set up the Web site was $2.5 million. Worse, it would take more than six months to set up the Web site and get it running smoothly. Such a delay would evaporate the lead Samsung had over its competitors. So Samsung found a better way—it

employed an application service provider. Instead of Samsung's resources being distracted to manage a half-dozen companies to configure the e-commerce software, set up the network, design business processes, and manage the system day-to-day, the application service provider did all these tasks. And instead of building or buying the e-commerce software, Samsung rents it. As a result, the Web site was launched in a matter of weeks. Using an ASP put Samsung even further ahead of its competition.

Discussion Questions

1. What advantages did Samsung gain through use of an application service provider?
2. Are there any potential risks with this approach for Samsung?

Critical Thinking Questions

3. Given that Samsung already has a lead on its competitors and that they are reluctant to sell their products over the Web, does the use of an application service provider to build a Web site represent an unfair advantage over the competition?
4. How would you address complaints from members of the information systems organization that use of an application service provider was reducing the amount of work they had to do?

Sources: Adapted from "Samsung Develops High-Speed SDRAM for Graphics," February 14, 2000, accessed at the Samsung Web page on March 1, 2000, at http://samsung electronics.com/news/global/ns_378_en.html; John Madden, "Digex Hosts Apps, Securely," *PC Week*, January 3, 2000, p. 32; Matt Hicks, "Great ASP-irations," *PC Week*, September 8, 1999, pp. 67, 74; Timothy S. Mullaney and Peter Burrows, "Suddenly, Software Isn't a Product, It's a Service," *Business Week*, June 21, 1999, pp. 134–138; and Michael Moeller, "Again, a Trendsetter," *Business Week*, June 21, 1999, p. 138.

Graphics Programs

With today's graphics programs, it is easy to develop attractive graphs, illustrations, and drawings. Graphics programs can be used to develop advertising brochures, announcements, and full-color presentations. If you are asked to make a presentation at school or work, you can use a graphics program to develop and display slides while you are giving your talk. A graphics program can be used to help you make a presentation, a drawing, or an illustration (see Figure 2.23 on page 75).

Software Suites

software suite

a collection of single-application software packages in a bundle

A **software suite** is a collection of single-application software packages in a bundle. Software suites can include word processors, spreadsheets, database management systems, graphics programs, communications tools, organizers, and more. There are a number of advantages to using a software suite. The software programs have been designed to work similarly, so that once you learn the

FIGURE 2.18

TurboTax

Tax preparation programs can save hours of work and are typically more accurate than doing a tax return by hand. Programs can check for potential problems and give you help and advice about what you may have forgotten to deduct.
(Source: Courtesy of Intuit)

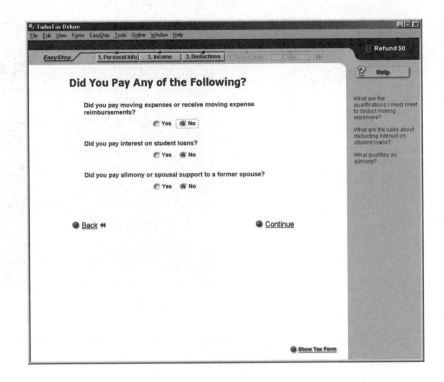

basics for one application, the other applications are easier to learn and use. Buying software in a bundled suite is cost-effective: the programs usually sell for a fraction of what they would cost individually.

Microsoft Office, Corel's WordPerfect Office, and Lotus SmartSuite are examples of popular general-purpose software suites for personal computer users (see Figure 2.24 on page 76). Each of these software suites includes a spreadsheet program, word processor, database program, and graphics package with the ability to move documents, data, and diagrams among them. Thus, a user can create a spreadsheet and then cut and paste that spreadsheet into a document created using the word processing application.

Since one or more applications in a suite may not be as desirable as the others, some people still prefer to buy separate packages. Another issue with the use of software suites is the large amount of main memory required to run them

FIGURE 2.19

Quicken

Off-the-shelf financial management programs are useful for paying bills and tracking expenses.
(Source: Courtesy of Intuit)

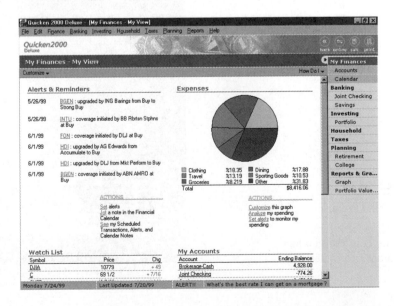

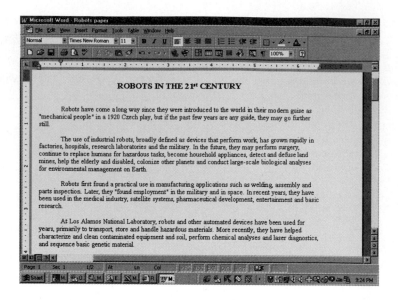

effectively. For example, many users find that they must spend hundreds of dollars for additional internal memory to upgrade their personal computer to be able to run a software suite.

Workgroup Application Software

Workgroup application software, often called *groupware*, helps groups of people work together more efficiently and effectively toward a common goal. Such software can support a team of managers working on the same production problem, letting them share their ideas and work via connected computer systems. Examples of such software include group scheduling software, electronic mail, and other software that enables people to share ideas. New York–based Swiss Reinsurance America deployed a claims processing application to more than 300 workers. By streamlining its work processes and using workgroup software, it was able to reduce the number of steps needed to process a claim from 18 to 7 and reduce the time it takes to process a claim from three days to one.

FIGURE 2.22

Database Program

Once entered into a database application, information can be manipulated and used to produce reports and documents.

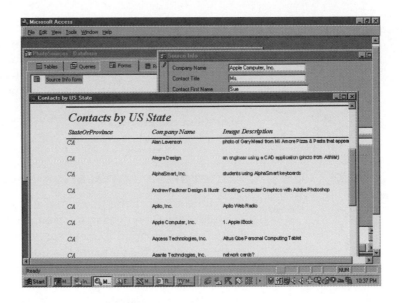

Lotus Notes and Domino

Lotus has defined *knowledge management* as the ability to provide individuals and groups of users with a method to find, access, and deliver valuable information in a coherent fashion. Its Lotus Notes product is an attempt to provide this ability. Lotus Notes gives companies the capability of using one software package, and one user interface, to integrate many business processes. For example, it can allow a global team to work together from a shared set of documents, have electronic discussions using common threads of discussion, and schedule team meetings. As Lotus Notes matured, Lotus added services to it and renamed it Domino, and now an entire market has emerged to build collaborative software based on Domino. For example, Domino.Doc is a Domino-based document management application with built-in workflow and archiving capabilities. Its "life cycle" feature tracks a document through the review, approval, publishing, and archiving processes. Similarly, the workflow integration adds support for multiple roles, log tracking, and distributed approval.

FIGURE 2.23

Graphics Program

Graphics programs can help you make a presentation at school or work. They can also be used to develop attractive brochures, illustrations, drawings, and maps.
(Source: Courtesy of Adobe Systems, Inc.)

FIGURE 2.24

Software Suite

A software suite, such as Microsoft Office, offers a collection of powerful programs, including word processing, spreadsheet, database, graphics, and other programs. The programs in a software suite are designed to be used together. In addition, the commands, icons, and procedures are the same for all programs in the suite.
(Source: Courtesy of Microsoft Corporation)

Group Scheduling

Group scheduling is another form of workgroup software, but not all software schedulers approach their tasks the same way. Some schedulers, known as personal information managers (PIMs), tend to focus on personal schedules and lists, as opposed to coordinating the schedules and meetings of a team or group. Schedulers do not suit everyone's needs, and if they are not truly required, they could impede efficiency. The "Three Cs" rule for successful implementation of groupware is summarized in Table 2.10.

Enterprise Application Software

Software that benefits the entire organization can also be developed or purchased. A fast-food chain, for example, might develop a materials ordering and distribution program to make sure that each fast-food franchise gets the necessary raw materials and supplies during the week. This materials ordering and distribution program can be developed internally using staff and resources in the IS department or purchased from an external software company. Table 2.11 lists a number of applications that can be addressed with enterprise software.

Many organizations are moving to integrated enterprise software that supports supply chain management (movement of raw materials from suppliers through shipment of finished goods to customers), as shown in Figure 2.25.

Organizations can no longer respond to market changes using nonintegrated information systems based on overnight processing of yesterday's business transactions, conflicting data models, and obsolete technology. As a result, many corporations are turning to *enterprise resource planning (ERP)* software, a set of integrated programs that manage a company's vital business operations for an entire multisite, global organization. An ERP system must be able to support multiple legal entities, multiple languages, and multiple currencies. Although the scope of an ERP system varies from vendor to vendor, most ERP systems provide integrated software to support manufacturing and finance. In addition to these core business processes, some ERP systems support additional business functions such as human resources, sales, and distribution. Software vendors that provide integrated enterprise software are listed in Table 2.12 on page 78.

Most ERP vendors specialize in software that addresses the needs of well-defined markets, such as automotive, semiconductor, petrochemical, and food/beverage manufacturers, with solutions targeted to meet their specific needs. Increased global competition, new executive management needs for control over the total cost and product flow through their enterprises, and more customer interactions are driving the demand for enterprisewide access to current business information. ERP offers integrated software from a single vendor that helps meet those needs. The primary benefits of implementing ERP include eliminating inefficient systems, easing adoption of improved work processes, improving access to data for operational decision making, standardizing technology vendors and equipment, and enabling the implementation of supply chain management.

Now that we have discussed the software that businesses use to accomplish their daily tasks, we turn to a brief discussion of programming languages—the

TABLE 2.10

Ernst & Young's "Three Cs" Rule for Groupware

Convenient	If it's too hard to use, it doesn't get used; it should be as easy to use as the telephone.
Content	It must provide a constant stream of rich, relevant, and personalized content.
Coverage	If it isn't close to everything you need, it may never get used.

TABLE 2.11

Examples of Enterprise
Application Software

Accounts receivable	Sales ordering
Accounts payable	Order entry
Airline industry operations	Payroll
Automatic teller systems	Human resource management
Cash-flow analysis	Check processing
Credit and charge card administration	Tax planning and preparation
Manufacturing control	Receiving
Distribution control	Restaurant management
General ledger	Retail operations
Stock and bond management	Invoicing
Savings and time deposits	Shipping
Inventory control	Fixed asset accounting

programs that IS professionals—and even everyday users today—can use to write new applications.

PROGRAMMING LANGUAGES

programming languages

coding schemes used to write both
systems and application software

FIGURE 2.25

Use of Integrated Supply Chain
Management Software

Both systems and application software are written in coding schemes called **programming languages**. The primary function of a programming language is to provide instructions to the computer system so that it can perform a processing activity. IS professionals work with programming languages, which are sets of symbols and rules used to write program code. Programming involves translating what a user wants to accomplish into a code that the computer can understand and execute. Like writing a report or a paper in English, writing a computer program in a programming language requires that the programmer follow a set of rules. Each programming language uses a set of symbols that have special meaning. Each language also has its own set of rules, called the

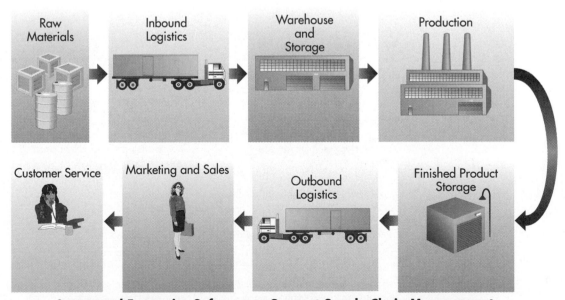

Integrated Enterprise Software to Support Supply Chain Management

TABLE 2.12

Selected Enterprise Resource
Planning Vendors

SAP	Baan
Oracle	SSA
PeopleSoft	Marcam
Dun & Bradstreet	QAD
JD Edwards	Ross Systems

syntax

a set of rules associated with a
programming language

syntax of the language. The language syntax dictates how the symbols should be combined into statements capable of conveying meaningful instructions to the CPU.

Programming languages were developed to help solve particular problems. Since they were each designed for different problems, they contain different attributes. Each of the attributes in Table 2.13 represents two extremes, with most languages falling somewhere between these extremes.

The Evolution of Programming Languages

The desire to use the power of information processing efficiently in problem solving has pushed development of newer programming languages. The evolution of programming languages is typically discussed in terms of generations of languages.

First-Generation Languages

The first generation of programming languages is *machine language,* which required the use of binary symbols (0s and 1s). As this is the language of the CPU, text files translated into binary sets can be read by almost every computer system platform.

Second-Generation Languages

Developers of programming languages attempted to overcome some of the difficulties inherent in machine language by replacing the binary digits with symbols programmers could more easily understand. Assembly languages use codes like A for add, MVC for move, and so on. This second-generation language was

TABLE 2.13

Programming Language
Attributes

Extreme 1	Extreme 2
Supports programming of batch processing systems with data collected into a set and processed at one time.	Supports programming of real-time systems with each data transaction processed when it occurs.
Requires programmer to write procedure-oriented code, describing step by step each action the computer must take.	Enables a programmer to write nonprocedure-oriented code, describing the end result desired without having to specify how to accomplish it.
Supports business applications that require the ability to store, retrieve, and manipulate alphanumeric data and process large files.	Supports sophisticated scientific computations.
Programmers write code with a relatively high level of errors.	Programmers write code with a relatively low level of errors.
Programmers are less productive and able to create only a small amount of code per unit time.	Programmers are more productive and are able to create a large amount of code per unit time.

termed *assembly language,* after the system programs, called assemblers, used to translate it into machine code. Systems software programs such as operating systems and utility programs are often written in an assembly language.

Third-Generation Languages

Third-generation languages continued the trend toward greater use of symbolic code and away from specifically instructing the computer how to complete an operation. BASIC, COBOL, C, and FORTRAN are examples of third-generation languages that use Englishlike statements and commands. This type of language is easier to learn and use than machine and assembly languages because it more closely resembles everyday human communication and understanding. With third-generation programming languages, each statement in the language translates into several instructions in machine language. In addition, third-generation languages take the programmer one step further from directing the actual operation of the computer. Although easier to program, third-generation languages are not as efficient in terms of speed and memory.

Fourth-Generation Languages

Fourth-generation programming languages emphasize what output results are desired rather than how programming statements are to be written. As a result, many managers and executives with little or no training in computers and programming are using fourth-generation languages (4GLs). Prime examples include Visual C++, Visual Basic, PowerBuilder, Delphi, Forte, Focus, Powerhouse, and SAS. Another popular fourth-generation language is called *structured query language (SQL),* which is often used to perform database queries and manipulations.

Fifth-Generation Languages

With fifth-generation languages, you do not tell the computer how to do a job, but what you want it to do, and it figures out what you need. SunSoft is currently testing Java Studio, a connect-the-dots style of visual development toolset that uses components known as Java Beans. Java Studio is so simple to use that you can easily develop a working program without any previous programming experience.

Object-Oriented Programming Languages

The preceding programming languages separate data elements from the procedures or actions that will be performed on them, but object-oriented programming languages tie them together into objects. Thus, an object consists of data and the actions that can be performed on the data. For example, an object could be data about an employee and all the operations (such as payroll calculations) that might be performed on the data.

Building programs and applications using object-oriented programming languages is like constructing a building using prefabricated modules or parts. The object containing the data, instructions, and procedures is a programming building block. The same objects (modules or parts) can be used repeatedly. An object can relate to data on a product, an input routine, or an order-processing routine. An object can even direct a computer to execute other programs or to retrieve and manipulate data. One of the primary advantages of an object is that it contains *reusable code.* In other words, the instruction code within that object can be reused in different programs for a variety of applications, just as the same basic prefabricated door can be used in two different houses. Thus, a sorting routine developed for a payroll application could be used in both a billing program and an inventory control program. By reusing program code, programmers can write programs for specific application problems more quickly (see Figure 2.26). By combining existing program objects with new ones, programmers can easily and efficiently develop new object-oriented programs to accomplish organizational goals.

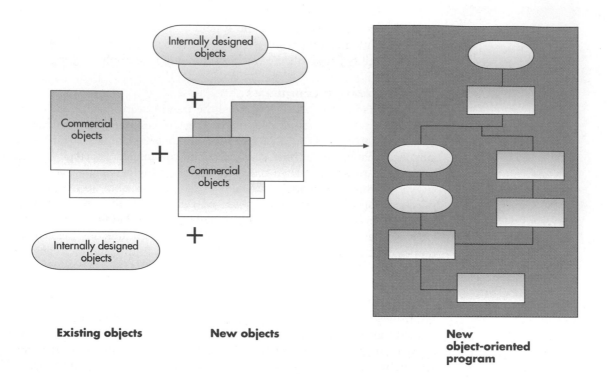

Existing objects   **New objects**   **New
                                                                object-oriented
                                                                program**

FIGURE 2.26

Reusable Code in Object-Oriented Programming

By combining existing program objects with new ones, programmers can easily and efficiently develop new object-oriented programs to accomplish organizational goals. Note that these objects can be either commercially available or designed internally.

There are several object-oriented programming languages; some of the most popular include Smalltalk, C++, and Java. One of the key advantages of Java is the ability of a Java application to run on a variety of computers and operating systems, including Unix, Windows, and Macintosh operating systems. An increasing number of colleges in the United States are using Java as their first programming language.

Visual Programming Languages

Visual programming languages use a visual environment including a mouse, icons, and pull-down menus to make programming easier and more intuitive. Visual Basic, PC COBOL, and Visual C++ are examples of visual programming languages.

The various languages have certain characteristics that make them appropriate for particular types of problems or applications. Among the third-generation languages, COBOL has excellent file-handling and database-handling capabilities for manipulating large volumes of business data, while FORTRAN is better suited for scientific applications. Java is an obvious choice for Web developers. End users will choose one of the fourth- or fifth-generation languages to develop programs. Although many programming languages are used to write new business applications, more lines of code are written in COBOL in existing business applications than any other programming language.

language translator

systems software that converts a programmer's source code into its equivalent machine language

source code

high-level program code written by the programmer

object code

machine language code

Language Translators

Because machine language programming is extremely difficult, very few programs are actually written in machine language. However, machine language is the only language capable of directly instructing the CPU. Thus, every non-machine-language program instruction must be translated into machine language prior to its execution. This is done by systems software called **language translators**. A language translator converts a programmer's source code into its equivalent in machine language. The high-level program code is referred to as the **source code**, whereas the machine language code is referred to as the **object code**.

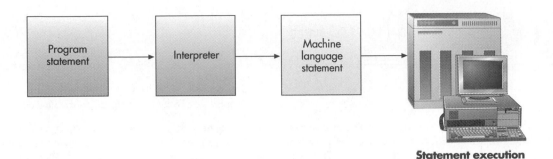

Statement execution

FIGURE 2.27

How an Interpreter Works

An interpreter translates each program statement or instruction in sequence. The CPU then executes the statement, erases it from memory, and translates another statement. An interpreter does not produce a complete machine-readable version of a program.

interpreter

a language translator that translates one program statement at a time into machine code

compiler

a language translator that converts a complete program into a machine language to produce a program that the computer can process in its entirety

There are two types of language translators—interpreters and compilers. An **interpreter** translates one program statement at a time, as the program is running. It displays on screen any errors it finds in the statement (Figure 2.27). This line-by-line translation makes interpreters ideal for those who are learning programming, but it does slow down the execution process.

A **compiler** is a language translator that converts a complete program into a machine language to produce a program that the computer can process in its entirety (Figure 2.28). Once the compiler has translated a complete program into machine language, the computer can run the machine language program as many times as needed. A compiler creates a two-stage process for program execution. First, it translates the program into a machine language; second, the CPU executes that program.

Software Bugs

To detect computer program logic errors, programmers usually must run through test data and check the results of the program against the results of running the same data by hand or calculator. For example, if you made the logic error $A = B + C$ instead of the correct $A = B - C$, then you tell the computer

Stage 1:   **Convert program**

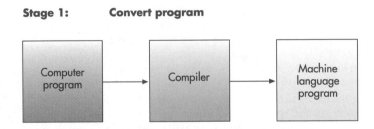

Stage 2:   **Execute program**

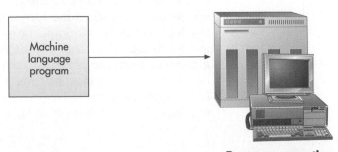

Program execution

FIGURE 2.28

How a Compiler Works

A compiler translates a complete program into a complete set of machine instructions (Stage 1). Once this is done, the CPU can execute the converted program in its entirety (Stage 2).

INFORMATION SYSTEMS IN ACTION
The Microprocessor—A Brief History

Dr. James E. LaBarre
Department of MIS, University of Wisconsin–Eau Claire

The microprocessor was first used in a calculator in the early 1970s. The Intel 4004 microprocessor enabled the inventors to store mathematical functions. This first calculator allowed other developers to see the future of the microprocessor. Later the Intel 8008 was used in the development of an electronic typewriter containing a video display. Shortly afterward, the 8080 microprocessor was used in a computer hobbyist kit. This first computer was named the Altair and was built by people interested in electronics.

By 1977, the MOS Technology 6502 microprocessor was being used in the first Apple computer. In 1978 IBM started using the Intel 8086-8088 in its first personal computer. This event gave the whole microcomputer era the momentum that has carried into the twenty-first century.

Increasing software needs required chip manufacturers to develop faster processors. Early in 1981 Apple Computer introduced the LISA. Using the Motorola MC68000 microprocessor, this computer provided the opportunity to produce a graphical user interface (GUI), and its microprocessor became the standard that was used in the Macintosh and succeeding Apple computers throughout the 1980s. In 1982 Intel developed the 80286. This processor created a de facto standard that has had an enormous effect on computing. Most microcomputer manufacturers attempted to match IBM's processor by building an IBM compatible microcomputer. With the exception of the Apple, most computers are still referred to as "IBM compatible" today.

In 1985, the Intel 386 provided the capability to do "multitasking," or to complete functions in the background. People could print while working on a different word processing document. This microprocessor also spurred the movement to Microsoft Windows. By 1989 the introduction of the Intel 486 made point-and-click computing easier. The increased speed of the processor provided options such as colored graphics and a math coprocessor.

The Pentium processor, developed in 1993, was often dubbed the "586." Although plagued by manufacturing errors in the beginning, it has emerged as the chip that has transformed computing. This chip was enhanced with the introduction of the Pentium Pro in 1995, adding features such as sound, character recognition, graphics, and networking. Recognizing the need to compete in this marketplace, Apple started using the 603 microprocessor, which provided the various versions of the Mac with basically the same capabilities.

By 1997 small and large businesses were looking for a microcomputer-based server that would meet their needs. Intel further enhanced the Pentium processor with the Pentium II MMX processor. Music and video graphic features were greatly enhanced. Shortly thereafter the Pentium II Xeon processor was introduced. This processor enabled computer manufacturers to build high-end servers that greatly facilitated local area networks.

The Pentium III processor was introduced in 1999. This high-end processor greatly enhanced the speed of computing for video display and voice activation. Computers operating at speeds of 1 gigahertz or faster are being used for video streaming, high-speed Internet access, and database queries in business and industry. Again Apple Computer met this challenge by introducing the PowerPC 750 at about the same time.

Intel and Motorola are not the only developers of microprocessors. Texas Instruments, Advance Micro Devices, Digital Equipment Corporation, and IBM are just a few of the other companies also producing microprocessors that have been or continue to be used in microcomputers.

Sources: "AMD-K6 Chipset and BIOS Vendors," AMD Web site at http://www.amd.com/K6/k6mbl/bios.html; Apple History Web site at http://www.apple-history.com; Yahoo Web site at http://dir.yahoo.com/Computers_and_Internet/Hardware/Components/Microprocessors/Motorola_68K/; "A History of the Microprocessor," Intel Web site at http://www.intel.com/intel/museum/25anniv/index.htm; and "History of Computers," Community Learning Network Web site at http://www.cln.org/themes/computer_history.html.

that B equals 42 and C equals 13 and ask it to find A, you'll quickly find the erroneous + sign when the answer comes out 55 instead of 29. Unfortunately, few logic errors are so simple or so easily detected. Most computer programs involve thousands of lines of code, so it can take years, even for teams of programmers, to debug programs such as those used to control emergency shutdown systems on nuclear reactors. According to the Pentagon and the Software Engineering Institute at Carnegie Mellon University, there are typically 5 to 15 bugs in every 1,000 lines of code.

● SUMMARY

PRINCIPLE • Information system users must work closely with information system professionals to define business needs, evaluate options, and select the hardware and software that provide a cost-effective solution to those needs.

Hardware devices work together to perform input, processing, data storage, and output. Processing is performed by an interplay between the central processing unit (CPU) and memory. Primary storage, or memory, provides working storage for program instructions and data to be processed and provides them to the CPU. Together, a CPU and memory process data and execute instructions.

Processing that uses several processing units is called multiprocessing. One form of multiprocessing uses coprocessors; coprocessors execute one type of instruction while the CPU works on others. Parallel processing involves linking several processors to work together to solve complex problems.

Computer systems can store larger amounts of data and instructions in secondary storage, which is less volatile and has greater capacity than memory. Storage media can be either sequential access or direct access. Common forms of secondary storage include magnetic tape, magnetic disk, optical disk storage, and PC memory cards. Redundant array of independent/inexpensive disks (RAID) is a method of storing data that allows the system to more easily recover data in the event of a hardware failure. Storage area network (SAN) uses computer servers, distributed storage devices, and networks to provide fast and efficient storage.

Input and output devices allow users to provide data and instructions to the computer for processing and allow subsequent storage and output. These devices are part of a user interface through which humans interact with computer systems. Input and output devices vary widely, but they share common characteristics of speed and functionality.

Scanners are input devices that convert images and text into binary digits. Point-of-sale (POS) devices are terminals with scanners that read and enter codes into computer systems. Automatic teller machines (ATMs) are terminals with keyboards used for transactions.

Output devices provide information in different forms, from hard copy to sound to digital format. Display monitors are standard output devices; monitor quality is determined by size, number of colors that can be displayed, and resolution. Other output devices include printers and plotters.

* * *

The six computer system types are network computer, personal computer, workstation, midrange computer, mainframe computer, and supercomputer. The network computer is a diskless, inexpensive computer used for accessing server-based applications and the Internet. Personal computers (PCs) are small, inexpensive computer systems. Two major types of PCs are desktop and laptop computers. Workstations are advanced PCs with greater memory, processing, and graphics abilities. Filing cabinet–size minicomputers have greater secondary storage and support transaction processing. Even larger mainframes have higher processing capabilities, while supercomputers are extremely fast computers used to solve the most intensive computing problems.

* * *

There are two main categories of software: systems software and application software. Systems software is a collection of programs that act as a buffer between hardware and application software. Application software enables people to solve problems and perform specific tasks. Application software may be proprietary or off-the-shelf. Although there are literally hundreds of computer applications that can help individuals at school, home, and work, the primary ones are word processing, spreadsheet analysis, database, graphics, and on-line services.

* * *

An operating system (OS) is a set of computer programs that controls the computer hardware to support users' computing needs. OS hardware functions convert an instruction from an application into a set of instructions needed by the hardware. The OS also serves as an intermediary between application programs and hardware, allowing hardware independence. Memory management involves controlling storage access and use by converting logical requests into physical locations and by placing data in the best storage space. Task management allocates computer resources through multitasking and time-sharing. With multitasking, users can run more than one application at a time. Time-sharing allows more than one person to use a computer system at the same time. An OS also provides a user interface, which allows users to access and command the computer.

Over the years, several popular operating systems have been developed. These include several proprietary operating systems used primarily on mainframes. MS-DOS is an early OS for IBM-compatibles. Older Windows operating systems are GUIs used with DOS. Newer versions, such as Windows 95, Windows 98, and Windows NT, are fully functional operating systems that do not need DOS. Apple computers typically use a proprietary operating system, the Mac OS. Unix is the leading portable operating system, usable on many computer system types and platforms.

PRINCIPLE • Organizations do not develop proprietary application software unless doing so will meet a compelling business need that can provide a competitive advantage.

Application software applies the power of the computer to solve problems and perform specific tasks. Application software can support individuals, groups, and organizations. User software, or personal productivity software, includes general-purpose programs that enable users to improve their personal effectiveness, increasing the amount of work and its quality. Software that helps groups work together is often referred to as groupware. Enterprise software that benefits the entire organization can also be developed or purchased.

• • •

Three approaches organizations use to obtain applications are as follows: (1) build proprietary application software, (2) buy existing programs off the shelf, or (3) use a combination of customized and off-the-shelf application software.

PRINCIPLE • End users and IS professionals use a programming language whose functional characteristics are appropriate to the task at hand.

There are five generations of programming languages, plus object-oriented and visual programming languages. End users learn and typically use fourth-generation programming languages such as Visual C++, Visual Basic, PowerBuilder, Delphi, Forte, Focus, Powerhouse, SAS, and SQL. These fourth-generation languages are less procedural and more English like than the first three generations of programming languages, typically used by IS professionals.

Object-oriented programming languages, such as C++, Smalltalk, and Java, use groups of related data, instructions, and procedures called objects, which serve as reusable modules in various programs. These languages can reduce program development and testing time. Java can be used to develop applications on the Internet.

● REVIEW QUESTIONS

1. Why is it said that the components of all information systems are interdependent?
2. Explain the two-phase process for executing instructions.
3. What is Moore's Law?
4. Describe the various types of memory.
5. Describe various types of secondary storage media in terms of access method, capacity, and portability.
6. What are the computer system types? How do these types differ?
7. What is a computer system platform? Give two examples.
8. Give four examples of personal productivity software.
9. What is meant by scalability? Why is it important?

10. Name four operating systems that support the personal, workgroup, and enterprise spheres of influence.
11. Identify the two primary sources for acquiring application software.
12. What is an application service provider? What issues arise in considering the use of one?
13. What is an operating system? What are the key activities that it performs?
14. Discuss the advantages and disadvantages of using a set of individual personal productivity applications from multiple vendors versus an integrated set of applications in a software suite.
15. Define *enterprise resource planning (ERP) system*. What functions does such a system perform?

● DISCUSSION QUESTIONS

1. What are the implications of Moore's Law—continuing the trend of increased computing power at lower costs? Use Moore's Law to forecast the computing power that could be available in three years. What sort of applications could benefit from that level of computer power?

2. What functions and capabilities is it reasonable to expect palmtop computers to perform? Are there functions that they should be able to perform that a desktop computer cannot perform?

3. Imagine that you are the business manager for your university. What type of computer would you recommend for broad deployment in the university's computer labs—a standard desktop personal computer or a network computer? Why?

4. If cost were not an issue, describe the characteristics of your ideal laptop computer.

5. Assume that you must take a computer programming course next semester. What language do you think would be best for you to study? Why? Do you think that a professional programmer needs to know more than one programming language? Why or why not?

6. Imagine that you have been assigned to choose the software suite to be adopted by your organization. On what basis would you make this decision? How would you do a comparison of the leading contenders—Microsoft Office, Corel WordPerfect Office, and Lotus SmartSuite? How much consideration would you place on the availability of good user documentation?

7. How can application software improve the effectiveness of a small workgroup? What are some of the benefits associated with implementation of groupware? What are some of the issues that may arise that could keep the use of groupware from being successful?

8. Many organizations have mandated that off-the-shelf application software be the first choice when implementing new software and that the software not be customized at all. What are the advantages and disadvantages of such an approach? What would you recommend for your firm if you were the manager in charge of application development? If you were a senior vice president of a business area?

● PROBLEM-SOLVING EXERCISES

1. Choose a fourth-generation programming language of interest to you and develop a slide presentation of its history, current level of usage, typical applications, ease of use, and so on.

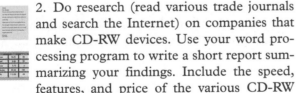

2. Do research (read various trade journals and search the Internet) on companies that make CD-RW devices. Use your word processing program to write a short report summarizing your findings. Include the speed, features, and price of the various CD-RW devices. Also summarize internal and external CD-RW devices. Develop a simple spreadsheet to compare the features and costs of the CD-RW devices you found.

3. Over the upcoming year, your department expects to add eight people to its staff. You will need to acquire eight personal computer systems and two additional printers for the new employees to share. Standard office computers have a Pentium (600 MHz) processor with 64 MB of RAM, SVGA color monitor, and a minimum 3-GB hard disk drive. At least four of the new people will use their computers more than three hours per day. You would like to provide larger monitors and special ergonomic keyboards for these people—if it fits within your budget. You are not sure whether you want to upgrade the machines to 96 MB of RAM and 6-GB hard drives.

Your department budget allows a maximum of $20,000 for computer hardware purchases this year, and you want to select only one vendor for all of the hardware. A price list from three vendors appears in the accompanying table, with prices for a single unit of each component. Use a spreadsheet to find the department's best solution; write a short memo explaining your rationale. Specify which vendor to choose and which items to order as well as the total cost.

4. Choose a graphics package and track data of interest to you. Produce a simple line chart that shows at least ten values for each of two variables over a two-week period—high and low stock price of a particular company, high and low temperature at your favorite vacation location, and so on. The graph must have a title, and the two line graphs must be labeled with the names of the variables. Use a word processing package to write a paragraph summarizing the data. Cut and paste the graph into the word processing document.

Component	Expert Solutions Ltd.	Business Processing Enterprises	Super Systems Inc.
600 MHz Pentium with 64 MB RAM 4.8-GB hard drive	$1,245	$1,275	$1,200
Upgrade to 128 MB RAM	250	225	245
Upgrade to 10-GB hard drive	190	215	205
15-inch .28-dpi SVGA monitor	350	330	340
17-inch .28-dpi SVGA monitor	625	600	615
Ergonomic keyboard	55	50	50
8-ppm color ink-jet printer	325	325	320
Surge protector/ power strip	35	32	35
Three-year warranty (parts and labor)	340	300	320

● TEAM ACTIVITIES

1. With two of your classmates, visit a major computer retail store (for example, CompUSA, MicroCenter). Spend a couple of hours during which each of you concentrates on identifying the latest developments in processing, input, and output devices. Write a brief report summarizing your findings.
2. Identify a manager of a computer center and seek his or her permission to tour the facility. Make a list of various types of computers and input, output, and secondary storage devices that you see during your tour. Does the center have more than one class of computer? If so, why? What types of jobs are run on each class of computer?
3. Form a group of three or four classmates. Find articles from business periodicals, search the Internet, or interview people on the topic of software bugs. How frequently do they occur, and how serious are they? What can software users do to encourage defect-free software? Compile your results for an in-class presentation or a written report.

● WEB EXERCISES

1. Search the Web for companies that make secondary storage devices and systems, including disk, tape, RAID, SAN, and others. Summarize your findings using your word processing program.
2. Visit the Web sites of three PC makers. Your search can include Dell, Gateway, IBM, Compaq, and others. From each of the three Web sites, summarize your ideal PC system. Develop a spreadsheet that compares features and prices. Which system and manufacturer do you prefer?
3. Do research on the Web and develop a two-page report on object-oriented programming.

● CASES

 Electronic Ink

One early prediction of the computer revolution was the elimination of paper. Forecasters thought that offices and homes would not require it; everything would be done electronically, saving trees, if not entire forests. This prediction, however, did not come true. Even with all the technological breakthroughs, it seems that paper is being used to a greater extent today in homes and offices than ever before. If the racks of paper at Office Depot, Staples, and Office Max are any indication, it seems that paper is here to stay. Although the trend to using more paper is likely to remain, some advances are making paper less of a necessity. One of these is a result of a new printing technology, using what some call "electronic ink."

E-ink was first conceived in the 1970s by Xerox, the same company that pioneered copiers and the graphical user interface that is typical in Apple computers and PCs using Windows. All of these ideas came out of the famous Palo Alto Research Center (PARC). The initial work was done using microscopic balls that were half white and half black. When an electrical charge was applied, the balls would turn to show either white or black.

The idea of printing using electronic ink instead of traditional printing on paper was intriguing. Yet, Xerox decided to concentrate on copiers, and electronic ink was put on hold. Although Xerox would restart its research into electronic ink in the 1990s, other companies would also investigate the technology.

In the mid-1990s, a young physicist at the MIT Media Lab hired some students to look into developing an electronic book. One of the students, Barrett Comiskey, experimented with several approaches, including balls painted half white and half black as at PARC. After many failures, he tried the idea of all white balls in a mixture of oil and dark dye. When the white balls were on the surface, you could see the white color. When the balls were submerged, you would see black. These tests led to a workable solution, and Comiskey and a few others formed a new company, E-Ink Corporation.

Today, E-Ink is testing its electronic ink in stores such as J.C. Penney. A four-foot-by-four-foot foam board from E-Ink is being tested in the athletic-wear department. The new technology allows J.C. Penney to display messages such as, "Hello down there! You've got to try the Nike Air Quest. It's super comfortable and lightweight too." With a few keystrokes, the message can be changed. The success of the trials has resulted in financial backing from such companies as IBM and Motorola.

Discussion Questions

1. What other applications would be good for the printing technology developed by E-Ink?
2. What applications would still require paper?

Critical Thinking Questions

3. A similar idea is to develop electronic books. Do you think books that display text and figures electronically will ever replace today's textbooks and novels?
4. Compare electronic ink to a flat-panel display. What are the advantages and disadvantages of each?

Sources: Adapted from Alec Klein, "Will the Future Be Written in E-Ink?" *The Wall Street Journal*, January 4, 2000, p. B1; and "Showing Off," *The Economist*, October 30, 1999.

 CD-ROM Titles

Truck and auto repair shops can devote more space to manuals than parts. These large and sometimes confusing manuals can be as thick as large phone books and difficult to use. A mechanic can spend hours searching for the correct manual and part information. If a page is ripped out or missing, it can be very frustrating. In addition, new manuals can be very expensive. With new car and truck models coming out every year, the cost can add up quickly. As a result, many auto and truck companies are starting to put their repair manuals on CD-ROMs instead of printing traditional paper manuals.

The CD-ROM manuals offer a number of advantages. They are much cheaper to produce and send through the mail. Easy-to-use search routines can make it a snap for mechanics to find the information they need. Each year, the auto company can send a CD containing repair manuals for multiple car and truck models. Yet although the CD-ROM repair manuals offer numerous advantages over printed manuals, there are still problems and hassles. Car mechanic Charles Schultz experienced some headaches when he used CDs in his shop.

Schultz specializes in foreign cars, such as BMWs. It was a blessing when BMW started to convert more of its shop repair manuals to CD-ROM. Using a laptop computer, Schultz could quickly find the information he needed by inserting the CD into the drive in his laptop and using the built-in search routines. But Schultz also quickly realized that his growing library of CD repair manuals had its own problems. Often he would have to scrub his hands and arms of work-related grease before he inserted a new, clean CD into his laptop. Getting dirt and grime on a printed repair manual was no problem, but dirt and grime on a CD could destroy both the CD and the CD drive in his laptop.

To solve this problem, Schultz found an inexpensive software solution. For $35, Schultz purchased Virtual Drive 2000, which simplified the task of downloading portions of a CD-ROM to a hard disk. Now Schultz can load only the information he needs from multiple CD-ROMs onto his hard disk. In addition to the advantage of not having to wash his hands every few minutes to load another CD-ROM into his laptop, Schultz also found that he could retrieve the needed information much faster from his hard disk than from the CD-ROM drive.

Discussion Questions

1. How was Charles Schultz able to take advantage of the convenience of repair manuals on CD-ROM while eliminating some of the problems?
2. With what other applications could this approach be used?

Critical Thinking Questions

3. What other features would Schultz likely find useful with the software solution described in this case?
4. Could Schultz's solution be used in other applications?

Sources: Adapted from Mitt Jones, "Hassle-Free Access to CD-ROM Titles," *PC World*, February 2000, p. 43; and Brian Quinton, "Ready to Play," *Telephony*, November 15, 1999.

Flash Chips

Sweden is known for its world-class Nordic skiers, people with fair skin and blond hair, and a telecommunications giant, Ericsson. The company has quarterly results of about $430 million in profits on about $6 billion in revenues. The president of Ericsson is a Harley-Davidson motorcycle fan and a hard-driving executive. He has been quoted as saying, "I will be here until they kick me out, I retire, or I die."

Not only is Ericsson a leader in networks, it is also a leader in cell phones and related equipment. Cell phone usage is expected to explode worldwide as technology improves and the cost of cellular service continues to decline. Today, there are about 275 million mobile phones in use worldwide, and the growth rate is expected to be about 30 percent per year for the next several years.

A key component of any cell phone is the memory device, which is used to store phone numbers and internal operating instructions for the phone. To get quality memory devices for its cell phones, Ericsson turned to Intel, the leading hardware maker of chips and memory devices. The move will guarantee that flash memory chips will be available for Ericsson when the demand for these memory devices skyrockets.

In a recently announced deal, Ericsson agreed to purchase about $1.5 billion in flash memory chips from Intel over a three-year period. Unlike the standard volatile memory devices used in computers, flash chips are not volatile. They retain their contents even when the power is turned off. In addition to computers and mobile phones, flash chips are used in handheld personal organizers, digital cameras, and music players. According to Benny Ginman, Intel's Director of European sales and marketing, "If you look at the growth in cell phones, you can see why we want to be a major supplier to this industry."

To provide high-quality, low-cost chips, Intel continues to innovate. Using new manufacturing and fabrication techniques, Intel has been able to streamline its operations and reduce costs. By mid- or late 2000, the company is expected to migrate from more traditional .25-micron to .18-micron flash chips. This change should allow Intel to decrease its cost per megabit by about 30 percent annually. To meet increased demand, Intel is also acquiring additional manufacturing facilities. In a separate deal, Intel agreed to purchase two older chip factories from Rockwell International. Intel plans to completely renovate the chip factories for a total cost of about $1.5 billion.

In addition to flash memory, Intel continues to be a leader in Pentium processors used in personal computers. Annual sales for Intel are approximately $30 billion. About 80 percent of these revenues comes from processors, such as the Pentium. The remaining 20 percent comes from flash memory and other types of chips used in a variety of operations.

Discussion Questions

1. Describe the characteristics of flash memory compared with normal memory.
2. What types of devices are ideal for flash memory?

Critical Thinking

3. What are the advantages to Ericsson of making a flash memory deal with Intel?
4. Will flash memory become more or less important in the future?

● NOTES

Sources for the opening vignette on p. 39: Adapted from Bryan Larsen, "Home Depot Strives for IT Simplicity," *Enterprise Development*, March 2000, pp. 10–19; Home Depot Web page, "Company Info and Financial Info," http://www.homedepot.com, accessed March 28, 2000; and Craig Stedman, "Java Fuels Home Depot Expansion," *Computerworld*, August 23, 1999, p. 34.

1. Kathleen Melymuka, "CEO Meg Whitman Powers eBay," *Computerworld*, January 10, 2000.
2. "Desktops Hit 750 MHz," *PC Magazine*, January 18, 2000, p. 11.
3. Cade Metz, "A Cool 1,000 MHz," *PC Magazine*, February 22, 2000, p. 50.
4. Michael Miller, "Moore's Law Will Continue to Drive Computing," *PC Magazine*, June 22, 1999, p. 146.
5. Mitch Wagner, "SAN Backs Up Financial Assets," *Internet Week*, April 26, 1999, p. 19.
6. Alfred Poor, "Optical-Drive Compatibility," *PC Magazine*, January 18, 2000, p. 124.
7. Gary Anthes, "Experiment with Voice Recognition," *Computerworld*, January 3, 2000, p. S20.
8. Stephen Manes, "Compute While You're Driving," *Forbes*, January 11, 1999, p. 112.
9. David Grotta, "New CD Writes Twice," *PC Magazine*, September 1, 1999, p. 30.
10. Unisys Web site, http://www.unisys.com, accessed February 17, 2000.
11. Andy Reinhardt, "Can Steve Jobs Keep His Mojo Working?" *Business Week*, August 2, 1999, p. 32.
12. Dominique Deckmyn, "Intel Technology Pushes Laptops to Desktop Speeds," *Computerworld*, January 17, 2000, p. 12.
13. Matt Hamblen, "Handhelds Help Boeing Boost Quality Inspections," *Computerworld*, November 8, 1999, p. 38.
14. Dean Takahasi, "Intel Moves from Windows with Line of Web Devices," *The Wall Street Journal*, January 5, 2000, p. B6.
15. Stephen Wildstrom, "Age of the E-Pliance," *Business Week*, January 17, 2000, p. 20.
16. Mary Jo Foley and Steven J. Vaughan-Nicols, "Microsoft Trims Windows 2000," *PC Week*, September 20, 1999, p. 18.
17. "Mac OS 9 Special Report," *The Macintosh News Network*, November 17, 1999, at http://www.macnn.com/reports/os9.shtml, accessed March 4, 2000.
18. Stephanie Neil, "Testing the Linux Waters," *PC Week*, March 15, 1999, pp. 81–82.
19. Aaron Ricadela, "Linux Comes Alive," *Information Week*, January 24, 2000, pp. 47–63.
20. Kevin Young, "Inside Windows 2000: A Guide to Implementation," *PC Week*, January 31, 2000, p. 47.
21. Steve Hamm, Peter Burrows, and Andy Reinhardt, "Is Windows Ready to Run E-Business?" *Business Week*, January 24, 2000, pp. 154–162.
22. Steve Hamm, Peter Burrows, and Andy Reinhardt, "Is Windows Ready to Run E-Business?" *Business Week*, January 24, 2000, pp. 154–162.
23. Steve Hamm, Peter Burrows, and Andy Reinhardt, "Is Windows Ready to Run E-Business?" *Business Week*, January 24, 2000, pp. 154–160.
24. Peter Burrows, "How Sun Became the Eddie and Scottie Show," *Business Week Online*, December 3, 1999, at http://www.businessweek.com/search.htm, accessed March 5, 2000.
25. "Sun Microsystems Upgrades Solaris," *Information Week Online NewsFlash*, January 24, 2000, at http://www.informationweek.com/story/IWK20000124S0001, accessed March 5, 2000.
26. Russell Kay, "A First Peek at the Newest Windows," *Computerworld*, March 20, 2000, p. 72.
27. Carmen Nobel, "Transmeta's Link to Linux," *PC Week*, January 31, 2000, p. 37.
28. David Penn, "Ericsson's Screen Phone Runs Linux," *Linux Journal*, February 29, 2000, at http://www2.linuxjournal.com/cgi-bin/frames.pl/index.html, accessed March 2, 2000.

CHAPTER 3

Organizing Data and Information

Organizations that want their employees to make proactive, informed decisions—decisions that provide a real competitive edge—are turning to business intelligence solutions.

— Samuel Greengard, business and technology writer

Principles	Learning Objectives
The database approach to data management provides significant advantages over the traditional file-based approach.	• *Define general data management concepts and terms, highlighting the advantages and disadvantages of the database approach to data management.* • *Name three database models and outline their basic features, advantages, and disadvantages.*
A well-designed and well-managed database is an extremely valuable tool in supporting decision making.	• *Identify the common functions performed by all database management systems and identify three popular end-user database management systems.*
Further improvements in the use of database technology will continue to evolve and yield real business benefits.	• *Identify and briefly discuss recent database developments.*

Catalina Marketing Corporation

Providing Data Services

Catalina Marketing Corporation is a leading supplier of in-store electronic scanner-activated consumer promotions. It provides marketing services to such consumer goods companies as Bristol-Myers Squibb, Campbell, Coca-Cola, and General Mills. Its Catalina Marketing Network provides customized communications and promotions based on customer purchases that reach shoppers every week in supermarkets that include Kroger, Meijer, Ralphs, and Winn-Dixie. Catalina uses actual purchase behavior to target the future buying behavior of more people than newspaper coupons and direct mail.

The shopping patterns of more than 150 million shoppers each week in 11,000 supermarkets nationwide are captured and stored in an enormous Catalina Marketing database that has grown to more than 2 TB. The database has 18 billion rows of data, containing consumer purchases for the past 65 weeks. Information is captured every time there's a transaction at one of the supermarkets that subscribes to the Catalina Network. Buy a gallon of milk, and the store gets a scanned transaction including what time of day, how it was paid for, and whether you bought bread or cereal with the milk. By itself, this snapshot means little. But combining the millions of transactions at that supermarket and at the entire supermarket chain over a day, a month, or even longer results in a complete profile of customer behavior that consumer goods manufacturers and supermarket retailers can use to better target their promotions and boost sales.

Each retailer that subscribes to the Catalina Network program has a PC in the supermarket. Catalina polls this computer nightly via a data network, and more than 70 million rows of data concerning that day's purchases are loaded into the database. During this polling process, Catalina also downloads to the store PC the information about targeted promotions scheduled for the next day.

With the data from its database, Catalina develops promotions for manufacturers that reach tens of millions of consumers. As groceries are scanned at the checkout counter, the PC flags any item—or the shopper, if the store has a loyalty program—eligible for a promotion. A printer located in each checkout lane prints out a targeted coupon, rebate, in-store game, or other incentive for the consumer. Two people buying the same product following each other in the checkout line can get different offers. The occasional user might get a coupon, while a user of a competitive product may get a free sample. The result of such a targeted campaign is a much higher redemption rate than can be achieved with traditional coupons. Magazine and newspaper coupon redemption rates are around 0.6 percent, and direct mail's redemption rate is around 4.3 percent. In contrast, Catalina Marketing claims that its scanner-based coupon system has an 8.9 percent redemption rate.

As you read this chapter, consider the following:

- How can databases be used to support critical business objectives?

- What are some of the issues associated with compiling and managing massive amounts of data?

database management system (DBMS)

a group of programs that manipulate the database and provide an interface between the database and the user of the database and other application programs

As we saw in Chapter 1, a database is a collection of data organized to meet users' needs. Throughout your career, you will be directly or indirectly accessing a variety of databases, ranging from a simple roster of departmental employees to a fully integrated corporatewide database. You will probably access these databases using software called a **database management system (DBMS)**. A DBMS consists of a group of programs that manipulate the database and provide an interface between the database and the user of the database and other application programs. A database, a DBMS, and the application programs that use the data in the database make up a database environment. Understanding basic database system concepts can enhance your ability to use the power of a computerized database system to support organizational goals and advance your career.

The bane of modern business is too much data and not enough information. Computers are everywhere, accumulating gigabytes galore. Yet it only seems to get harder to find the forest for the trees—that is, to extract significance from the blizzard of numbers, facts, and statistics. Like other components of a computer-based information system, the overall objective of a database is to help an organization achieve its goals. A database can contribute to organizational success in a number of ways, including the ability to provide managers and decision makers with timely, accurate, and relevant information based on data. As we saw in the case of Catalina Marketing, a database can help companies organize data to learn from this valuable resource. Databases also help companies generate information to reduce costs, increase profits, track past business activities, and open new market opportunities. In fact, the ability of an organization to gather data, interpret it, and act on it quickly can distinguish winners from losers in a highly competitive marketplace. It is critical to the success of the organization that database capabilities be aligned with the company's goals. Because data is so critical to an organization's success, many firms develop databases to help them access data more efficiently and use it more effectively. In this chapter, we will investigate the development and use of different types of databases.

DATA MANAGEMENT

Without data and the ability to process it, an organization would not be able to complete most business activities successfully. It could not pay employees, send out bills, order new inventory, or produce information to assist managers in decision making. As you recall, data consists of raw facts, such as employee numbers and sales figures. For data to be transformed into useful information, it must first be organized meaningfully.

The Hierarchy of Data

character

basic building block of information, consisting of uppercase letters, lowercase letters, numeric digits, or special symbols

field

typically a name, number, or combination of characters that describes an aspect of a business object or activity

record

a collection of related data fields

Data is generally organized in a hierarchy that begins with the smallest piece of data used by computers (a bit) and progresses through the hierarchy to a database. As discussed in Chapter 2, a bit (a binary digit) represents a circuit that is either on or off. Bits can be organized into units called bytes. A byte is typically eight bits. Each byte represents a **character**, which is the basic building block of information. A character may consist of uppercase letters (A, B, C, . . . , Z), lowercase letters (a, b, c, . . . , z), numeric digits (0, 1, 2, . . . , 9), or special symbols (.![+][-]/ . . .).

Characters are put together to form a **field**. A field is typically a name, number, or combination of characters that describes an aspect of a business object (e.g., an employee, a location, a truck) or activity (e.g., a sale). A collection of related data fields is a **record**. By combining descriptions of various aspects of an object or activity, a more complete description of the object or activity is obtained. For instance, an employee record is a collection of fields about one

FIGURE 3.1

The Hierarchy of Data

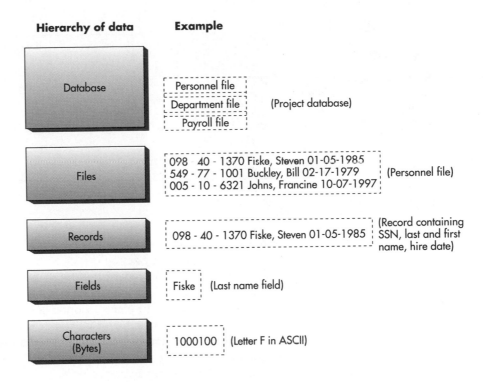

Hierarchy of data

Example

Database — Personnel file / Department file / Payroll file (Project database)

Files —
098 - 40 - 1370 Fiske, Steven 01-05-1985
549 - 77 - 1001 Buckley, Bill 02-17-1979
005 - 10 - 6321 Johns, Francine 10-07-1997
(Personnel file)

Records — 098 - 40 - 1370 Fiske, Steven 01-05-1985 (Record containing SSN, last and first name, hire date)

Fields — Fiske (Last name field)

Characters (Bytes) — 1000100 (Letter F in ASCII)

employee. One field would be the employee's name, another her address, and still others her phone number, pay rate, earnings made to date, and so forth. A collection of related records is a **file**—for example, an employee file is a collection of all company employee records. Likewise, an inventory file is a collection of all inventory records for a particular company or organization. PC database software often refers to files as tables.

At the highest level of this hierarchy is a *database*, a collection of integrated and related files. Together, bits, characters, fields, records, files, and databases form the **hierarchy of data** (Figure 3.1). Characters are combined to make a field, fields are combined to make a record, records are combined to make a file, and files are combined to make a database. A database houses not only all these levels of data but the relationships among them.

Data Entities, Attributes, and Keys

Entities, attributes, and keys are important database concepts. An **entity** is a generalized class of people, places, or things (objects) for which data is collected, stored, and maintained. Examples of entities include employees, inventory, and customers. An **attribute** is a characteristic of an entity. For example, employee number, last name, first name, hire date, and department number are attributes for an employee (Figure 3.2). Inventory number, description, number of units on hand, and the location of the inventory item in the warehouse are examples of attributes for items in inventory. Customer number, name, address, phone number, credit rating, and contact person are examples of attributes for customers. Attributes are usually selected to capture the relevant characteristics of entities such as employees or customers. The specific value of an attribute, called a **data item**, can be found in the fields of the record describing an entity.

As we mentioned, a collection of fields about a specific object is a record. A **key** is a field or set of fields in a record that is used to identify the record. A **primary key** is a field or set of fields that uniquely identifies the record. No other record can have the same value for its primary key. The primary key is used to distinguish records so that they can be accessed, organized, and manipulated. For an

file

a collection of related records

hierarchy of data

bits, characters, fields, records, files, and databases

entity

generalized class of people, places, or things for which data is collected, stored, and maintained

attribute

a characteristic of an entity

data item

the specific value of an attribute

key

a field or set of fields in a record that is used to identify the record

primary key

a field or set of fields that uniquely identifies the record

FIGURE 3.2

Keys and Attributes

The key field is the employee number. The attributes include last name, first name, hire date, and department number.

Employee #	Last name	First name	Hire date	Dept. number	
005-10-6321	Johns	Francine	10-07-1997	257	
549-77-1001	Buckley	Bill	02-17-1979	632	} Entities (records)
098-40-1370	Fiske	Steven	01-05-1985	598	

Key field

Attributes (fields)

employee record such as the one shown in Figure 3.2, the employee number is an example of a primary key.

Locating a particular record that meets a specific set of criteria may require the use of a combination of secondary keys. For example, a customer might call a mail-order company to place an order for clothes. If the customer does not know his primary key (such as a customer number), a secondary key (such as last name) can be used. In this case, the order clerk enters the last name, such as Adams. If there are several customers with a last name of Adams, the clerk can check other fields, such as address, first name, and so on, to find the correct customer record. Once the correct customer record is obtained, the order can be completed and the clothing items shipped to the customer.

The Traditional Approach versus the Database Approach

Traditionally, organizations collected data within each department of a business. Customer order data was maintained within the sales and order fulfillment department, invoicing data was kept in the billing department, and tax information was logged in the accounting department. Today, organizations have begun to tie functions together to streamline their information systems and avoid unnecessary duplication. Let's look at both the traditional and database approaches to information systems.

The Traditional Approach

One of the most basic ways to manage data is via files. Because a file is a collection of related records, all records associated with a particular application (and therefore related by the application) could be collected and managed together in an application-specific file. At one time, most organizations had numerous application-specific data files; for example, customer records often were maintained in separate files, with each file relating to a specific process completed by the company, such as shipping or billing. This approach to data management, in which separate data files are created and stored for each application program, is called the **traditional approach**. For each particular application, one or more data files is created (Figure 3.3).

One of the flaws in this traditional file-oriented approach to data management is that much of the data—for example, customer name and address—is duplicated in two or more files. This duplication of data in separate files is known as **data redundancy**. The problem with data redundancy is that changes to the data (e.g., a new customer address) might be made in one file and not the other. The order-processing department might have updated its file to the new address, but the billing department is still sending bills to the old address. Data redundancy, therefore, conflicts with **data integrity**—the degree

traditional approach to data management

an approach whereby separate data files are created and stored for each application program

data redundancy

duplication of data in separate files

data integrity

the degree to which the data in any one file is accurate

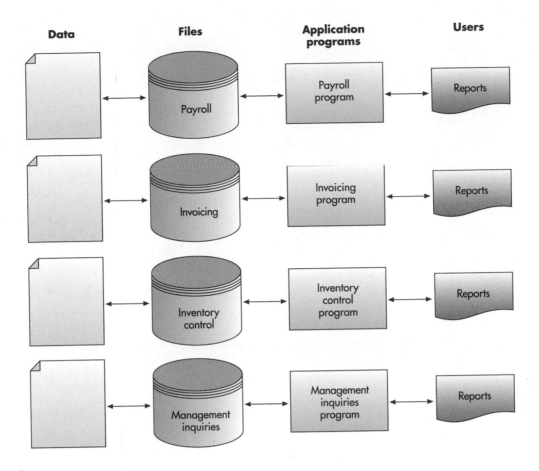

Data · Files · Application programs · Users

FIGURE 3.3

The Traditional Approach to Data Management

With the traditional approach, one or more data files is created and used for every application. For example, the inventory control program would have one or more files containing inventory data, such as the inventory item, number on hand, and item description. Likewise, the invoicing program can have files on customers, inventory items being shipped, and so on. With the traditional approach to data management, it is possible to have the same data, such as inventory items, in several different files used by different applications.

database approach to data management

an approach whereby a pool of related data is shared by multiple application programs

to which the data in any one file is accurate and consistent. Data integrity is established by controlling or eliminating data redundancy. Keeping a customer's address in only one file decreases the possibility that the customer will have two different addresses stored in different locations. The efficient operation of a business requires a high degree of data integrity.

Despite the drawbacks of using the traditional file approach in computerized database systems, some organizations continue to use it. For these firms, the cost of converting to another other approach is too high.

The Database Approach

Because of the problems associated with the traditional approach to data management, many managers wanted a more efficient and effective means of organizing data. The result was the **database approach** to data management. In a database approach, a pool of related data is shared by multiple application programs. Rather than having separate data files, each application uses a collection of data that is either joined or related in the database.

The database approach offers significant advantages over the traditional file-based approach. For one, by controlling data redundancy, the database approach can use storage space more efficiently and increase data integrity. The database approach can also increase an organization's flexibility in the use of data. Because data once kept in two files is now located in the same database, it is easier to locate and request data for many types of processing, and departments can share data and information resources. This flexibility can be critical when coordinating organization-wide responses across diverse functional areas of a corporation. However, some consistency needs to be established among software programs.

To use the database approach to data management, additional software—a database management system (DBMS)—is required. As previously discussed,

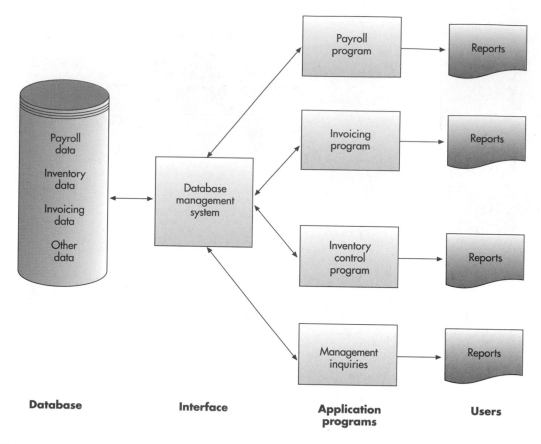

Database   Interface   Application   Users
programs

FIGURE 3.4

The Database Approach to
Data Management

An enterprisewide database
enables Allina Health System
to reduce healthcare costs and
provide patients with the best
treatment possible. The
database pulls cost information
from its hospital and health
plans, comparing best practices
in treatments and matching
cost to level of service.
(Source: Stone/Frank Siteman)

a DBMS consists of a group of programs that can be used as an interface
between a database and the user or the database and application programs.
Typically, this software acts as a buffer between the application programs and
the database itself. Figure 3.4 illustrates the database approach.

The database approach to data management involves a combination of hardware and software. Tables 3.1 and 3.2 list some of the primary advantages and
disadvantages of the database approach and explore some of these issues.

Because of the many advantages of the database approach, most businesses
use databases to store data on customers, orders, inventory, employees, and
suppliers. This data is used as the input to the various information systems
throughout an organization. For example, the transaction processing system
can use the data to support daily business processes such as billing, inventory
tracking, and ordering. This same data can be processed by a management
information system to create reports or a decision support system to provide
information to aid managerial decision making.

Many modern databases are enterprisewide, encompassing much of the data
of the entire organization. Often, distinct yet related databases are linked to
provide enterprisewide databases. Much planning and organization go into the
development of enterprisewide databases. Allina Health System, a healthcare
system that owns and manages 19 hospitals and 50 clinics and provides health
plans to more than one million customers, uses an enterprisewide database.
Allina must try to reduce healthcare costs and hold the line on insurance premiums while providing its patients with the best treatment possible. It must also
integrate the information systems and business practices of the organizations it
has acquired through mergers. To address these issues, management called for
an information systems strategy that included the development of enterprisewide databases spanning all its hospitals, clinics, and patients. The goal
was to enable Allina to gather information and integrate it in ways never done

Advantages	Explanation
Improved strategic use of corporate data	Accurate, complete, up-to-date data can be made available to decision makers where, when, and in the form they need it.
Reduced data redundancy	The database approach can reduce or eliminate data redundancy. Data is organized by the DBMS and stored in only one location. This results in more efficient utilization of system storage space.
Improved data integrity	With the traditional approach, some changes to data were not reflected in all copies of the data kept in separate files. This is prevented with the database approach because there are no separate files that contain copies of the same piece of data.
Easier modification and updating	With the database approach, the DBMS coordinates updates and data modifications. Programmers and users do not have to know where the data is physically stored. Data is stored and modified once. Modification and updating is also easier because the data is stored at only one location in most cases.
Data and program independence	The DBMS organizes the data independently of the application program. With the database approach, the application program is not affected by the location or type of data. Introduction of new data types not relevant to a particular application does not require the rewriting of that application to maintain compatibility with the data file.
Better access to data and information	Most DBMSs have software that makes it easy to access and retrieve data from a database. In most cases, simple commands can be given to get important information. Relationships between records can be more easily investigated and exploited, and applications can be more easily combined.
Standardization of data access	A primary feature of the database approach is a standardized, uniform approach to database access. This means that the same overall procedures are used by all application programs to retrieve data and information.
A framework for program development	Standardized database access procedures can mean more standardization of program development. Because programs go through the DBMS to gain access to data in the database, standardized database access can provide a consistent framework for program development. In addition, each application program need address only the DBMS, not the actual data files, reducing application development time.
Better overall protection of the data	The use of and access to centrally located data are easier to monitor and control. Security codes and passwords can ensure that only authorized people have access to particular data and information in the database, thus ensuring privacy.
Shared data and information resources	The cost of hardware, software, and personnel can be spread over a large number of applications and users. This is a primary feature of a DBMS.

TABLE 3.1

Advantages of the Database Approach

before—for example, pulling cost information from the hospital and the health plans, comparing best practices in treatments, and matching cost to level of service.[1]

DATA MODELING AND DATABASE MODELS

Because there are so many elements in today's businesses, it is critical to keep data organized so that it can be used effectively. A database should be designed to store all data relevant to the business and provide quick access and easy modification. In addition, it must reflect the business processes of the organization. When building a database, careful consideration must be given to these questions:

- Content: What data should be collected and at what cost?
- Access: What data should be provided to which users and when?
- Logical structure: How should data be arranged so that it makes sense to a given user?
- Physical organization: Where should data be physically located?

Disadvantages	Explanation
Relatively high cost of purchasing and operating a DBMS in a mainframe operating environment	Some mainframe DBMSs can cost hundreds of thousands of dollars.
Increased cost of specialized staff	Additional specialized staff and operating personnel may be needed to implement and coordinate the use of the database. However, some organizations have been able to implement the database approach with no additional personnel.
Increased vulnerability	Even though databases offer better security because security measures can be concentrated on one system, they also make more data accessible to the trespasser if security is breached. In addition, if for some reason there is a failure in the DBMS, multiple application programs are affected.

TABLE 3.2

Disadvantages of the Database Approach

planned data redundancy

a way of organizing data in which the logical database design is altered so that certain data entities are combined, summary totals are carried in the data records rather than calculated from elemental data, and some data attributes are repeated in more than one data entity to improve database performance

data model

a diagram of data entities and their relationships

enterprise data modeling

data modeling done at the level of the entire enterprise

entity-relationship (ER) diagrams

a data model that uses basic graphical symbols to show the organization of and relationships between data

Data Modeling

Key considerations in organizing data in a database include determining what data is to be collected, who will have access to it, and how they might use it. Based on these determinations, a database can then be created. Building a database requires two different types of designs: a logical design and a physical design. The logical design of a database shows an abstract model of how the data should be structured and arranged to meet an organization's information needs. The logical design of a database involves identifying relationships among the different data items and grouping them in an orderly fashion. Because databases provide both input and output for information systems throughout a business, users from all functional areas should assist in creating the logical design to ensure that their needs are identified and addressed. Physical database design starts from the logical database design and fine-tunes it for performance and cost considerations (e.g., improved response time, reduced storage space, lower operating cost). The person identified to fine-tune the physical design must have an in-depth knowledge of the DBMS to implement the database. For example, the logical database design may need to be altered so that certain data entities are combined, summary totals are carried in the data records rather than recalculated, and some data attributes are repeated in more than one data entity. These are examples of **planned data redundancy**. It is done to improve the system performance so that user reports or queries can be created more quickly.

One of the tools database designers use to show the logical relationships among data is a data model. A **data model** is a diagram of entities and their relationships. Data modeling usually involves understanding a specific business problem and analyzing the data and information needed to deliver a solution. When done at the level of the entire organization, this is called **enterprise data modeling**. Enterprise data modeling is an approach that starts by investigating general data and information needs of the organization at the strategic level and then examining more specific data and information needs for the various functional areas and departments within the organization. Various models have been developed to help managers and database designers analyze data and information needs. An entity-relationship diagram is an example of such a data model.

Entity-relationship (ER) diagrams use basic graphical symbols to show the organization of and relationships between data. In most cases, boxes are used in ER diagrams to indicate data items or entities, and diamonds show relationships between data items and entities.

Figure 3.5 shows an ER diagram for a customer who places orders. ER diagrams can show a number of relationships. For example, one customer can place many orders. This is an example of a one-to-many relationship, as shown by the one-to-many symbol (1:N) used in Figure 3.5. Each order can include

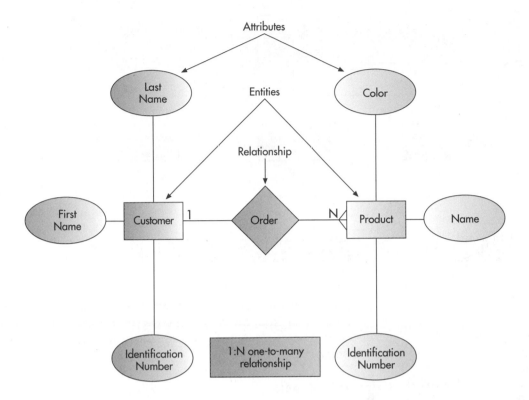

FIGURE 3.5

An Entity-Relationship (ER) Diagram for a Customer Ordering Database

Development of this type of diagram helps ensure the logical structuring of application programs that are able to serve users' needs and are consistent with the data relationships in the database.

one or more line items, where a line item specifies the product identification and quantity ordered. One-to-one, many-to-many, and other relationships can also be revealed using ER diagrams. ER diagrams help ensure that the relationships among the data entities in a database are logically structured so that application programs can be developed to serve user needs. In addition, ER diagrams can be used as reference documents once a database is in use. If changes are to be made in the database, ER diagrams can help design them.

Database Models

The structure of the relationships in most databases follows one of three logical database models: hierarchical, network, and relational. Hierarchical and network models are still being used today, but relational models are the most popular. It is important to remember that the records represented in the models are actually linked or related logically to one another. These links dictate the way users can access data with application programs. Because the different models involve different links between data, each model has unique advantages and disadvantages.

Hierarchical (Tree) Models

In many situations, data follows a hierarchical, or treelike, structure. In a **hierarchical database model**, the data is organized in a top-down, or inverted tree, structure. For example, data about a project for a company can follow this type of model, as shown in Figure 3.6. The hierarchical model is best suited to situations in which the logical relationships between data can be properly represented with the one-to-many approach.

Network Models

In the **network model** there is an owner-member relationship in which a member may have many owners (Figure 3.7). Thus the network model is capable of supporting many-to-many relationships.

hierarchical database model

a data model in which data is organized in a top-down, or inverted tree, structure

network model

an expansion of the hierarchical database model with an owner-member relationship in which a member may have many owners

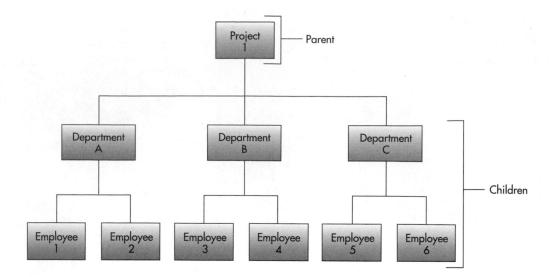

FIGURE 3.6

A Hierarchial Database Model

Project 1 is the top, or root, element. Departments A, B, and C are under this element, with Employees 1 through 6 beneath them as follows: Employees 1 and 2 under Department A, Employees 3 and 4 under Department B, and Employees 5 and 6 under Department C. Thus, there is a one-to-many relationship among the elements of this model.

relational model

a database model that describes data in which all data elements are placed in two-dimensional tables, called relations, that are the logical equivalent of files

domain

the allowable values for data attributes

selecting

data manipulation that chooses rows according to certain criteria

Databases structured according to either the hierarchical model or the network model suffer from the same deficiency: once the relationships are established between data elements, it is difficult to modify them or to create new relationships.

Relational Models

Relational models have become the most popular database models, and their use will increase in the future. The relational model describes data using a standard tabular format. In a database structured according to the relational model, all data elements are placed in two-dimensional tables, called relations, that are the logical equivalent of files. The tables in relational databases organize data in rows and columns, simplifying data access and manipulation. It is easier for managers to understand the relational model (Figure 3.8) than hierarchical and network models.

In the relational model, each row of a table represents a data entity, with the columns of the table representing attributes. Each attribute can take on only certain values. The allowable values for these attributes are called the **domain**. The domain for a particular attribute indicates what values can be placed in each of the columns of the relational table. For instance, the domain for an attribute such as gender would be limited to male or female. A domain for pay rate would not include negative numbers. Defining a domain can increase data accuracy. For example, a pay rate of -$5.00 could not be entered into the database because it is a negative number and not in the domain for pay rate.

Once data has been placed into a relational database, users can make inquiries and analyze data. Basic data manipulations include selecting, projecting, and joining. **Selecting** involves choosing rows according to certain criteria. Suppose a project table contains the project number, description, and department number for all projects being performed by a company. The president of

FIGURE 3.7

A Network Database Model

In this network model, two projects are at the top. Departments A, B, and C are under Project 1; Departments B and C are under Project 2. Thus, the elements of this model represent a many-to-many relationship.

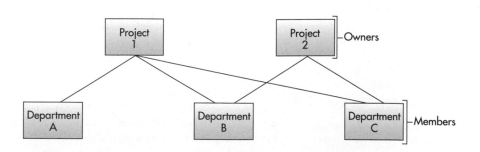

FIGURE 3.8

A Relational Database Model

In the relational model, all data elements are placed in two-dimensional tables, or relations. As long as they share at least one common element, these relations can be linked to output useful information.

Data table 1: Project table

Project number	Description	Dept. number
155	Payroll	257
498	Widgets	632
226	Sales Manual	598

Data table 2: Department table

Dept. number	Dept. name	Manager SSN
257	Accounting	005-10-6321
632	Manufacturing	549-77-1001
598	Marketing	098-40-1370

Data table 3: Manager table

SSN	Last name	First name	Hire date	Dept. number
005-10-6321	Johns	Francine	10-07-1997	257
549-77-1001	Buckley	Bill	02-17-1979	632
098-40-1370	Fiske	Steven	01-05-1985	598

the company might want to find the department number for Project 226, a sales manual project. Using selection, the president can choose the row for Project 226 and see that the department number for the department completing the sales manual project is 598.

Projecting involves choosing columns in a table. For example, we might have a department table that contains the department number, department name, and social security number (SSN) of the manager in charge of the project. The sales manager might want to create a new table with only the department number and the social security number of the manager in charge of the sales manual project. Projection can be used to create a new table containing only department number and SSN.

Joining involves combining two or more tables. For example, we can combine the project table and the department table to get a new table with the project number, project description, department number, department name, and social security number for the manager in charge of the project.

As long as the tables share at least one common data attribute, the tables in a relational database can be **linked** to provide useful information and reports (see Figure 3.9). Being able to link tables to each other through common data attributes is one of the keys to the flexibility and power of relational databases. Suppose the president of a company wants to find out the name of the manager of the sales manual project and how long the manager has been with the company. The president would make the inquiry to the database, perhaps via a desktop personal computer. The DBMS would start with the project description and search the project table to find out the project's department

projecting

data manipulation that chooses columns in a table

joining

data manipulation that combines two or more tables

linking

data manipulation that combines two or more tables using common data attributes to form a new table with only the unique data attributes

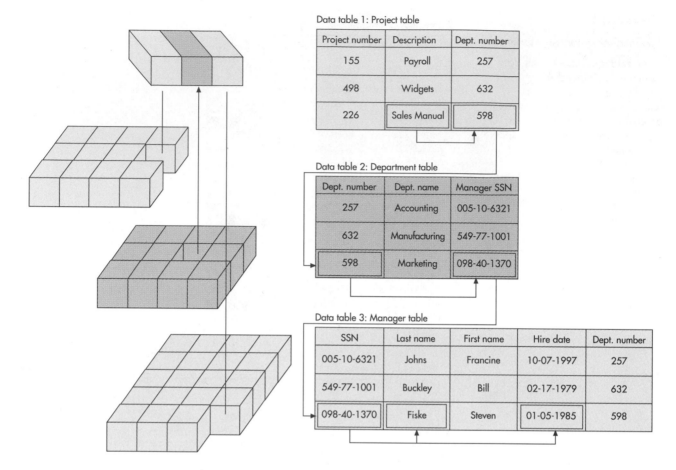

Data table 1: Project table

Project number	Description	Dept. number
155	Payroll	257
498	Widgets	632
226	Sales Manual	598

Data table 2: Department table

Dept. number	Dept. name	Manager SSN
257	Accounting	005-10-6321
632	Manufacturing	549-77-1001
598	Marketing	098-40-1370

Data table 3: Manager table

SSN	Last name	First name	Hire date	Dept. number
005-10-6321	Johns	Francine	10-07-1997	257
549-77-1001	Buckley	Bill	02-17-1979	632
098-40-1370	Fiske	Steven	01-05-1985	598

FIGURE 3.9

Linking Data Tables to Answer an Inquiry

In finding the name and hire date of the manager working on the sales manual project, the president needs three tables: project, department, and manager. The project description (Sales Manual) leads to the department number (598) in the project table, which leads to the manager's SSN (098-40-1370) in the department table, which leads to the managers' name (Fiske) and hire date (01-05-1985) in the manager table.

number. It would then use the department number to search the department table for the department manager's social security number. The department number is also in the department table and is the common element that allows the project table and the department table to be linked. The DBMS then uses the manager's social security number to search the manager table for the manager's hire date. The manager's social security number is the common element between the department table and the manager table. The final result: the manager's name and hire date are presented to the president as a response to the inquiry.

The relational database model is by far the most widely used. It is easier to control, more flexible, and more intuitive than the others because it organizes data in tables. As seen in Figure 3.10, a relational database management system, such as Access, provides a number of tips and tools for building and using database tables. This figure shows the database displaying information about data types and indicating that additional help is available.

DATABASE MANAGEMENT SYSTEMS (DBMSs)

Creating and implementing the right database system ensures that the database will support an organization's business activities and goals. But how do we actually create, implement, use, and update a database? The answer is found in the database management system. DBMSs are classified by the type of database model they support. For example, a relational database management system follows the relational model. Access by Microsoft is a popular relational DBMS for personal computers. Popular mainframe relational DBMSs include DB2 by

FIGURE 3.10

Building and Modifying a
Relational Database

Relational databases provide
many tools, tips, and tricks to
simplify the process of creating
and modifying a database.

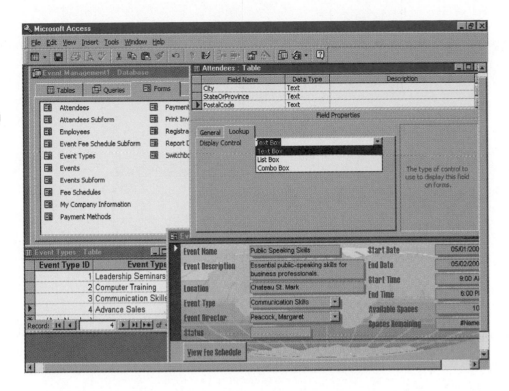

FIGURE 3.10

Building and Modifying a
Relational Database

Relational databases provide
many tools, tips, and tricks to
simplify the process of creating
and modifying a database.

IBM, Oracle, Sybase, and Informix. All DBMSs share some common functions: providing a user view, physically storing and retrieving data in a database, allowing for database modification, manipulating data, and generating reports.

Providing a User View

Because the DBMS is responsible for access to a database, one of the first steps in installing and using a database involves telling the DBMS the logical and physical structure of the data and relationships among the data in the database. This description is called a **schema** (as in schematic diagram). A schema can be part of the database or a separate schema file. The DBMS can reference a schema to find where to access the requested data in relation to another piece of data.

A DBMS also acts as a user interface by providing a view of the database. A user view is the portion of the database a user can access. To create different user views, subschemas are developed. A **subschema** is a file that contains a description of a subset of the database and identifies which users can modify the data items in that subset. While a schema is a description of the entire database, a subschema shows only some of the records and their relationships in the database. For example, a sales representative might need only data describing customers in her region, not the sales data for the entire nation. A subschema could be used to limit her view to data from her region. With subschemas, the underlying structure of the database can change, but the view the user sees might not change. For example, even if all the data on the southern region changed, the northeast region sales representative's view would not change if she accessed data on her region.

A number of subschemas can be developed for different managers or users and the various application programs. Typically, the database user or application will access the subschema, which then accesses the schema (Figure 3.11). Subschemas can also provide additional security because programmers, managers, and other users are typically allowed to view only certain parts of the database.

schema

a description of the entire database

subschema

a file that contains a description of a subset of the database and identifies which users can view and modify the data items in the subset

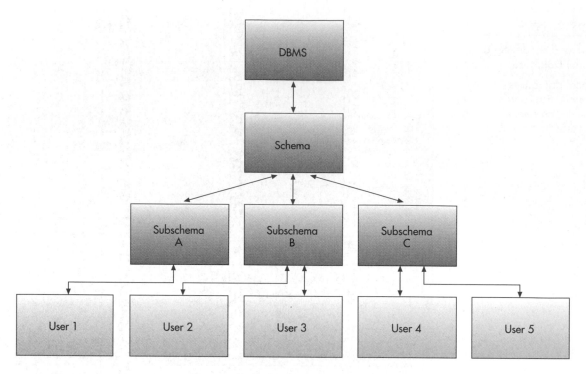

FIGURE 3.11

The Use of Schemas and Subschemas

data definition language (DDL)

a collection of instructions and commands used to define and describe data and data relationships in a specific database

data dictionary

a detailed description of all the data stored in the database

Creating and Modifying the Database

Schemas and subschemas are entered into the DBMS (usually by database personnel) via a data definition language. A **data definition language (DDL)** is a collection of instructions and commands used to define and describe data and data relationships in a specific database. A DDL allows the database's creator to describe the data and the data relationships that are to be contained in the schema and the many subschemas. In general, a DDL describes logical access paths and logical records in the database.

Another important step in creating a database is to establish a **data dictionary**, a detailed description of all data used in the database. The data dictionary contains the name of the data item, aliases or other names that may be used to describe the item, the range of values that can be used, the type of data (such as alphanumeric or numeric), the length of the data item in bytes, a notation of the person responsible for updating it and the various users who can access it, and a list of reports that use the data item. Following are some of the typical uses of a data dictionary.

- *Provide a standard definition of terms and data elements.* Standardization can help in programming by providing consistent terms and variables to be used for all programs. Programmers know what data elements are already "captured" in the database and how they relate to other data elements.
- *Assist programmers in designing and writing programs.* Programmers do not need to know which storage devices are used to store needed data. Using the data dictionary, programmers specify the required data elements. The DBMS locates the necessary data. More important, programmers can use the data dictionary to see which programs already use a piece of data and, if appropriate, copy the relevant section of the program code into their new program, thus eliminating duplicate programming efforts.
- *Simplify database modification.* If for any reason a data element needs to be changed or deleted, the data dictionary would point to specific programs that use the data element that may need modification.

A data dictionary helps achieve the advantages of the database approach in these ways:

- *Reduced data redundancy.* With standard definitions of all data, it is less likely that the same data item will be stored in different places under different names. For example, a data dictionary would reduce the likelihood that the same part number would be stored as two different items, such as PTNO and PARTNO.
- *Increased data reliability.* A data dictionary and the database approach reduce the chance that data will be destroyed or lost. In addition, it is more difficult for unauthorized people to gain access to sensitive data and information.
- *Faster program development.* With a data dictionary, programmers can develop programs faster. They don't have to develop names for data items because the data dictionary does that for them.
- *Easier modification of data and information.* The data dictionary and the database approach make modifications to data easier because users do not need to know where the data is stored. The person making the change indicates the new value of the variable or item, such as part number, that is to be changed. The database system locates the data and makes the necessary change.

Storing and Retrieving Data

When an application program needs data, it requests that data through the DBMS. Suppose that to calculate the total price of a new car, an auto dealer pricing program needs price data on the engine option—six cylinders instead of the standard four cylinders. The application program requests this data from the DBMS. In doing so, the application program follows a logical access path. Next, the DBMS, working in conjunction with various system software programs, accesses a storage device, such as disk or tape, where the data is stored. When the DBMS goes to this storage device to retrieve the data, it follows a path to the physical location (physical access path) where the price of this option is stored. In the pricing example, the DBMS might go to a disk drive to retrieve the price data for six-cylinder engines. This relationship is shown in Figure 3.12.

This same process is used if a user wants to get information from the database. First, the user requests the data from the DBMS. For example, a user might give a command, such as LIST ALL OPTIONS FOR WHICH PRICE IS GREATER THAN 200 DOLLARS. This is the logical access path (LAP). Then the DBMS might go to the options price sector of a disk to get the information for the user. This is the physical access path (PAP).

When two or more people or programs attempt to access the same record in the same database at the same time, there can be a problem. For example, an inventory control program might attempt to reduce the inventory level for a product by ten units because ten units were just shipped to a customer. At the same time, a purchasing program might attempt to increase the inventory level for the same product by 200 units because more inventory was just received. Without proper database control, one of the inventory updates may not be correctly made, resulting in an inaccurate inventory level for the product. **Concurrency control** can be used to avoid this potential problem. One approach is to lock out all other application programs from access to a record if the record is being updated or used by another program.

Users increasingly need to be able to access and update databases via the Internet. Many database vendors are incorporating this capability into their products, including Microsoft, Allaire, Inline Internet Systems, Netscape Communications, EveryWhere Development, and StormCloud Development. Such databases allow companies to create an Internet-accessible catalog, which is nothing more than a database of items, descriptions, and prices.

concurrency control

a method of dealing with a situation in which two or more people need to access the same record in a database at the same time

FIGURE 3.12

Logical and Physical
Access Paths

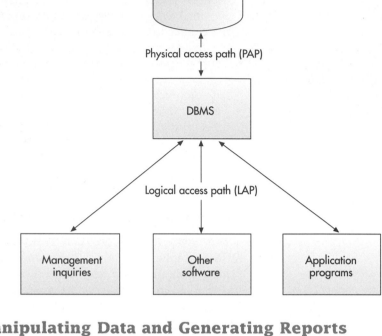

Manipulating Data and Generating Reports

Once a DBMS has been installed, the system can be used by all levels of employees via specific commands in various programming languages. For example, COBOL commands can be used in simple programs that will access or manipulate certain pieces of data in the database. Here's another example of a DBMS query: SELECT * FROM EMPLOYEE WHERE JOB_ CLASSIFICATION = "C2". The * tells the program to include all columns from the EMPLOYEE table for just those employees with a C2 job classification. In general, the commands that are used to manipulate the database are part of the **data manipulation language (DML)**. This specific language, provided with the DBMS, allows managers and other database users to access, modify, and make queries about data contained in the database to generate reports. Again, the application programs go through subschemas, schemas, and the DBMS before actually getting to the physically stored data on a device such as a disk.

As discussed in Chapter 2, SQL lets programmers learn one powerful query language and use it on systems ranging from PCs to the largest mainframe computers (Figure 3.13). Programmers and database users also find SQL valuable because SQL statements can be embedded into many programming languages, such as the widely used COBOL. Because SQL uses standardized and simplified procedures for retrieving, storing, and manipulating data in a database system, the popular database query language can be easy to understand and use.

Once a database has been set up and loaded with data, it can produce desired reports, documents, and other outputs (Figure 3.14). These outputs usually appear in screen displays or hard-copy printouts. The output-control features of a database program allow you to select the records and fields to appear in reports. You can also complete calculations specifically for the report by manipulating database fields. Formatting controls and organization options (such as report headings) help you customize reports and create flexible,

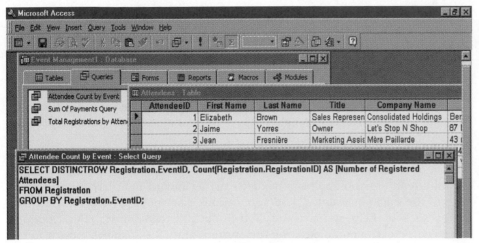

FIGURE 3.13

Structured Query Language

SQL has become an integral part of most relational database packages, as shown by this screen from Microsoft Access.

convenient, and powerful information-handling tools.

A database program can produce a wide variety of documents, reports, and other outputs that can help organizations achieve their goals. The most common reports select and organize data to present summary information about some aspect of company operations. For example, accounting reports often summarize financial data such as current and past-due accounts. Many companies base their routine operating decisions on regular status reports that show the progress of specific orders toward completion and delivery. Increasingly, companies are using databases to provide improved customer services.

Exception, scheduled, and demand reports highlight events that require urgent management attention. Exception reports are produced only upon the occurrence of some predefined exception condition—e.g., a sales change by +/-10%. Scheduled reports are produced according to a predetermined time schedule—e.g., the fourth work night of the month. Demand reports are produced only upon the explicit request of the user. Database programs can produce documents and reports. A few examples include these:

• Form letters with address labels
• Payroll checks and reports
• Invoices
• Orders for materials and supplies
• A variety of financial performance reports

FIGURE 3.14

Database Output

A database application offers sophisticated formatting and organization options to produce the right information in the right format.

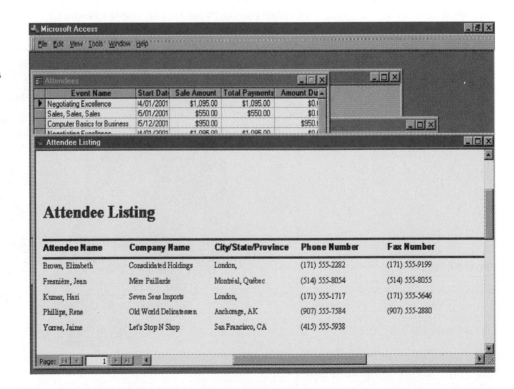

Popular Database Management Systems

The latest generation of database management systems makes it possible for end users to build their own database applications. End users are using these tools to address everyday problems such as how to manage a mounting pile of information on employees, customers, inventory, and sales or how to organize wine lists, CD collections, and video libraries. These database management systems are an important personal productivity tool along with word processing, spreadsheet, and graphics software.

A key to making DBMSs more usable for some databases is the incorporation of "wizards" that walk you through how to build customized databases, modify ready-to-run applications, use existing record templates, and quickly locate the data you want. These applications also include powerful new features such as help systems and Web-publishing capabilities. For example, users can create a complete inventory system and then instantly post it to the Web, where it does double duty as an electronic catalog. Some of the more popular DBMSs for end users include Microsoft Access, Lotus Approach, and Inprise's dBASE.

The complete database management software market encompasses software used by professional programmers and that runs on midrange, mainframe, and supercomputers. The entire market generates $10 billion per year in revenue, with Oracle, IBM, Microsoft, Informix, and Sybase the leaders (Figure 3.15).[2] Microsoft rules in desktop PC software, but Oracle has the largest share of database software on mainframe computers.

Selecting a Database Management System

Selecting the best database management system begins by analyzing database needs and characteristics. The information needs of the organization affect what type of data is collected and what type of database management system is used. Important characteristics of databases include the size of the database, number of concurrent users, performance, the ability of the DBMS to integrate with other systems, the features of the DBMS, vendor considerations, and the cost of the database management system.

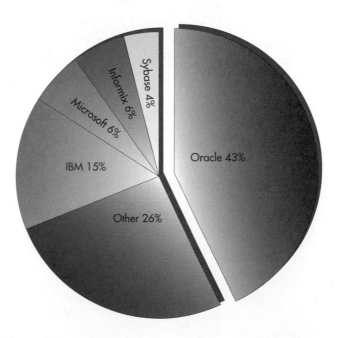

FIGURE 3.15

Worldwide Database Market
Share, 1998
(Source: Data from Julie Pitta, "Squeeze
Play: Databases Get Ugly," *Forbes,*
February 22, 1999, pp. 50–51.)

Database Size

The database size depends on the number of records or files in the database. The size determines the overall storage requirement for the database. Most database management systems can handle relatively small databases of less than 100 million bytes; fewer can manage terabyte-size databases.

Number of Concurrent Users

The number of simultaneous users that can access the contents of the database is also an important factor. Clearly, a database that is meant to be used in a large workgroup must be able to support a number of concurrent users; if it cannot, then the efficiency of the members of the workgroup will be lowered. The term *scalability* is sometimes used to describe how well a database performs as the size of the database or the number of concurrent users is increased. A highly scalable database management system is desirable to provide flexibility. Unfortunately, many companies make a poor DBMS choice in this regard and then later are forced to convert to a new DBMS when the original does not meet expectations.[3]

Performance

How fast the database can update records can be the most important performance criterion for some organizations. Credit card and airline companies, for example, must have database systems that can update customer records and check credit or make a plane reservation in seconds, not minutes. Other applications, such as payroll, can be done once a week or less frequently and do not require immediate processing. If an application demands immediacy, it also demands rapid recovery facilities in the event the computer system shuts down temporarily. Other performance considerations include the number of concurrent users that can be supported and how much main memory is required to execute the database management program.

Integration

A key aspect of any database management system is its ability to integrate with other applications and databases. A key determinant here is what operating systems it can run under—such as Unix, Windows NT, and Windows 2000. Some companies use several databases for different applications at different locations. A manufacturing company with four plants in three different states might have a separate database at each location. The ability of a database program to import data from and export data to other databases and applications can be a critical consideration.

Features

The features of the database management system can also make a big difference. Most database programs come with security procedures, privacy protection, and a variety of tools. Other features can include how easy the database package is to use and the availability of manuals and documentation that can help the organization get the most from the database package. Additional features such as wizards and ready-to-use templates help improve the product's ease of use and are very important.

The Vendor

The size, reputation, and financial stability of the vendor should also be considered in making any database purchase. Some vendors are well respected in the information systems industry and have a large staff of support personnel to give assistance, if needed. A well-established and financially secure database company is more likely to remain in business than others.

Cost

Database packages for personal computers can cost a few hundred dollars, while large database systems for mainframe computers can cost hundreds of thousands of dollars. In addition to the initial cost of the database package, monthly operating costs should be considered. Some database companies rent or lease their database software. Monthly rental or lease costs, maintenance costs, additional hardware and software costs, and personnel costs can be substantial.

Databases continually evolve to include new capabilities and make work processes more efficient. In addition, they allow organizations to perform functions that were unheard of only a few years ago. We explore some of these new developments next.

DATABASE DEVELOPMENTS

The types of data and information that managers need change as business processes change. A number of developments in the use of databases and database management systems can help managers meet their needs, including setting up data warehouses and marts, using the object-oriented approach in database development, and searching for and using unstructured data such as graphics and video.

Data Warehouses, Data Marts, and Data Mining

The raw data necessary to make sound business decisions is stored in a variety of locations and formats—hierarchical databases, network databases, flat files, and spreadsheets, to name a few. This data is initially captured, stored, and managed by transaction processing systems that are designed to support the day-to-day operations of the organization. For decades, organizations have collected operational, sales, and financial data with their on-line transaction processing (OLTP) systems.

Traditional OLTP systems are designed to put data into databases very quickly, reliably, and efficiently, but they are not good at supporting meaningful analysis of the data. In fact, tuning a system to provide speedy performance for OLTP often sacrifices data analysis capabilities. Also, data stored in OLTP databases is inconsistent and constantly changing. The database contains the current transactions required to operate the business, including errors, duplicate entries, and reverse transactions, which get in the way of a business analyst, who needs stable data. Historical data is missing from the OLTP database, which makes trend analysis impossible. Because of the application orientation of the data, the variety of non-integrated data sources, and the lack of historical data, companies were limited in their ability to access and use the data for other purposes. So, although the data collected by OLTP systems doubles every two years, it does not meet the needs of the business decision maker—those systems are data rich but information poor.

Data Warehouses

data warehouse

a database that collects business information from many sources in the enterprise, covering all aspects of the company's processes, products, and customers

A **data warehouse** is a database that collects business information from many sources in the enterprise, covering all aspects of the company's processes, products, and customers.[4] The data warehouse provides business users with a multi-dimensional view of the data they need to analyze business conditions. Data warehousing is designed specifically to support management decision making, not to meet the needs of transaction processing systems. The data warehouse provides a specialized decision support database that manages the flow of information from existing corporate databases and external sources to end-user decision support applications. A data warehouse stores historical data that has been extracted from operational systems and external data sources (Figure 3.16). This operational and external data is "cleaned up" to remove inconsistencies and integrated to create a new information database that is more suitable for business analysis.

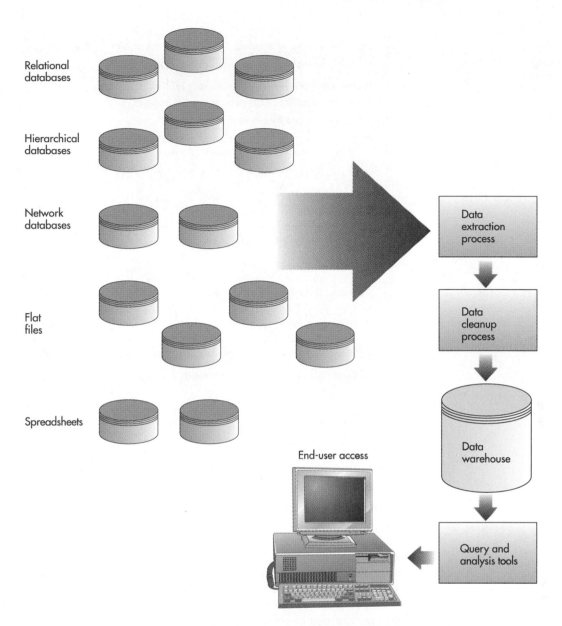

FIGURE 3.16

Elements of a Data Warehouse

Data warehouses typically start out as very large databases containing millions and even hundreds of millions of data records. As this data is collected from the various production systems, a historical database is built that business analysts can use. To remain fresh and accurate, the data warehouse receives regular updates, and old data that is no longer needed is purged. Updating the data warehouse must be fast, efficient, and automated, or its ultimate value is sacrificed. It is common for a data warehouse to contain 3–10 years of current and historical data. Data cleaning tools can merge data from many sources into one database, automate data collection and verification, delete unwanted data, and maintain data in a database management system. Wal-Mart has one of the largest data warehouses in the world—a whopping 101 TB of data representing two years of detailed sales data that goes down to the level of individual receipts.[5] This data is used to make purchasing, pricing, and promotion decisions.

The primary advantage of data warehousing is the ability to relate data in new, innovative ways. However, a data warehouse can be extremely difficult to establish, with the average cost of building one estimated at $2.2 million.[6] Table 3.3 provides some advice on how to create a data warehouse.

TABLE 3.3

How to Design a Customer
Data Warehouse
(Source: Adapted from Anne Field,
"Precision Marketing," *Inc.*, vol. 18,
no. 9, September, 1996, p. 54.)

Sharply define your goals and objectives before you build the warehouse.

- Are you looking to increase your customer base?
- Do you want to double sales of a product line?
- Would you like to encourage repeat business?

Choose the software that best fits your goals.

- If you need a system that is tightly tied to sales rep activity in the field, choose a contact-management program with database capability.
- If you do a lot of telemarketing, you may need a telemarketing program.

Determine who should be in the database.

- First, figure out what types of customers have the most potential.
- Then construct the database, using everything from salespeople's contacts to lists bought from outside suppliers.

Develop a plan.

- Only after your objectives are laid out and your database is constructed is it time to devise a marketing program.
- Generally, it will fall into one of three categories: direct marketing, rewards for repeat purchases, and relationship-building promotions for long-time customers that generate profits.

Measure results.

- Generate periodic status reports in which you determine items like cost per contact and cost per sale.
- If you want to be really careful, do not launch your program at full tilt. Start with a small prototype; if you like what you see, expand to include the entire database.

Data Marts

data mart

a subset of a data warehouse

A **data mart** is a subset of a data warehouse. Data marts bring the data warehouse concept—on-line analysis of sales, inventory, and other vital business data that has been gathered from transaction processing systems—to small and medium-size businesses and to departments within larger companies. Rather than store all enterprise data in one monolithic database, data marts contain a subset of the data for a single aspect of a company's business—for example, finance, inventory, or personnel. In fact, there may even be more detailed data for a specific area in a data mart than a data warehouse would provide.

Data marts are most useful for smaller groups who want to access detailed data. A warehouse is used for summary data for the rest of the company. Because data marts typically contain tens of gigabytes of data, as opposed to the hundreds of gigabytes in data warehouses, they can be deployed on less powerful hardware with smaller disks, delivering significant savings to an organization. Although any database software can be used to set up a data mart, some vendors deliver specialized software designed and priced specifically for data marts. Already, companies such as Sybase, Software AG, and Microsoft have announced products and services that make it easier and cheaper to deploy these scaled-down data warehouses. The selling point is that data marts put targeted business information into the hands of more decision makers.

Data Mining

data mining

an information analysis tool that involves the automated discovery of patterns and relationships in a data warehouse

Data mining is an information analysis tool that involves the automated discovery of patterns and relationships in a data warehouse. Data mining represents the next step in the evolution of decision support systems. It makes use of advanced statistical techniques and machine learning to discover facts in a large database, including databases on the Internet. Unlike query tools, which require users to formulate and test a specific hypothesis, data mining uses built-in analysis tools to automatically generate a hypothesis about the patterns and anomalies found in the data and then from the hypothesis predicts future behavior.

Data mining's objective is to extract patterns, trends, and rules from data warehouses to evaluate (i.e., predict or score) proposed business strategies, which in turn will improve competitiveness, improve profits, and transform business processes. It is used extensively in marketing to improve customer retention; cross-selling opportunities; campaign management; market, channel, and pricing analysis; and customer segmentation analysis (especially one-to-one marketing). In short, data mining tools help end users find answers to questions they never even thought to ask.

There are thousands of data mining applications. Bell Canada uses data mining to identify patterns, group customers with similar characteristics, and create predictive target models to determine which customers should receive a particular offer.[7] Fingerhut uses data mining to create specialized catalogs and optimize mailings to secure the highest possible revenue from the more than one million catalogs it mails each day.[8] Credit card issuers and insurers mine their data warehouses for subtle patterns within thousands of customer transactions to identify fraud, often just as it happens. One U.S. cellular phone company is using Silicon Graphics MineSet software to dig through mountains of call data and pinpoint illegally cloned cell phone ID numbers. Manufacturers mine data collected from factory floor sensors to identify where an intermittent assembly line error is causing a defect that will show up only months after an appliance goes into use. Read the "E-Commerce" box to learn about an interesting data mining application to introduce state-of-the-art technology into everyday life.

E-commerce presents another major opportunity for effective use of data mining. Attracting customers to on-line Web sites is tough; keeping them can be next to impossible. For example, when on-line retail Web sites launch deep-discount sales, they cannot easily figure out how many first-time customers are likely to come back and buy again. Nor do they have a way of understanding which customers acquired during the sale are price sensitive and more likely to jump on future sales. As a result, companies are gathering data on user traffic through their Web sites. This data is then analyzed using data mining techniques to personalize the Web site and develop sales promotions targeted at specific customers.

Faced with nagging questions about how best to push products and understand customers, Reel.com (an on-line provider of movies to rent, buy, and sell; a movie guide; and new-movie information) decided to use data mining techniques to analyze its Web site traffic data to better understand its customers. It employed consumer behavior-tracking software called LifeTime from Verbind. The software crunches through 6 to 12 months of on-line transaction data to build behavior maps—digital blueprints of a customer's Web site activity. The maps reveal a shopper's "velocity"—what he or she bought, how much was spent, and how often purchases were made. By viewing a customer's past behavior, Reel.com can predict that customer's average transactional amount and develop plans to target offers to that specific customer. Verbind is just one of an array of sophisticated consumer tracking tools being deployed by e-commerce sites. Analytic profilers such as DataSage, E-piphany, and Personify all mimic Verbind's capabilities.[9]

Reel.com employed data mining to map individual shoppers' habits. Armed with a customer's data, Reel.com can develop plans to target offers to that specific customer.

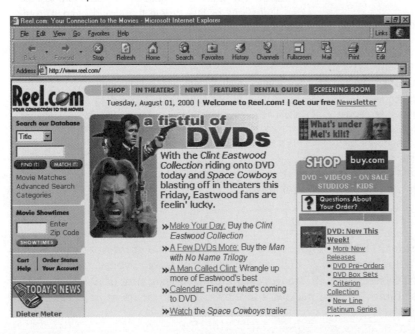

E-COMMERCE
Safeway Customers Involved in E-Commerce Experiment

Basingstoke, England, isn't normally considered a cutting-edge kind of place. The quintessential British suburb lies 60 miles southwest of London, in the Hampshire countryside. "Boringstoke," as it is sometimes called, has long been the butt of jokes for the likes of Gilbert and Sullivan and Monty Python. Lately, however, 500 residents of Basingstoke have been participating in Easi-Order, a shopping experiment at the Safeway supermarket on Worting Road. They use a Palm III PDA equipped with a bar-code scanner, a modem, and software from IBM's Institute for Advanced Commerce to do their shopping from home. The PDA is provided free of charge to the test shoppers.

Here's how it works. Safeway built a data warehouse of every item bought from its inventory of 22,000 products by 10 million British shoppers over the past four years—some 3 TB of grocery-buying intelligence. Data mining software written for Safeway by IBM determines what groceries a family needs based on how long it has been since they visited the store and details from their past orders. The software is programmed to suggest additional items as well. For example, it may suggest that the family try Oracle toothpaste—Safeway's own brand—or promote baby products to new mothers based on birth notices obtained from a government health agency.

A suggested order is transmitted to the family's Palm III PDA whenever they connect by telephone to Safeway. Of course, families sometimes want to buy something they haven't ordered before or something they bought so long ago the Safeway computer figures they've lost interest in it. If the family has an empty box or wrapper for the item, someone swipes the bar code with a scanner built into the PDA, and the item is added to the electronic order. If the family doesn't have one of the items on hand or if it has no bar code, a member just describes the item in a free-format field that turns into e-mail to Safeway: "One quart of strawberries."

When the family finishes editing the order, the person attaches the PDA to the telephone and transmits the order to an IBM server. This midrange computer is a Java-based intranet server that connects to Safeway's S/390 mainframe computer. In addition to the order, the server receives a message saying when the family will pick up its groceries.

The following morning, a Safeway Easi-Order specialist arrives at the Basingstoke store, logs on to the server and prints out all the orders that are scheduled for pickup that day. Then sometime before the scheduled pickup, the Easi-Order specialist goes up and down the store's aisles filling a shopping cart with each order. The specialist logs each item as it is put into the basket by scanning its bar code with a handheld scanner.

When the specialist completes the order, it is brought to a holding area at the front of the store. The scanner is plugged into a docking station that reads the order and holds the information until the family comes. When the family arrives, a member swipes its Safeway account card at the same station, and the system matches the order data with the customer data and sends both to the server, and from there they go to the 3-TB database at Safeway's data center. The order information will rest in the database until the family next connects its PDA to Safeway's computer and obtains a new suggested order.

Most of the test shoppers use the service weekly. Some prefer to buy a month's worth of nonperishable items with Easi-Order and shop weekly for other items in the conventional way. Of course, people have certain items they prefer to pick on their own, such as fresh fruit and vegetables. The big advantage that shoppers appreciate is the time saved using the electronic ordering service—they can go in and out of the store in less than 15 minutes. Some shoppers think they have joined an elite class of shoppers because they no longer have to line up at the checkout counter.

Discussion Questions

1. Identify specific costs and benefits for Safeway associated with setting up this novel approach to shopping. Do you think that Safeway can cost-justify this shopping experiment? Why or why not?
2. What specific pieces of data are contained in the Safeway data warehouse for each family?

Critical Thinking Questions

3. What do you think are some of the potential issues and limitations of this system that could keep it from being successful?
4. Would you consider shopping in this manner? Why or why not?

Sources: Adapted from Gary H. Anthes, "Easi-Order," *Computerworld*, March 30, 2000, pp. 46–48; Chip Bayers, "Capitalist Econstruction," *Wired*, March 2000, accessed at http://www.wired.com/wired/archive/8.03/markets.html; Safeway UK Web site at http://www.safeway.co.uk, accessed April 11, 2000; and Pervasive Computing at IBM Web site at http://www-3.ibm.com/pvc/tech/safeway.shtml., accessed April 11, 2000 (this site includes the capability to demo the Safeway system).

Traditional DBMS vendors are well aware of the great potential of data mining. Thus, companies such as Oracle, Informix, Sybase, Tandem, and Red Brick Systems are all incorporating data mining functionality into their products. Table 3.4 summarizes a few of the most frequent applications for data mining.

Application	Description
Market segmentation	Identifies the common characteristics of customers who buy the same products from your company.
Customer churn	Predicts which customers are likely to leave your company and go to a competitor.
Fraud detection	Identifies which transactions are most likely to be fraudulent.
Direct marketing	Identifies which prospects should be included in a mailing list to obtain the highest response rate.
Market basket analysis	Identifies what products or services are commonly purchased together (e.g. beer and diapers).
Trend analysis	Reveals the difference between a typical customer this month versus last month.

TABLE 3.4

Common Data Mining Applications
(Source: Vance McCarthy, "Strike It Rich," *Datamation*, February 1997, pp. 44–50.)

on-line analytical processing (OLAP)

software that allows users to explore data from a number of different perspectives

On-Line Analytical Processing (OLAP)

Most industry surveys today show that the majority of data warehouse users rely on spreadsheets, reporting and analysis tools, or their own custom applications to retrieve data from warehouses and format it into business reports and charts. In general, these approaches work fine for questions that can be answered when the amount of data involved is relatively modest and can be accessed with a simple table lookup.

For nearly two decades, multidimensional databases and their analytical information display systems have provided flashy sales presentations and trade show demonstrations. All you have to do is ask where a certain product is selling well, for example, and a colorful table showing sales performance by region, product type, and time frame automatically pops up on the screen. Called **on-line analytical processing (OLAP)**, these programs are now being used to store and deliver data warehouse information. OLAP allows users to explore corporate data from a number of different perspectives.

OLAP servers and desktop tools support high-speed analysis of data involving complex relationships, such as combinations of a company's products, regions, channels of distribution, reporting units, and time periods. Speed is essential as businesses grow and accumulate more and more data in their operational systems and data warehouses. Long popular with financial planners, OLAP is now being put in the hands of other professionals. The leading OLAP software vendors include Cognos, Comshare, Hyperion Solutions, Oracle, MineShare, WhiteLight, and Microsoft.

Access to data in multidimensional databases can be very quick because they store the data in structures optimized for speed, and they avoid SQL and index processing. But multidimensional databases can take a great deal of time to update; in very large databases, update times can be so great that they force updates to be made only on weekends. Despite this flaw, multidimensional databases have continued to prosper because of their great retrieval speed. Some software providers are attempting to counteract this flaw through the use of partitioning and calculations-on-the-fly capabilities.

Consumer goods companies use OLAP to analyze the millions of consumer purchase records captured by scanners at the checkout stand. This data is used to spot trends in purchases and to relate sales volume to promotions and store conditions, such as displays, and even the weather. OLAP tools let managers analyze business data using multiple dimensions, such as product, geography, time, and salesperson. The data in these dimensions, called measures, is generally aggregated—for example, total or average sales in dollars or units, or budget dollars or sales forecast numbers. Rarely is the data studied in its raw, unaggregated form. Each dimension also can contain some hierarchy. For example, in the time dimension, users may examine data by year, by quarter, by month, by week, and even by day. A geographic dimension may compile data from city, state, region, country, and even hemisphere.

Resort Condominiums International (RCI) is a wholly owned subsidiary of Cendant Corporation and a market leader in vacation exchange services, with more than 3,500 affiliated time-share resorts and over 2.5 million member families worldwide. RCI uses Applix's iTM1 real-time OLAP solution to enhance its budgeting, forecasting, inventory, sales reporting, and analysis for locations in the United States and Europe. As in many companies, this information resides in various locations—such as databases, flat files, and spreadsheets. RCI developed a centralized data warehouse to hold the data. Business analysts use iTM1 to access the data warehouse and quickly develop models incorporating this voluminous and complex data. The OLAP features of iTM1 enable RCI to make better business decisions because they can easily integrate and analyze both operational and financial data.[10]

The value of data ultimately lies in the decisions it enables. Powerful information-analysis tools in areas such as OLAP and data mining, when incorporated into a data warehousing architecture, bring market conditions into sharper focus and help organizations deliver greater competitive value. OLAP provides top-down, query-driven data analysis; data mining provides bottom-up, discovery-driven analysis. OLAP requires repetitive testing of user-originated theories; data mining requires no assumptions and instead identifies facts and conclusions based on patterns discovered. OLAP, or multidimensional analysis, requires a great deal of human ingenuity and interaction with the database to find information in the database. A user of a data mining tool does not need to figure out what questions to ask; instead, the approach is, "Here's the data, tell me what interesting patterns emerge." For example, a data mining tool in a credit card company's customer database can construct a profile of fraudulent activity from historical information. Then, this profile can be applied to all incoming transaction data to identify and stop fraudulent behavior, which may otherwise go undetected.

Table 3.5 compares an OLTP database to a data warehouse. Table 3.6 compares the OLAP and data mining approaches to data analysis.

TABLE 3.5

Comparison of OLTP and Data Warehousing

Characteristic	OLTP Database	Data Warehousing
Purpose	Support transaction processing	Support decision support
Source of data	Business transactions	Multiple files, databases—data internal and external to the firm
Data access allowed users	Read and write	Read only
Primary data access mode	Simple database update and query	Simple and complex database queries with increasing use of data mining to recognize patterns in the data
Primary database model employed	Relational	Relational
Level of detail	Detailed transactions	Often summarized data
Availability of historical data	Very limited—typically a few weeks or months	Multiple years
Update process	On-line, ongoing process as transactions are captured	Periodic process, once per week or once per month
Ease of update	Routine and easy	Complex, must combine data from many sources; data must go through a data cleanup process
Data integrity issues	Each individual transaction must be closely edited	Major effort to "clean" and integrate data from multiple sources

Characteristic	OLAP	Data mining
Purpose	Supports data analysis and decision making	Supports data analysis and decision making
Type of analysis supported	Top-down, query-driven data analysis	Bottom-up, discovery-driven data analysis
Skills required of user	Must be very knowledgeable of the data and its business context	Must trust in data mining tools to uncover valid and worthwhile hypothesis

TABLE 3.6

Comparison of OLAP and Data Mining

open database connectivity (ODBC)

standards that ensure that software written to comply with these standards can be used with any ODBC-compliant database

Open Database Connectivity (ODBC)

To help with database integration, many companies rely on **open database connectivity (ODBC)** standards. Software written to comply with these standards can be used with any ODBC-compliant database, making it easier to transfer and access data among different databases. For example, a manager might want to take several tables from one database and incorporate them into another database that uses a different database management system. Or, a manager might want to transfer one or more database tables into a spreadsheet program. If all this software meets ODBC standards, the data can be imported, exported, or linked to other applications (see Figure 3.17). For example, a table in an Access database can be exported to a Paradox database or a spreadsheet. Tables and data can also be imported using ODBC. For example, a table in a dBASE database or an Excel spreadsheet can be imported into an Access database. Linking allows an application to use data or an object stored in another application without actually importing the data or object into the application. The Access database, for example, can link to a table in the Lotus 1-2-3 spreadsheet or the Sybase database. Applications that follow the ODBC standard can use these powerful ODBC features to share data between different applications stored in different formats.

FIGURE 3.17

Advantages of ODBC

ODBC can be used to export, import, or link tables between different applications.

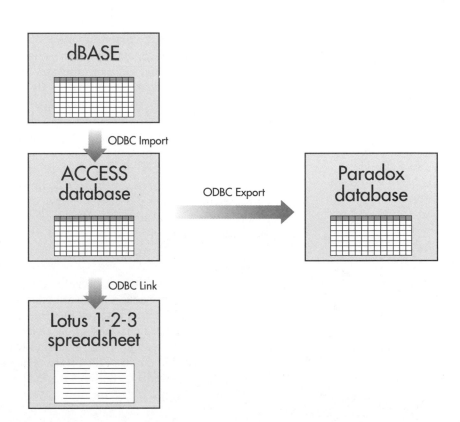

Object-Relational Database Management Systems

Many of today's newer application programs require the ability to manipulate audio, video, and graphical data. Conventional database management systems are not well suited for this, because these types of data cannot easily be stored in rows or tables. Manipulation of such data requires extensive programming so that the DBMS can translate data relationships. An **object-relational database management system (ORDBMS)** provides a complete set of relational database capabilities plus the ability for third parties to add new data types and operations to the database. These new data types can be audio, images, unstructured text, spatial data, or time-series data that require new indexing, optimization, and retrieval features.

In such a database, these types of data are stored as objects, which contain both the data and the processing instructions needed to complete the database transaction. The objects can be retrieved and related by an ORDBMS. Businesses can then mix and match these elements in their daily search for clues and information. For example, by clicking on a picture of a red Corvette, a market analyst at General Motors might be able to call up a profile of red Corvette buyers. If he wants to break that up by geographic region, he might circle and click on a map. All in the same motion, he might view a GM sales-training film to see whether the sales pitch is appropriate, given the most recent market trends. Chances are the analyst will do all this using an Internet site tied into a database. As another example, MasterCard is interested in object-oriented technology to combine transactional data with cardholder fingerprints to prevent fraud.

Each of the vendors offering ORDBMS facilities provides a set of application programming interfaces to allow users to attach external data definitions and methods associated with those definitions into the database system. They are essentially offering a standard socket into which users can plug special instructions. DataBlades, Cartridges, and Extenders are the names applied by Informix, Oracle, and IBM to describe the plug-ins to their respective products. Other plug-ins serve as interfaces to Web servers.

Web-based applications increasingly require complex object support to link graphical and other media components back to the database. These systems make sense for developers of systems that are highly dependent on complex data types, particularly Web and multimedia applications. Because it supports so many applications, an ORDBMS is also called a universal database server.

An increasing amount of data that organizations use is in the form of images, which can be stored in object-relational databases. Credit card companies, for example, input pictures of charge slips into an image database using a scanner. The images can be stored in the database and later sorted by customer, printed, and sent to customers along with their monthly statements. Image databases are also used by physicians to store X rays and transmit them to clinics away from the main hospital. Financial services, insurance companies, and government branches are also using image databases to store vital records and replace paper documents.

Hypertext

The object-relational database provides greater flexibility in defining relationships between data. The previously discussed relational, hierarchical, and network models were designed for data that can be organized into fixed-length data fields and records with structured relationships. With hypertext, users can search and manipulate alphanumeric data in an unstructured way. In an object-relational database, text data is placed in chunks called *nodes*. The user then establishes links between the nodes. The relationships between the data can be created to user specifications, instead of following one of the more structured database models. Suppose a

---

object-relational database management system (ORDBMS)

a DBMS capable of manipulating audio, video, and graphical data doctor is treating a new patient with symptoms similar to those of two other patients. The notes on each patient are placed in nodes, and the three nodes are linked. The doctor can use hypertext to retrieve and cross-reference these three cases when treating future symptoms of the new patient.

Hypermedia

Hypermedia is an extension of hypertext. Hypermedia allows businesses to search and manipulate multimedia forms of data—graphics, sound, video, and alphanumeric data. A marketing manager, for example, might store notes about the competition and new trends in the marketplace. These notes could include written material on new markets, images of products and advertising brochures, and TV commercials used by competitors. Using a hypermedia database management system, the marketing manager could organize this data into nodes and define the relationships among them. For example, she could link all TV commercials and written brochures about new products from several competitors. With the hypermedia database approach, many types of data can be organized into a web of nodes connected by links established by the user.

Spatial Data Technology

Spatial data technology involves the use of an object-relational database to store and access data according to the locations it describes and to permit spatial queries and analysis. NASA has set up a massive environmental monitoring program that calls for satellites and earth stations to deliver more than a terabyte of data every day to databases. The cumulative data will eventually multiply to a petabyte, which is 1,000 times a terabyte. Once the data is edited, it will be loaded into other databases for analysis. The primary database will offer satellite photo images and measurements. If environmentalists wanted to compare carbon dioxide levels within a certain latitude, longitude, and altitude with another polygon of space, they could easily do so. Implications of spatial data technology are potentially significant. For example, the data derived from the NASA project and others like it could help find oil and mineral reserves or identify sources of pollutants. Builders and insurance companies can use spatial data to make decisions related to natural hazards. Spatial data can even be used to improve financial risk management with information stored by derivative type, currency type, interest rates, and time.

The painstaking process of ensuring accuracy, completeness, and currency of the data that's been extracted from operational databases and formatted for analysis by a variety of departments is key to the success of any data warehouse. Evolutionary Technologies International, Platinum Technology, Prism Solutions, and other companies sell products for extracting and cleaning data. Also, a new class of tools is emerging for managing **metadata**, the data that describes the contents of a database. Metadata tells users when a piece of data was last updated, its format, where the data originated, and its intended uses. That information can guide users through databases and help them understand the meaning and context of financial data, customer records, and business transactions.[11] But these products address only part of the problem—the harder part is having procedures in place to manage data as it is extracted from operational systems or imported from data providers and transformed into meaningful information. Often data problems can be traced back to a transaction processing error that occurred during an update of an operational database. Warehouse implementers should expect to spend four to five times more time and effort on data cleaning activities than they would at first estimate.

---

metadata

the data that describes the contents of a database

Spatial data technology is used by NASA to store data from satellites and earth stations. Location-specific information can be accessed and compared. (Source: Courtesy of NASA.)

Business Intelligence

business intelligence

the process of getting enough of the right information in a timely manner and usable form and analyzing it so that it can have a positive impact on business strategy, tactics, or operations

Business intelligence is the process of getting enough of the right information in a timely manner and usable form and analyzing it so that it can have a positive impact on business strategy, tactics, or operations. Business intelligence involves turning data into useful information that is then distributed throughout an enterprise. Companies use this information to make improved strategic decisions about which markets to enter, how to select and manage key customer relationships, and how to select and effectively promote products to increase profitability and market share.

Boise Cascade Office Products uses Aonix's NOMAD mainframe reporting software to provide business intelligence to key employees in marketing, accounting, product planning, logistics, and distribution centers across the country. Users can select data from more than 72 databases and, using the power of the IBM S/390 mainframe computer, perform reporting tasks based on any combination of parameters, including market segment, customer size, date range, and product category. Employees can obtain the data needed to zero in on problem areas, obtain a detailed picture of the profitability of any customer, and see which products are selling and which are not. If sales decline, they can track those declines to specific offices or even to individual sales reps to pinpoint problems and take immediate steps to remedy them.[12]

competitive intelligence

a continuous process involving the legal and ethical collection of information, analysis, and controlled dissemination of information to decision makers

Competitive intelligence is one aspect of business intelligence and is limited to information about competitors and how that knowledge affects strategy, tactics, and operations. Effective **competitive intelligence** is a continuous process involving the legal and ethical collection of information, analysis that doesn't avoid unwelcome conclusions, and controlled dissemination of that information to decision makers. Are you ahead of your competitors or are you racing to catch up? To stay ahead in the marketplace, you must be able to integrate competitive intelligence into your company's strategic plans and decisions. Competitive intelligence is a critical part of your company's ability to see and respond quickly and appropriately to the changing marketplace.

Competitive intelligence is not espionage—the use of illegal means to gather information. In fact, almost all the information a competitive intelligence professional needs can be collected by examining published information sources, conducting interviews, and using other legal, ethical methods. Using a variety of analytical tools, a skilled competitive intelligence professional can by deduction fill the gaps in information already gathered. Read the "Ethical and Societal Issues" box to see how one company is using competitive intelligence.

Larger businesses often have staff resources who can do most of the data gathering and analysis. They will often benefit, though, by having an outside perspective during analysis. Smaller businesses almost never have the internal staff resources to do effective business intelligence. The exception to this is if there is a person (often the CEO) who does the business intelligence work because he or she likes it. Smaller businesses and independent business professionals can usually benefit from outside help with information gathering and with analysis and recommendation.

counterintelligence

the steps an organization takes to protect information sought by "hostile" intelligence gathers

The term **counterintelligence** describes the steps an organization takes to protect information sought by "hostile" intelligence gatherers. One of the most effective counterintelligence measures is to define "trade secret" information relevant to the company and control its dissemination.

ETHICAL AND SOCIETAL ISSUES
United Technologies Gathers Competitive Intelligence

United Technologies (UTC) provides high-technology products to the aerospace and building systems industries throughout the world. UTC's business units include Pratt & Whitney (aircraft engines), Carrier (air conditioning), Otis (elevators), International Fuel Cells (fuel cell technology), Hamilton Sundstrand (aerospace systems), and Sikorsky (helicopters). Recent annual income was $3 billion from revenues of $24 billion.

UTC gathers a lot of competitive intelligence. Each UTC business unit has long had its own competitive intelligence (CI) function, and UTC recently added a corporate CI unit to offer a big-picture perspective. The unit resides within UTC's Research Center, which monitors and develops technology for the company. In aerospace, consolidation has reduced the number of competitors, while in the worldwide air conditioning and ventilation market, the challenge is to keep an eye on a growing number of rivals. UTC constantly monitors what companies are up to, how they use suppliers, how their distribution chains function, and who their customers are.

UTC's intelligence gathering supports its strong acquisition strategy of buying companies that possess complementary technologies. Three major acquisitions occurred during 1999. Sundstrand joined Hamilton Standard to more than double the size of UTC's aerospace systems activities and presence. International Comfort Products brought $750 million of residential product sales to Carrier's North American cooling and heating businesses. LG Industrial Systems' elevator business added more than $500 million to Otis's sales, mostly in Korea, where LG is the leading elevator company.

To begin researching a competitor, UTC gets as much information as possible from public sources, such as financial reports, Web site information, patents, market activity, and alliances with universities. All that information is brought to UTC's CI group to evaluate the competitor's technology—does it complement UTC's, could UTC improve the competitor's technology, does the competitor seem to be a good value? This intelligence is complemented with information from unconventional sources to complete the picture. For example, UTC staff might question a magazine writer about information on the company that didn't make it into the article, or staffers might contact potential CI sources at a trade association meeting. UTC might also interview customers and suppliers of the company in question.

UTC's CI staff is told to adhere to strict legal and ethical codes in gathering information. They must always identify who they are, where they're from, and their purpose for seeking information.

Discussion Questions

1. Who are the primary sources of CI for UTC? Why? What kinds of information can they provide?
2. Generate a list of three questions you would ask of a supplier to a UTC competitor to gain competitive intelligence.

Critical Thinking Questions

3. Do you think that adherence to its strict legal and ethical code helps or hinders the UTC staff in obtaining information? Why?
4. What additional means of gathering competitive intelligence can you identify? Which of these might be outside the limits of the UTC code of ethics?

Sources: Adapted from "Letter to Shareholders" and "Year in Review," United Technologies 1999 annual report, found at http://www.utc.com/annual99/, accessed April 4, 2000; and Gary Abramson, "All Along the Watchtower," *CIO Enterprise*, July 15, 1999, Section 2, pp. 25–34.

---

knowledge management

the process of capturing a company's collective expertise wherever it resides—in computers, on paper, in people's heads—and distributing it wherever it can help produce the biggest payoff

Knowledge management is the process of capturing a company's collective expertise wherever it resides—in computers, on paper, or in people's heads—and distributing it wherever it can help produce the biggest payoff. The goal of knowledge management is to get people to record knowledge (as opposed to data) and then share it. Although a variety of technologies can support it, knowledge management is really about changing people's behavior to make their experience and expertise available to others.[13] Knowledge management had its start in large consulting firms and has expanded to nearly every industry. Dow Chemical was one of the first manufacturing companies to employ a disciplined process in managing its knowledge, notably its previously underexploited portfolio of more than 29,000 patents. Intellectual Asset Management is now a formal, six-stage process, whose mission is to capitalize on the revenue potential of Dow's patents through new product development or licensing.[14]

INFORMATION SYSTEMS IN ACTION
Grandma and Limp Bizkit, Diapers and Beer

Roger McHaney,
Kansas State University

What do diapers have in common with beer? Or Snoop Doggy Dogg and Limp Bizkit CDs with elderly senior citizens? Common sense would shout out, "Nothing!" However, new technology is quickly proving that common sense isn't always the best indicator of how and why products are being purchased.

Data mining, the process of conducting computer-based searches through mountains of corporate transaction data, is yielding all sorts of unexpected, statistically significant connections between different product purchases and consumer groups. Although the idea has been around for years, it is only recently that an increase in data warehouses, availability of off-the-shelf statistical software packages, a drop in secondary storage prices, and a plethora of consumer data from the Internet have come together in a way that has made data mining a value-adding activity.

SAS Institute is one of many organizations offering advanced tools to support data mining efforts. SAS's Enterprise Miner provides an integrated suite of data mining tools for businesses that need to conduct comprehensive analyses of customer data. These tools can help uncover previously unknown patterns of data that reveal customers' buying habits and provide a greater understanding of underlying motivation.

Data mining, however, is more than tools. Stuart Young, director of analytics at ICOM, has developed four rules that move data mining from theory into business practice.

According to Young, the following keys to success exist:

1. Keep data close to the customer. Data directly from the customer will be more recent and of higher quality.
2. Understand your customer. The more you know about your customers, the more likely you are to determine motives for their behavior.
3. Use past behavior to predict future actions. Statisticians can use a variety of techniques to project expected actions.
4. Rely on your team. Turning information into business value takes teamwork and discipline.

By following Young's suggestions and by using tools such as SAS Institute's Enterprise Miner, business analysts gain new insight into their marketing efforts. Thanks to data mining, it's no longer a secret that men sent out to buy diapers between 6 and 8 P.M. are also likely to pick up a six-pack of beer. Analysts have also discovered that although seniors buy Snoop Doggy Dogg and Limp Bizkit CDs, an effort to sell concert tickets at retirement homes would probably fail. Instead, a targeted marketing campaign emphasizing discounted music might sell more to seniors on fixed incomes buying presents for their grandchildren.

Sources: K Haegele, "Getting Gold from a Data Mine," *Target Marketing*, no. 4, April 2000, p. 23; T. Wasserman, "Of Diapers and Beer," *Brandweek*, no. 9, February 28, 2000, p. 34; T. Wasserman, G. Khermouch, and J. Green, "Mining Is Everyone's Business," *Brandweek*, no. 9, February 28, 2000, pp. 32–36, 41; and SAS Web site, http://www.sas.com.

● SUMMARY

PRINCIPLE • **The database approach to data management provides significant advantages over the traditional file-based approach.**

Data is one of the most valuable resources a firm possesses. It is organized into a hierarchy that builds from the smallest element to the largest: bit, byte, field, record, file, and database.

An entity is a generalized class of objects for which data is collected, stored, and maintained. An attribute is a characteristic of an entity. Specific values of attributes—called data items—can be found in the fields of the record describing an entity. A data key is a field within a record that is used to identify the record. A primary key uniquely identifies a record, while a secondary key is a field in a record that does not uniquely identify the record.

The traditional approach to data management has been from a file perspective. Separate files are created for each application. This approach can create problems over time: as more files are created for new applications, data that is common to the individual files becomes redundant. Also, if data is changed in one file, those changes might not be made to other files, reducing data integrity.

To address problems of traditional file-based data management, the database approach was developed. Benefits of this approach include reduced data redundancy, improved data consistency and integrity, easier modification and updating, standardization of data access, and more efficient program development.

Potential disadvantages of the database approach include the relatively high cost of purchasing and operating a DBMS in a mainframe operating environment; specialized staff required to implement and coordinate the use of the database; and increased vulnerability if security is breached and there is a failure in the DBMS.

•　•　•

When building a database, careful consideration must be given to content and access, logical structure, and physical organization. One of the tools database designers use to show the relationships among data is a data model that shows data entities and their relationships. Enterprise data modeling involves analyzing the data and information needs of the entire organization. Entity-relationship (ER) diagrams can be employed to show the relationships between entities in the organization.

Databases typically use one of three common models: hierarchical (tree), network, and relational. The relational model, the most widely used database model, is easier to control, more flexible, and more intuitive than the other models because it organizes data in tables.

PRINCIPLE • A well-designed and well-managed database is an extremely valuable tool in supporting decision making.

A DBMS is a group of programs used as an interface between a database and application programs. When an application program requests data from the database, it follows a logical access path. The actual retrieval of the data follows a physical access path. Records can be considered in the same way: a logical record is what the record contains; a physical record is where the record is stored on storage devices. Schemas are used to describe the entire database, its record types, and their relationships to the DBMS.

A database management system provides four basic functions: providing user views, creating and modifying the database, storing and retrieving data, and manipulating data and generating reports.

Subschemas are used to define a user view, the portion of the database a user can access and/or manipulate. Schemas and subschemas are entered into the computer via a data definition language, which describes the data and relationships in a specific database. Another tool used in database management is the data dictionary, which contains detailed descriptions of all data in the database.

Once a DBMS has been installed, the database may be accessed, modified, and queried via a data manipulation language. SQL is used in several popular database packages today and can be installed on PCs and mainframes.

Popular end-user DBMSs include Microsoft Access, Lotus Approach, and Inprise's dBASE. Oracle, IBM, Microsoft, Sybase, and Informix are the leading DBMS vendors.

PRINCIPLE • Further improvements in the use of database technology will continue to evolve and yield real business benefits.

Organizations are building data warehouses, which are relational database management systems specifically designed to support management decision making.

Multidimensional databases and on-line analytical processing (OLAP) programs are being used to store data and allow users to explore the data from a number of different perspectives.

Data mining, which is the automated discovery of patterns and relationships in a data warehouse, is emerging as a practical approach to generate a hypothesis about the patterns and anomalies in the data that can be used to predict future behavior.

An object-relational database management system (ORDBMS) provides a complete set of relational database capabilities, plus the ability for third parties to add new data types and operations to the database. These new data types can be audio, images, unstructured text, spatial data, or time series data that require new indexing, optimization, and retrieval features.

Business intelligence is the process of getting enough of the right information in a timely manner and usable form and analyzing it so that it can have a positive impact on business strategy, tactics, or operations. Competitive intelligence is one aspect of business intelligence limited to information about competitors and how that information affects strategy, tactics, and operations. Counter-intelligence describes the steps an organization takes to protect information sought by "hostile" intelligence gatherers. Knowledge management is the process of capturing a company's collective expertise wherever it resides—in computers, on paper, or in people's heads—and distributing it wherever it can help produce the biggest payoff. The goal of knowledge management is to get people to record knowledge (as opposed to data) and then share it.

● REVIEW QUESTIONS

1. Describe the hierarchy of data.
2. Define the term *database*. How is it different from a database management system?
3. What are the advantages of the database approach to data management, as opposed to the traditional file-based approach?
4. Describe the following three types of database models: hierarchical model, network model, and relational model.
5. How is a database schema used?
6. Identify important characteristics in selecting a database management system.
7. What is the difference between a data definition language (DDL) and a data manipulation language (DML)?

8. What advantages does the open database connectivity (ODBC) standard offer?
9. What is a data warehouse, and how is it different from a traditional database used to support OLTP?
10. What is OLAP? Does OLAP imply the use of a relational database?
11. What is data mining? How is it different from OLAP?
12. What is an ORDBMS? What kind of data can it handle?
13. What is business intelligence? How is it used?
14. What is competitive intelligence?

● DISCUSSION QUESTIONS

1. You have been selected to represent the student body on a project to develop a new student database for your school. What actions might you take to fulfill this responsibility to ensure that the project meets the needs of students and is successful?
2. Your company is building a new manufacturing facility to handle the increase in sales volume expected from the introduction of a new line of products. Imagine that you are the manager of the new plant. What counterintelligence initiatives might you undertake?
3. What is a data model and what is data modeling? Why is data modeling an important part of strategic planning?
4. You are going to design a database for your wine-tasting club to track its inventory of wines. Identify the database characteristics most important to you in choosing a DBMS to implement this system. Which of the database management systems described in this chapter would you choose? Why? Is it important for you to know what sort of computer the database will run on? Why or why not?

5. How would you distinguish a data mart from a data warehouse? Would you recommend a strategy of developing numerous data marts or a single data warehouse to meet the decision support needs of a large multidivisional organization? Why?
6. Make a list of the databases in which data about you exists. How is the data in each database captured? Who updates each database and how often? Is it possible for you to request a printout of the contents of your data record from each database?
7. Develop a list of conditions under which you would use the OLAP approach to analyze data in a data warehouse. Develop a list of conditions that favor the data mining approach.
8. You are the vice president of information technology for a large commercial bank. You are to make a presentation to the board of directors recommending the investment of $5 million to establish a competitive intelligence organization including people, data gathering services, and software tools. What are your key points in favor of this investment? What counterpoints can you anticipate others might argue?

● PROBLEM-SOLVING EXERCISES

1. Use an end-user database management system to design and implement a system to record your personal items and home/apartment furnishings so that you have a log of all valuable items for insurance purposes in case of theft, fire, or natural disaster. For each item, you should record a complete description, date of purchase, cost, and any identifying numbers such as a serial number or registration number. What will be the unique key for the records in your database?

2. A video movie rental store is using a relational database to store information on movie rentals to answer customer questions. Each entry in the database contains the following items: Movie ID No. (primary key), Movie Title, Year Made, Movie Type, MPAA Rating, Number of Copies on Hand, and Quantity Owned. Movie types are comedy, family, drama, horror, science fiction, and western. MPAA ratings are G, PG, PG-13, R, X, and NR (not rated). Use an end-user database management system to build a data entry screen to enter this data. Build a small database with at least ten entries.

3. To improve service to their customers, the salespeople at the video rental store have proposed a list of changes being considered for the database in the previous exercise. From this list, choose two database modifications and implement them.

Proposed changes:
a. Add the date that the movie was first available to help locate the newest releases.
b. Add the director's name.
c. Add the names of three actors in the movie.
d. Add a critic rating of one, two, three, or four stars.
e. Add the number of Academy Award nominations.

● TEAM ACTIVITIES

1. In a group of three or four classmates, do an analysis of the leading database management system vendors. Which one is currently the market leader in terms of annual revenue? In terms of market share? Which one do you believe has the best products? Why?
2. As a team of three or four classmates, interview business managers from three different businesses that use databases to help them in their work. What data entities and data attributes are contained in each database? How do they access the database to perform analysis? Have they received training in any query or reporting tools? What do they like about their database and what could be improved? Do any of them use data mining or OLAP techniques? Weighing the information obtained, select one of these databases as being most strategic for the firm and briefly present your selection and the rationale for the selection to the class.

3. Imagine that you and your classmates are a research team developing an improved process for evaluating college applicants. The goal of the research is to predict which students will be most successful in their college career. Those who score well on the profile will be accepted, those who score exceptionally well will be considered for scholarships. Prepare a brief report for your instructor addressing these questions:
a. What data do you need for each college applicant?
b. What data might you need that is not typically requested on the college application form?
c. From where might you get this data?

Take a first cut at designing a database for this application. Identify the data entities about which data must be captured. For each data entity, list the data attributes that you believe are necessary for this database. Identify the primary key for each data entity.

● WEB EXERCISES

1. Use a Web search engine to find information on one of the following topics: data warehouse, data mining, or OLAP. Find a definition of the term, an example of a company using the technology, and three companies that provide such software. Cut graphics and text material from the Web pages and paste them into a word processing document to create a two-page report on your selected topic. At the home page of each software company, request further information from the company about its products.

2. Use a Web search engine to find three companies that provide competitive intelligence services. How are the services that they provide similar? How are they different? Which companies seem to be the most ethical? Why?

● CASES

 ### Lockheed Martin Implements OLAP System

Lockheed Martin is a diversified technology company that researches, designs, manufactures, and integrates advanced technology products and services for government and commercial customers. Annual sales exceed $25 billion, and the firm employs 160,000 people worldwide. Its products and services include systems integration, military aircraft, launch vehicles, and defense systems.

The Tactical Aircraft Systems (TAS) division, located in Fort Worth, Texas, designs, develops, and produces state-of-the-art tactical military aircraft systems. Its current products include the T-16 Fighting Falcon, a small, agile, low-cost, high-capability fighter with outstanding mission success; the F-22 Fighter, designated by the U.S. Air Force as the next-generation air superiority fighter; and the Joint Strike Fighter, the next-generation multirole fighter. TAS is also working with Japan to help design, develop, and manufacture its next fighter, called the F-2 Fighter.

When you design and build high-performance fighter aircraft, gathering information about processes, parts, and procedures is critical to producing a superior airplane while controlling costs. TAS decided to implement a data warehouse and provide OLAP tools to create a single, consolidated view of the business information it needed for better decision making. The Lockheed division wanted to provide its employees with access to program and business-management information, staffing information, risk and technical measurements, sales and cost forecasting, and overhead analysis.

The division began implementing OLAP in 1998 as part of a long-term strategy to bolster business intelligence capabilities. OLAP lets users from engineering, purchasing, manufacturing, and other areas in the company examine data on aircraft design and manufacturing from many perspectives. By the end of 1999, the defense contractor had deployed more than 20 OLAP applications. One application, for example, is the Manpower Forecast Module used to analyze staffing requirements, employee skills, and skill requirements for various contracts. This application allows product team leaders to input their staffing needs and identify individuals who match the needs down to the skill level. Another application, Hours Per Unit, measures the cost of aircraft parts in terms of manufacturing hours. This information enables analysis of alternative designs and helps the company evaluate whether specific processes should be outsourced.

In implementing this system, the company needed strong protection for sensitive business data. Considerable effort was devoted to ensuring that security measures were in place, including user access permissions and user authentication. As a result, the division's system administrators can control what users are able to do with the OLAP applications and can manage how different user groups can interact with the data.

Discussion Questions

1. What benefits does Lockheed Martin gain from the use of OLAP and a data warehouse that were not possible through the use of more traditional database technology?

2. Briefly describe how the Manpower Forecast Module might work.

Critical Thinking Questions

3. Why does Lockheed consider the data in the warehouse to be sensitive? Who else could use the data to their benefit? How might they use this data?

4. Is there a high potential for application of this OLAP solution to other Lockheed Martin divisions? Why or why not?

Source: James Cope, "New Tools Help Lockheed Martin Prepare for Takeoff," *Computerworld*, March 27, 2000, p. 46; Tatilia Baron, "OLAP Goes Online," *Information Week*, September 20, 1999, pp. 90–92; Lockheed Martin Tactical Aircraft Systems home page, http://www.lmco.com/careers/tactical_aircraft_fortworth_tx.html, accessed December 6, 1999; "About Lockheed Martin," Lockheed Martin Web site, http://www.lmco.com/about/index.htm, accessed December 6, 1999.

 ## Allina Health System Implements a Data Warehouse

Minneapolis-based Allina Health System is a not-for-profit healthcare system serving one million people living in Minnesota, Wisconsin, North Dakota, and South Dakota. The vertically integrated healthcare system includes 13,000 physicians and 22,000 employees who own and manage 19 hospitals, 57 clinics, and seven nursing homes. Allina provides people with a lifetime of healthcare options and a full continuum of care—from prevention and wellness services, such as health screenings and immunizations, to the highest-quality and most technologically advanced inpatient and outpatient services.

Allina is pressured from all sides to reduce healthcare costs and hold the line on premiums, while providing its patients with the best treatment possible. It must also integrate the information systems and business practices of the multiple organizations it has acquired through various mergers in the recent past. To meet these challenges, management developed an enterprise data warehouse strategy. The goal is to enable Allina to pull information together and integrate it in ways never done before—for example, extracting cost information from the hospital and the health plans, comparing best practices in treatments, and matching cost to level of service better.

Meeting this goal was a data warehouse design challenge. The data modeling effort had to ensure that each data attribute, each data mart, and the data warehouse could be tied together logically and physically. For example, all data attributes are assigned to one of four different data subtypes: people (organizations and patients, members, providers, etc.), places, things, and events.

There were three key elements to the success of this project. First, a multitier database was developed so that both summary and detailed information is available. Second, the large implementation team was divided into three specialized groups, including a data management group in charge of architecture, data modeling, and databases; a tools group focusing on management of data access and data query tools; and an analysis group of business analysts, project managers, and consultants to implement new warehouse subject areas and data marts. Third, the organization used pilot projects, including a data mart for administrative reporting for one hospital and a large-scale data warehouse on patient histories for all hospitals; when completed, the warehouse exceeded 70 GB.

Discussion Questions

1. In which group of the implementation team would you feel most comfortable participating? Why?

2. What business and technical challenges made this project difficult?

Critical Thinking Questions

3. Starting with the four data subtypes (people, places, things, and events), make a list of the data attributes of interest for each data subtype.

4. Imagine that you are on the board of directors for Allina and the IT manager has requested approval to market the data warehouse system to other healthcare systems across the United States. What would be your reaction? Why?

Sources: "About Allina Health System," Allina Health System Web site, http://www.allina.com, accessed April 3, 2000; and "Rx for Reducing Paperwork," IBM Web site, http://www.ibm.com/stories, accessed April 4, 2000.

3 Fifth Third Bank Invests in Internet Systems

Fifth Third Bank traces its origins to the Bank of the Ohio Valley, which opened its doors in Cincinnati in 1858. Today Fifth Third comprises 14 affiliate banks with 650 branches and 12,000 employees located in Ohio, Kentucky, Indiana, Illinois, Florida, Arizona, and Michigan. Fifth Third is headquartered in Cincinnati, Ohio, and has $41 billion in assets and total deposits of $26 billion. Fifth Third is better than most banks at increasing revenues faster than expenses, and that helps it achieve above-average earnings growth—it is coming off its 26th straight year of record profits.

For the past two years, Fifth Third has invested the majority of its $60 million information systems budget in Internet technology. It has aggressively pushed large corporate customers to give up faxes and move onto the Internet to access and exchange information and even trade stocks on-line. Fifth Third developed tools that allow merchants to access credit card information over the Internet. Using passwords and secure networks, retailers are allowed to view their credit card transactions, refund money to customers, and review information issued by credit card companies. Currently, more than $4 billion of e-commerce transactions is completed every year using Fifth Third's network.

Another of Fifth Third's value-added services is the development of a mobile workforce. The company is testing a program that enables 700 commercial loan officers, armed with laptops, to go to different banks and sell cash management products and credit lines on the spot. Similarly, in the business-to-consumer arena, mortgage loan officers will be able to walk into an open house and help potential home buyers apply for a home mortgage.

Although the industry average for on-line banking hovers at 4 percent of customers, more than 10 percent of Fifth Third's customers rely on the Internet to access its services. The company is successfully attracting more on-line customers by continually improving its ability to serve them over the Internet. For instance, the bank is rolling out a 401(k) retirement planning site that will allow customers to execute trades electronically. Fifth Third will also incorporate the natural language inquiry ability from the popular Web search engine Ask Jeeves. This service will allow customers to use plain-English questions rather than keywords to search for answers regarding retirement planning.

Fifth Third is also using Internet technologies to build its own intranet and put over 95 percent of the company's manuals on-line. This move will create a more productive workforce.

Discussion Questions

1. Which of these many information system initiatives will increase bank revenue? Which will decrease bank expenses?
2. Fifth Third has a number of Internet-based services: providing merchants access to credit card information over the Internet; supporting mobile workers in selling cash management, credit lines, and home mortgages; developing a 401(k) retirement planning site; and providing on-line access to company manuals. Briefly summarize the database implications for each of these applications—the type of database technology required, size of the database, and contents of the database.

Critical Thinking Questions

3. Which of these new information system services do you think will be the most successful investment for Fifth Third? Why?
4. Is Fifth Third investing too much of its information systems budget in the Internet? If so, what other technologies should it be exploring? If not, why not?

Sources: Adapted from Jeff McKinney, "5/3 Sees Growth Opportunity in Chicago," *Cincinnati Enquirer*, February 13, 2000, pp. E1–E2; Anne Chen, "Fifth Third Places First on Fast@Track," *PC Week*, December 15, 1999, p. 90; and "Investor Relations," Fifth Third Web site, http://www.53.com, accessed February 15, 2000.

● NOTES

Sources for the opening vignette on p. 91: Adapted from "Catalina Marketing Corporation Reports Record Third-Quarter Results," company press release, *PRNewswire*, January 13, 2000, at http://www.prnewswire.com/cgi-bin/stories; and Catalina Marketing Corporation, "Investor Highlights," fact sheet, *PR Newswire*, http://www.prnewswire.com/cnoc/, accessed April 2, 2000.

1. "About Allina Health System" section of the Allina Health System Web site at http://www.allina.com, accessed on April 3, 2000.
2. Julie Pitta, "Squeeze Play: Databases Get Ugly," *Forbes*, February 22, 1999, pp. 50–51.
3. Stewart Deck, "SQL Users Turn to Oracle8 for Bulk," *Computerworld*, May 10, 1999, p. 4.
4. Amy Helen Johnson, "Data Warehousing," *Computerworld*, December 6, 1999, p. 75.
5. Craig Stedman, "Wal-Mart CIO Leaves Retailer and IT Leader," *Computerworld*, March 9, 2000, p. 4.
6. Gabrielle Gagnon, "Data Warehousing: An Overview," *PC Magazine*, March 19, 1999, pp. 245–246.
7. Candee Wilde, "Telcos Turn to Analytical Tools to Stay in Touch," *Information Week*, March 13, 2000, pp. 98–102.
8. James M. Connolly, "Mining Your Business," *Computerworld*, May 17, 1999, pp. 94–98.
9. Carol Pickering, "They're Watching You," *Business 2.0*, February 2000, pp. 135–137.
10. "RCI Implements Applix's Real-Time OLAP Solution to Enhance Sales Reporting & Analysis," August 5, 1999, Applix press releases at http://www.applix.com/releases/99-08-05_rci.cfm.
11. Craig Stedman, "Metadata," *Computerworld*, October 18, 1999, p. 74.
12. Samuel Greengard, "Business Intelligence," *Beyond Computing*, January/February 2000, pp. 18–23.
13. Rochelle Garner, "Knowledge Management," *Computerworld*, August 9, 1999, pp. 50–51.
14. James Allen, "The Process of Knowledge Management," presented at the Transforming Knowledge-Intensive Processes Conference, September 28, 1999 Boston, Massachusetts.

CHAPTER 4

Telecommunications, the Internet, Intranets, and Extranets

*T*he Internet changes everything.

— Larry Ellison, CEO of Oracle Corporation

Principles

The effective use of telecommunications and networks can turn a company into an agile, powerful, and creative organization, giving it a long-term competitive advantage.

The Internet is like many other new technologies—it provides a wide range of services, some of which are effective and practical for use today, others of which are still evolving, and still others of which will fade away from lack of use.

Learning Objectives

- *Define the term* telecommunications *and describe the components of a telecommunications system.*
- *Identify the benefits associated with a telecommunications network.*

- *Briefly describe how the Internet works, including alternatives for connecting to it and the role of Internet service providers.*
- *Identify and briefly describe the services associated with the Internet, including the World Wide Web.*
- *Define the terms* intranet *and* extranet *and discuss how organizations are using them.*
- *Identify several issues associated with the use of networks.*

FedEx

Sparring with UPS on the Internet

As more companies are setting up storefronts on the Internet, they must find reliable and cost-effective ways to ship their products to consumers. There are a number of shipping companies to choose from, but UPS and FedEx are the largest. UPS ships about 12 million packages daily. Its annual revenues are estimated to be about $26 billion, including roughly $14 billion from U.S. ground shipments, $5 billion from U.S. overnight deliveries, $3 billion from U.S. two-day deliveries, and $4 billion from international deliveries. UPS currently has about 326,000 employees. FedEx, on the other hand, has annual revenues of about $16 billion. Unlike UPS, FedEx's largest revenue component is U.S. overnight shipments, worth $7 billion annually. International shipments are the next largest contributor to total revenues, worth $4 billion. U.S. ground shipments represent $3 billion, and U.S. two-day air represents $2 billion of FedEx's total revenues. FedEx has about 141,000 employees. Clearly, UPS is doing more with ground transportation. The company has approximately 150,000 trucks, compared with FedEx's 43,500 trucks. The two companies are similar in terms of their air shipments. FedEx has a fleet of 637 planes, which is slightly larger than UPS's 610-plane fleet. The impressive size and success of these two companies can be directly traced to their Internet strategies.

FedEx was one of the first shipping companies to integrate the use of the Internet into its operations. According to Fred Smith, CEO of FedEx, "The information about the package is almost as important as the package itself." This total information approach was implemented using the convenience and power of the Internet. Being one of the first shipping companies to install elaborate scanning and tracking equipment, FedEx was able to tell its customers the location of their packages and shipments from the moment the deliveries left the customer's driveway. FedEx customers started to rely on FedEx when the package "absolutely, positively" had to be there on time. More recently, FedEx decided to invest an additional $100 million in technology and equipment that will allow it to deliver packages faster and with greater control.

Although UPS was late to implement sophisticated scanning and tracking systems, the company has now listed technology and the Internet as one of its primary strategic initiatives. During the last several years, UPS has invested about $1 million annually on its Internet site and technology in general. This investment has allowed UPS to compete with FedEx on reliable, overnight deliveries. In 1999, UPS shipped about $20 billion worth of goods from its Internet site, and that amount is expected to increase to $180 billion within five years. Both FedEx and UPS are integrating their Web sites with back-end computer systems and operations. In addition, both companies are looking to customize their shipping operations. Corporations such as Nike, Hewlett-Packard, and Cisco are using FedEx and UPS to streamline their shipping operations.

Even with huge investments in the Internet, these two companies are still locked in fierce competition to be the best. Clad in purple uniforms, FedEx employees are in a close race with the coffee-colored-uniformed UPS employees. Fortunately for both companies, the Internet has dramatically expanded the need for shipping. While these two companies are fighting it out to get a larger slice of the shipping pie, the pie is ever increasing.

As you read this chapter, consider the following:

- What impact has the Internet had on FedEx and UPS?

- In the next five years, what impact will the Internet have on businesses and the economy in general?

To speed communications and share information, businesses are linking formerly isolated employees, branch offices, and global operations via networks, whether they set up their own or use outside services. As seen with FedEx and UPS in the opening vignette, companies are using networks not only to speed communications but to compete for market share and profits. Telecommunications is just one aspect of the increasingly wired world we'll now explore.

AN OVERVIEW OF TELECOMMUNICATIONS AND NETWORKS

Telecommunications refers to the electronic transmission of signals for communications, and it has the potential to create profound changes in business because it lessens the barriers of time and distance. Telecommunications not only is changing the way businesses operate but also is altering the nature of commerce itself. As networks are connected with one another and information is transmitted more freely, a competitive marketplace is making excellent quality and service imperative for success.

Figure 4.1 shows a general model of telecommunications. The model starts with a sending unit (1), such as a person, a computer system, a terminal, or another device, that originates the message. The sending unit transmits a signal (2) to a telecommunications device (3). The telecommunications device performs a number of functions, which can include converting the signal into a different form or from one type to another. A telecommunications device is a hardware component that allows electronic communication to occur or to occur more efficiently. The telecommunications device then sends the signal through a medium (4). A **telecommunications medium** is anything that carries an electronic signal and interfaces between a sending device and a receiving device. The signal is received by another telecommunications device (5) that is connected to the receiving computer (6). In this chapter, we will explore the components of the telecommunications model shown in Figure 4.1. An important characteristic of telecommunications is the speed at which information is transmitted, measured in bits per second (bps). Common speeds are in the range of thousands of bits per second (Kbps) to millions of bits per second (Mbps).

Advances in telecommunications technology allow us to communicate rapidly with clients and co-workers almost anywhere in the world. Telecommunications also reduces the amount of time needed to transmit information that can drive and conclude business actions. A manufacturing sales representative, for example, can use telecommunications technology to get new product prices from the

telecommunications medium

anything that carries an electronic signal and interfaces between a sending device and a receiving device

FIGURE 4.1

Elements of a Telecommunications System

Telecommunications devices relay signals between computer systems and transmission media.

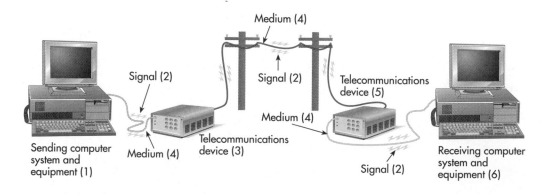

Sending computer system and equipment (1)
Signal (2)
Medium (4)
Telecommunications device (3)
Medium (4)
Signal (2)
Telecommunications device (5)
Medium (4)
Signal (2)
Receiving computer system and equipment (6)

Telecommunications technology enables businesspeople to communicate with coworkers and clients from remote locations.
(Source: Stone/Terry Vine.)

central sales office while working at a customer's location. This empowers the sales representative and often results in faster, higher-quality customer service. Telecommunications technology also helps businesses coordinate activities and integrate various departments to increase operational efficiency and support effective decision making. The far-reaching developments of telecommunications have a profound effect on business information systems and on society in general.

Telecommunications

The use of telecommunications can help businesses solve problems and maximize opportunities. Using telecommunications effectively requires careful analysis of telecommunications media, devices, and carriers and services.

Transmission Media

Various types of communications media are available. Each type exhibits its own characteristics, including cost, capacity, and speed. In developing a telecommunications system, the selection of media depends on the purpose of the overall information and organizational systems, the purpose of the telecommunications subsystems, and the characteristics of the media. As with other components, the media should be chosen to support the goals of the information and organizational systems at the least cost and to allow for possible modification of system goals over time. The proper media will help a company link subsystems to maximize effectiveness and efficiency. Various media types are summarized in Table 4.1, and common types of wiring and cabling are shown in Figures 4.2a through c.

TABLE 4.1

Media Types

Media Type	Description	Advantages	Disadvantages
Twisted-pair wire cable	Twisted pairs of copper wire, shielded or unshielded	Used for telephone service, widely available	Transmission speed and distance limitations
Coaxial cable	Inner conductor wire surrounded by insulation	Cleaner and faster data transmission than twisted-pair	More expensive than twisted-pair
Fiber-optic cable	Many extremely thin strands of glass bound together in a sheathing; uses light beams to transmit signals	Diameter of cable much smaller than coaxial, less distortion of signal, capable of high transmission rates	Expensive to purchase and install
Microwave transmission	High-frequency radio signal sent through atmosphere and space	Avoids cost and effort to lay cable or wires, capable of high-speed transmission	Must have unobstructed line of sight between sender and receiver; signal highly susceptible to interception
Cellular transmission	Divides coverage area into cells; each cell has mobile telephone subscriber office	Supports mobile users; costs are dropping	Signal highly susceptible to interception
Infrared transmission	Signals sent through air as light waves	Devices can be moved, removed, and installed without expensive wiring and network connections	Must have unobstructed line of sight between sender and receiver; transmission effective only for short distances

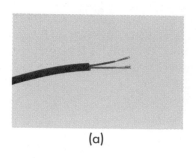
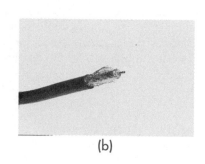
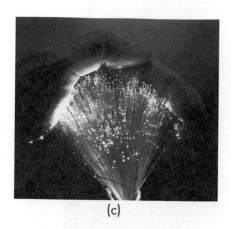

(a)   (b)

(c)

FIGURE 4.2

Common Wiring and
Cabling Types

(a) Twisted-pair wire, (b) coaxial
cable, (c) fiber-optic cable.
(Sources: a and b, Fred Bodin;
c, Stone/Greg Pease.)

common carriers

long-distance telephone companies

value-added carriers

companies that have developed
private telecommunications systems
and offer their services for a fee

switched line

a communications line that uses
switching equipment to allow one
transmission device to be connected
to other transmission devices

TABLE 4.2

Common Telecommunications
Devices

Telecommunications Devices

A telecommunications device is a hardware device that allows electronic communication to occur or to occur more efficiently. Almost every telecommunications system uses one or more of these devices to transmit or receive signals. Table 4.2 summarizes some of the more common telecommunications devices.

Carriers and Services

Telecommunications carriers provide the telephone lines, satellites, modems, and other communications technology used to transmit data from one location to another. They also provide many types of services. Telecommunications carriers are classified as either common carriers or special-purpose carriers. The **common carriers** are primarily the long-distance telephone companies. American Telephone & Telegraph (AT&T), one of the largest companies providing communications media and services, is a common carrier for long-distance service and a special-purpose carrier for other services. WorldCom, Sprint, and others make up a significant part of the telecommunications industry as well. **Value-added carriers** are companies that have developed private telecommunications systems and offer their services for a fee. Some value-added carriers that offer communications services include SprintNet, Telenet, and Tymnet.

Common carriers typically provide the use of standard telephone lines, called **switched lines**. These lines use switching equipment to allow one transmission device (e.g., your telephone) to be connected to other transmission devices (e.g., the telephones of your friends and relatives). A switch is a special-purpose circuit that directs messages along specific paths in a telecommunications system. When you make a phone call, the local telephone service provider's switching equipment connects your phone to the phone of the person you're calling.

Device	Function
Modem	Translates data from a digital form (as it is stored in the computer) into an analog signal that can be transmitted over ordinary telephone lines. This process is called modulation. Also performs a demodulation function to convert the analog signal received back into digital form.
Fax modem	Facsimile devices, commonly called fax devices, allow businesses to transmit text, graphs, photographs, and other digital files via standard telephone lines. A fax modem is a very popular device that combines a fax with a modem, giving users a powerful communications tool.
Multiplexer	Allows several telecommunications signals to be transmitted over a single communications medium at the same time.
PBX	A communications system that manages both voice and data transfer within a building and to outside lines. In a PBX system, switching equipment routes phone calls and messages within the building. PBXs can be used to connect hundreds of internal phone lines to a few phone company lines.

Telecommunications networks require state-of-the-art computer software technology to continuously monitor the flow of voice, data, and image transmission over billions of circuit miles worldwide. (Source: Stone/Roger Tully.)

dedicated line

a communications line that provides a constant connection between two points; no switching or dialing is needed, and the two devices are always connected

digital subscriber line (DSL)

a communications line that uses existing phone wires going into today's homes and businesses to provide transmission speeds exceeding 500 Kbps at a cost of $20 or more per month

A **dedicated line**, also called a leased line, provides a constant connection between two points. No switching or dialing is needed; the two devices are always connected. Many firms with high data transfer requirements between two points—say, an East Coast and a West Coast shared headquarters arrangement—use dedicated lines. The high initial cost of purchasing or leasing such a line is offset by eliminating long-distance charges incurred with a switched line.

Common carriers are providing more and more phone and dialing services to home and business users. Automatic number identification (ANI), or caller ID, equipment can be installed on a phone system to identify and display the number of an incoming call. In a business setting, ANI can be used to identify the caller and link that caller with information stored in a computer. For example, when a customer calls Federal Express, the customer service rep uses ANI to identify the name and address of the customer, thus saving time when handling a request for a pickup. ANI can be very useful in helping people screen calls before answering them. Common carriers offer even more services to extend the capabilities of the typical phone system, such as intelligent dialing (when a busy signal is received, the phone redials the number when your line and the line of the party you are trying to reach are both free) and access codes to screen out junk calls, wrong numbers, and unwanted phone calls.

A **digital subscriber line (DSL)** uses existing phone wires going into today's homes and businesses to provide transmission speeds exceeding 500 Kbps at a cost of $20 or more per month. This speed means faster Internet access and downloads compared with standard phone lines. A special modem costing a few hundred dollars is also required. DSL lines are not available everywhere; even so, DSL is growing rapidly in usage. The number of DSL modems installed in 2000 was estimated at about 6.2 million—nearly double the rate of a year earlier.[1] Figure 4.3 shows the expected growth of DSL compared with cable modems.[2] The problem for individuals and businesses is how to choose the best telecommunications option.[3] Each option has its own cost, speed, and reliability to consider. Table 4.3 shows some of the costs, advantages, and disadvantages of different lines and services offered by communications carriers.[4]

FIGURE 4.3

Growth of DSL versus Cable Modems
(Sources: DSL figures, Telechoice, Inc., Denver; cable figures, Kenetic Strategies, Inc., Phoenix, found in Matthew Hamblen, "Digital Subscriber Line," *Computerworld*, February 7, 2000, p. 69. Reprinted with permission of *Computerworld*.)

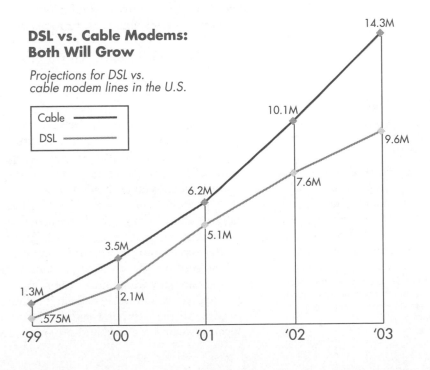

DSL vs. Cable Modems: Both Will Grow

Projections for DSL vs. cable modem lines in the U.S.

Cable ———
DSL ————

14.3M
10.1M
9.6M
7.6M
6.2M
5.1M
3.5M
2.1M
1.3M
.575M

'99   '00   '01   '02   '03

Line/Service	Speed	Cost per Month	Advantages	Disadvantages
Standard phone service	56 Kbps	$10–$30	Low cost and broadly available	Too slow for video and downloads of large files
ISDN	64–128 Kbps	$70–$120	Fast for video and other applications	Higher costs and not available everywhere
DSL	500 Kbps–1.544 Mbps	$20–$100 in addition to standard phone service	Fast, and the service comes over standard phone lines	Slightly higher costs and not available everywhere
Cable modem	Receive at up to 500 Kbps and send at speeds up to 1.544 Mbps	$20–$100	Fast and uses existing cable that comes into the home	Slightly higher costs and not available everywhere
T1	1.544 Mbps	$1,000	Very fast broadband service, typically used by corporations and universities	Very expensive, high installation fee, and users pay a monthly fee based on distance

TABLE 4.3

Costs, Advantages, and Disadvantages of Several Line and Service Types

computer network

the communications media, devices, and software needed to connect two or more computer systems and/or devices

centralized processing

processing alternative in which all processing occurs in a single location or facility

decentralized processing

processing alternative in which processing devices are placed at various remote locations

distributed processing

processing alternative in which computers are placed at remote locations but connected to each other via telecommunications devices

Networks and Distributed Processing

A **computer network** consists of communications media, devices, and software needed to connect two or more computer systems or devices. Once connected, computers can share data, information, and processing jobs. More and more businesses are linking computers in networks to streamline work processes and allow employees to collaborate on projects.

The effective use of networks can turn a company into an agile, powerful, and creative organization, giving it a long-term competitive advantage. Networks can be used to share hardware, programs, and databases across the organization. They can transmit and receive information to improve organizational effectiveness and efficiency. They enable geographically separated workgroups to share documents and opinions, which fosters teamwork, innovative ideas, and new business strategies. To take full advantage of networks, it is important to understand strategies, network concepts and considerations, network types, and related topics.[5]

Basic Data Processing Strategies

When an organization needs to use two or more computer systems, one of three basic data processing strategies may be followed: centralized, decentralized, or distributed. With **centralized processing**, all processing occurs in a single location or facility, providing the highest degree of control. For example, centralized processing is useful for financial institutions that require a high degree of security. With **decentralized processing**, processing devices are placed at various remote locations; however, individual computer systems are isolated and do not communicate with each other. Decentralized systems are suitable for companies that have independent operating divisions. Some drugstore chains, for example, operate each location as a completely separate entity; each store has its own computer system that works independently of the computers at other stores. With **distributed processing**, computers are placed at remote locations but are connected to each other via telecommunications devices. Consider a manufacturing company with plants in Milwaukee, Chicago, and Atlanta and a corporate headquarters in New York. Each location has its own computer system. By connecting all the computer systems into a distributed processing system, all the locations can share data and programs. Distributed processing also allows each

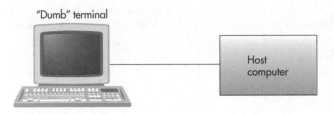

FIGURE 4.4

Terminal-to-Host Connection

terminal-to-host

an architecture in which the application and database reside on one host computer, and the user interacts with the application and data using a "dumb" terminal

file server

an architecture in which the application and database reside on one host computer, called the file server

client/server

an architecture in which multiple computer platforms are dedicated to special functions such as database management, printing, communications, and program execution

plant to perform its own processing (say, for example, inventory) while the New York computer system coordinates and processes other applications, such as payroll. Distribution of processing across the organizational system ensures that the right information is delivered to the right individuals, maximizing the capabilities of the overall information system by balancing the effectiveness and efficiency of each individual computer system.

Terminal-to-Host, File Server, and Client/Server Systems

If an organization chooses distributed information processing, it can connect computers in several ways, including terminal-to-host, file server, and client/server architecture. With **terminal-to-host** architecture, the application and database reside on one host computer, and the user interacts with the application and data using a "dumb" terminal. (Even if you use a PC to access the application, you run terminal emulation software on the PC to make it act as if it were a dumb terminal with no processing capacity.) Since a dumb terminal has no data processing capability, all computations, data accessing and formatting, and data display are done by an application that runs on the host computer (Figure 4.4).

In **file server** architecture, the application and database reside on one host computer, called the file server. The database management system runs on the end user's personal computer or workstation. If the user needs even a small subset of the data that resides on the file server, the file server sends the user the entire file that contains the data requested, including a lot of data the user does not want or need. The downloaded data can then be analyzed, manipulated, formatted, and displayed by a program that runs on the user's personal computer (Figure 4.5).

In **client/server** architecture, multiple computer platforms, called servers, are dedicated to special functions such as database management, data storage, printing, communications, network security, and program execution. Each server is accessible by all computers on the network. A server distributes programs and data files to the other computers (clients) on the network as they request them. The client requests services from the servers, provides a user interface, and presents results to the user. Once data is moved from a server to the client, the data may be processed on the client (Figure 4.6).

The type of application most appropriate for client/server architecture is one that uses large data files, requires fast response time, and needs strong security and recovery options. All these factors point to the kind of applications that are central to the operation and management of the business. On-line transaction processing and decision support applications are particularly good candidates for client/server computing.

FIGURE 4.5

File Server Connection

The file server sends the user the entire file that contains the data requested. The downloaded data can then be analyzed, manipulated, formatted, and displayed by a program that runs on the user's personal computer.

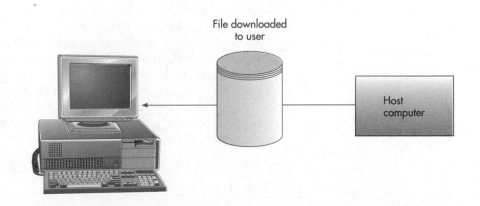

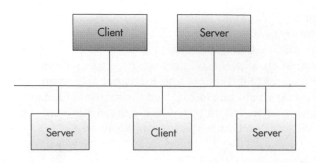

FIGURE 4.6

Client/Server Connection

Moving to client/server architecture can be a major two- to five-year conversion process. Over that time period, considerable costs are incurred for hardware, software, communications equipment and links, data conversion, and training. Also, controlling the client/server environment to prevent unauthorized use, invasion of privacy, and viruses is difficult. Implementing client/server architecture allows an organization to operate using hardware, software, and networks from multiple vendors with, in many cases, relatively new and untested products. Complex situations such as these make it likely that problems will arise, and often the problems are difficult to identify and isolate to a single vendor.

In spite of the drawbacks, the use of single-vendor environments and terminal-to-host architecture is fading fast as corporations move into the much more complex client/server environment. Adaptable systems are essential to implementing a client/server architecture so that organizations are free to choose clients and servers and be assured that their combinations can communicate with one another.

Network Types

Depending on the physical distance between nodes on a network and the communications and services provided by the network, networks can be classified as local area, wide area, or international. Local area networks tie together equipment in a building or local area; international networks are used to communicate between countries. Wide area networks operate over a broad geographic area.

local area network (LAN)

a network that connects computer systems and devices within the same geographic area

A network that connects computer systems and devices within the same geographic area is a **local area network (LAN)**. Typically, local area networks are wired into office buildings and factories (Figure 4.7). Although unshielded twisted-pair (UTP) wire cable is the most widely used medium with LANs, other media—including fiber-optic cable—are also popular. They can be built around powerful personal computers, minicomputers, or mainframe computers. When a personal computer is connected to a local area network, a network interface card (NIC) is usually required. A network interface card is a card that is placed in a computer's expansion slot to allow it to communicate with the network. A wire or connector from the network is then plugged directly into the network interface card.

wide area network (WAN)

a network that ties together large geographic regions using microwave and satellite transmission or telephone lines

A **wide area network (WAN)** ties together large geographic regions using microwave and satellite transmission or telephone lines. When you make a long-distance phone call, you are using a wide area network. AT&T, WorldCom, and others are examples of companies that offer WAN services to the public. Companies also design and implement WANs. These WANs usually consist of computer equipment owned by the user, together with data communications equipment provided by a common carrier. (See Figure 4.8.)

international network

a network that links systems between countries

Networks that link systems between countries are called **international networks**. In addition to requiring sophisticated equipment and software, such networks must meet specific national and international laws regulating the electronic flow of data across international boundaries, often called *transborder data flow*. Some countries have strict laws restricting the use of telecommunications and databases, making normal business transactions such as payroll costly, slow, or even impossible. Other countries, sometimes called *data havens*, have few laws restricting the use of telecommunications and databases. International networks in developing countries can have inadequate equipment and infrastructure that can cause problems and limit the usefulness of the network.

FIGURE 4.7

A Typical LAN

All network users within an office building can connect to each other's devices for rapid communication. For instance, a user in research and development could send a document from her computer to be printed at a printer located in the desktop publishing center.

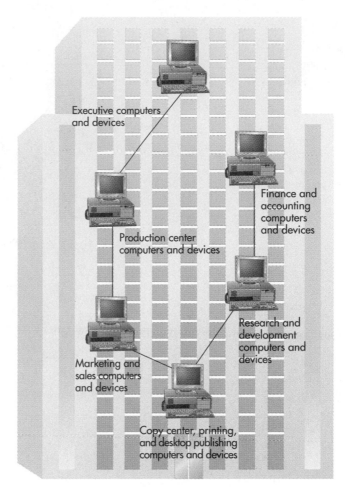

Executive computers and devices

Finance and accounting computers and devices

Production center computers and devices

Research and development computers and devices

Marketing and sales computers and devices

Copy center, printing, and desktop publishing computers and devices

Corporate headquarters

Despite the obstacles, numerous private and public international networks exist. United Parcel Service, for example, has invested in an international network, called UPSnet. UPSnet allows drivers to use handheld computers to send real-time information about pickups and deliveries to central data centers. The huge UPS network allows data to be retrieved by customers to track packages or to be used by the company for faster billing, better fleet planning, and improved customer service.[6]

With more people working at home, connecting home computing devices and equipment together into a unified network is on the rise. Small businesses are also starting to connect their systems and equipment. With a home or small-business network, computers, printers, scanners, and other devices can be connected. A person working on one computer, for example, can use data and programs stored on another computer's hard disk. In addition, a single printer can be shared by several computers on the network (Figure 4.9).

Communications Software and Protocols

communications software

software that provides a number of important functions in a network

Communications software provides a number of important functions in a network. Most communications software packages provide error checking and message formatting. In some cases, when there is a problem, the software can indicate what is wrong and suggest possible solutions. Communications software can also maintain a log listing all jobs and communications that have taken

FIGURE 4.8

A Wide Area Network

Wide area networks are the basic long-distance networks used by organizations and individuals around the world. The actual connections between sites, or nodes (shown by dashed lines), may be any combination of satellites, microwave, or cabling. When you make a long-distance telephone call, you are using a WAN.

North America

network operating system (NOS)

systems software that controls the computer systems and devices on a network and allows them to communicate with each other

network management software

software that enables a manager on a networked desktop to monitor the use of individual computers and shared hardware (such as printers), scan for viruses, and ensure compliance with software licenses

communications protocols

rules and standards that ensure communications among computers of different types and from different manufacturers

place over a specified period of time. In addition, data security and privacy techniques are built into most packages.

Consider a situation in which a computer is attached to a network that connects large disk drives, printers, and other equipment and devices. How does an application program request data from a disk drive on the network? The answer is through the network **operating system (NOS)**, which controls the computer systems and devices on a network and allows them to communicate with each other. An NOS performs the same types of functions for the network as operating system software does for a computer, such as memory and task management and coordination of hardware. When network equipment (such as printers, plotters, and disk drives) is required, the network operating system makes sure that these resources are correctly used.

With **network management software**, a manager on a networked desktop can monitor the use of individual computers and shared hardware (such as printers), scan for viruses, and ensure compliance with software licenses. Some of the many benefits of network management software include fewer hours spent on routine tasks (such as installing new software), faster response to problems, and greater overall network control.

Communications protocols are rules and standards that make communications possible. A number of communications protocols are used by companies and organizations of all sizes. Just as standards are important in building computer and database systems, established protocols help ensure communications among computers of different types and from different manufacturers. Several common protocols are summarized in Table 4.4.

Now that we've described the basics of telecommunications and networks, we turn to discussion of a network of global proportions—the Internet.

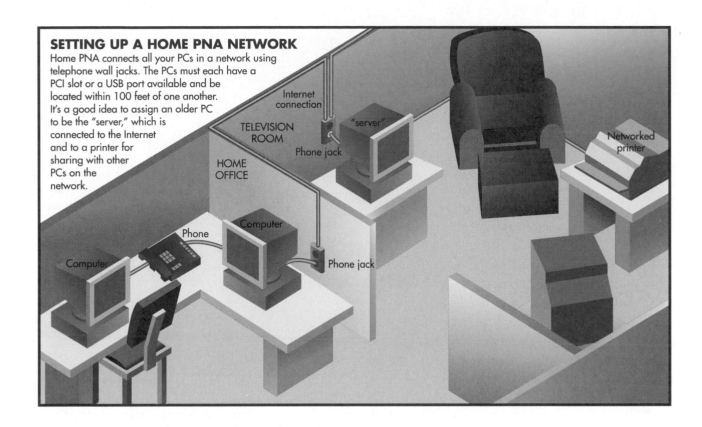

SETTING UP A HOME PNA NETWORK
Home PNA connects all your PCs in a network using telephone wall jacks. The PCs must each have a PCI slot or a USB port available and be located within 100 feet of one another. It's a good idea to assign an older PC to be the "server," which is connected to the Internet and to a printer for sharing with other PCs on the network.

Internet connection

"server"

TELEVISION ROOM

Phone jack

Networked printer

HOME OFFICE

Computer

Phone

Phone jack

Computer

FIGURE 4.9

Connecting Computing Devices Using a Home Network
(Source: Reprinted from "Network Your Home Painlessly," *PC Magazine*, April 4, 2000, p. 108. Copyright © 2000 Ziff Davis Media, Inc. All rights reserved.)

TABLE 4.4

Common Communications Protocols

USE AND FUNCTIONING OF THE INTERNET

It is hard to overestimate the impact the Internet has had on today's organizations and our daily lives. The Internet is the world's largest computer network. Actually, the Internet is a collection of interconnected networks, all freely exchanging information (see Figure 4.10). Research firms, colleges, and universities have long been part of the Internet, and now businesses, high schools, elementary schools, and other organizations are joining up as well. Nobody knows exactly how big the Internet is because it is a collection of separately run smaller computer networks with no single place where all the connections are registered.

Protocol	Description
Open Systems Interconnection (OSI)	This protocol divides data communications functions into seven distinct layers to simplify the development, operation, and maintenance of complex telecommunications networks.
Transport control protocol/Internet protocol (TCP/IP)	The primary communications protocol of the Internet, developed in the 1970s.
Systems Network Architecture (SNA)	Communications protocol used with IBM and IBM-compatible computers.
Ethernet	Popular communications protocol often used with local area networks that ensures compatibility among devices so that many people can attach to a common cable to share network facilities and resources.
X.400	A set of standards that are often used by businesses to process transactions electronically.
X.500	A set of standards for defining a network directory containing information on users—from names and e-mail addresses to job titles and resource-access privileges.

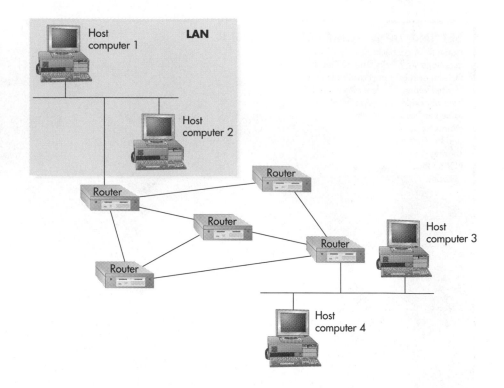

The Internet is truly international in scope, with users on every continent—including Antarctica. Although the United States still claims more Web activity than other countries, the Internet is expanding around the globe.[7] In Japan, for example, approximately 18 million people use the Internet.[8] In a few short years, the numbers are expected to swell to over 60 million Internet users. In Brazil, Latin America's biggest Internet market, a number of Internet companies are about to launch free Internet service.[9] These companies hope to make a profit from sales transactions and advertising fees. If they are successful, free Web access may become more common worldwide.

The ancestor of the Internet was **ARPANET**, a project started by the U.S. Department of Defense (DOD) in 1969. The ARPANET was both an experiment in reliable networking and a means to link DOD and military research contractors, including a large number of universities doing military-funded research. (*ARPA* stands for the Advanced Research Projects Agency, the branch of the DOD in charge of awarding grant money. The agency is now known as DARPA—the added *D* is for *Defense*.)

Unlike a corporate network with a centralized infrastructure, the Internet is nothing more than an ad hoc linkage of many networks that adhere to basic standards. Since these networks are constantly changing and being improved, the Internet itself is in a perpetual state of evolution. However, since the Internet is such a loose collection of networks, there is nothing to prevent some participants from using outdated or slow equipment.

To speed Internet access, a group of corporations and universities, called the University Corporation for Advanced Internet Development (UCAID) is working on a faster, new Internet.[10] Called Internet2 (I2), Next Generation Internet (NGI), and Abilene, depending on the universities or corporations involved, the new Internet offers the potential of faster Internet speeds, up to 2 Gbits per second or more. Although not related to the efforts of UCAID, Project Oxygen could also improve the speed of the Internet. The project is a global fiber-optic, undersea network that will connect almost 80 countries. The first phase of the project should be completed by 2003.

How the Internet Works

The Internet transmits data from one computer (called a host) to another (see Figure 4.10). If the receiving computer is on a network to which the first computer is directly connected, it can send the message directly. If the receiving computer is not on the same network as the sending computer, the sending computer relays the message to another computer that can forward it. The message may be sent through a *router* to reach the forwarding computer. The forwarding host, which presumably is attached to at least one other network, in turn delivers the message directly if it can or passes it to yet another forwarding host. It is quite common for a message to pass through a dozen or more forwarders on its way from one part of the Internet to another.

The various networks that are linked to form the Internet work pretty much the same way—they pass data around in chunks called *packets*, each of which carries the addresses of its sender and its receiver. The set of conventions used to pass packets from one host to another is known as the **Internet protocol (IP)**. Many other protocols are used in conjunction with IP. The best known is the **transport control protocol (TCP)**, which includes rules that computers on a network use to establish and break connections. Many people refer to TCP/IP, the combination of TCP and IP used by most Internet applications. Adhering to the same technical standards allows the more than 100,000 individual computer networks owned by governments, universities, nonprofit groups, and companies to constitute the Internet. Once a network following these standards links to a **backbone**—one of the Internet's high-speed, long-distance communications links—it becomes part of the worldwide Internet community.

Each computer on the Internet has an assigned address called its **uniform resource locator**, or *URL*, to identify it from other hosts. The URL gives those who provide information over the Internet a standard way to designate where Internet elements such as servers and documents can be found. Let's look at the URL for Course Technology, http://www.course.com/home.cfm.

The "http" specifies the access method and tells your software to access this particular file using the Hypertext Transfer Protocol. This is the primary method for interacting with the Internet. The "www" part of the address signifies that the address is associated with the World Wide Web service (discussed later). The "course.com/home.cfm" part of the address is the domain name that identifies the Internet host site. Domain names must adhere to strict rules. They always have at least two parts separated by dots (periods). For all countries except the United States, the rightmost part of the domain name is the country code (au for Australia, ca for Canada, dk for Denmark, fr for France, jp for Japan, etc.). Within the United States, the country code is replaced with a code denoting affiliation categories (see Table 4.5). The leftmost part of the domain name identifies the host network or host provider, which might be the name of a university or business.

Today, 20 companies, called *registrars,* can register domain names, and an additional 33 companies are accredited to register domain names. Another 30 companies are seeking accreditation to register domain names from the Internet Corporation for Assigned Names and Numbers (ICANN).[11] Some registrars are concentrating on large corporations, where the profit margins may be higher, compared with small businesses or individuals.

It has been estimated that there are over 270,000 registered domain names.[12] Some people, called cybersquatters, have registered domain names in the hope of selling the names to corporations at a later date. The domain name Business.com, for example, sold for $7.5 million. But some companies are fighting back. Ford Motor Company, for example, sued a person who tried to sell the domain names Ford-quality.com and Lincoln-quality.com on an Internet auction site.[13]

Internet protocol (IP)

a communication standard that enables traffic to be routed from one network to another as needed

transport control protocol (TCP)

a protocol that includes rules that computers on a network use to establish and break connections

backbone

one of the Internet's high-speed, long-distance communications links

uniform resource locator (URL)

an assigned address on the Internet for each computer

TABLE 4.5

U.S. Top-Level Domain
Affiliations

Affiliation ID	Affiliation
aero	air transport industry
biz	alternative to com
com	business organization
coop	non-profit cooperatives
edu	educational sites
firm	businesses and firms
gov	government sites
info	unrestricted use
mil	military sites
museum	museums
name	individuals
net	networking organizations
org	organizations
pro	accountants, lawyers, physicians

Accessing the Internet

There are three ways to connect to the Internet (Figure 4.11). Which method is chosen is determined by the size and capability of the organization or individual.

Connect via LAN Server

This approach requires the user to install on his or her PC a network adapter card and Open Datalink Interface (ODI) or Network Driver Interface Specification (NDIS) packet drivers. These drivers allow multiple protocols to run on one network card simultaneously. LAN servers are typically connected to the Internet at 56 Kbps or faster. Such speed makes for an exciting trip on the Internet but is

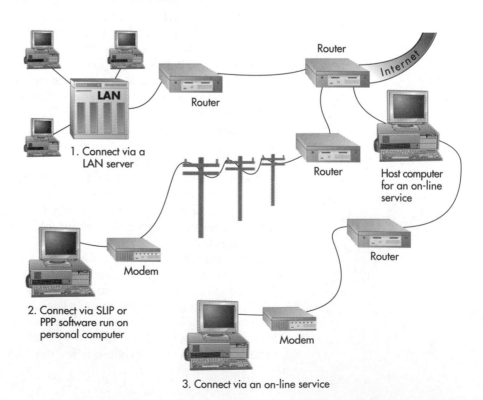

FIGURE 4.11

Three Ways to Access the Internet

There are three ways to access the Internet—using a LAN server, dialing into the Internet using SLIP or PPP, or using an on-line service with Internet access.

also very expensive—$2,000 or so a month! However, the cost of this connection can be shared among several dozen LAN users to get to a reasonable cost per user. Additional costs associated with a LAN connection to the Internet include the cost of the protocol software mentioned at the beginning of this section.

Connect via SLIP/PPP

This approach requires a modem and the TCP/IP protocol software plus **serial line internet protocol (SLIP)** or **point-to-point protocol (PPP)** software. SLIP and PPP are two communications protocols that transmit packets over telephone lines, allowing dial-up access to the Internet. If you are running Windows, you will also need Winsock. Users must also have an Internet service provider that lets them dial into a SLIP/PPP server. SLIP/PPP accounts can be purchased for $30 a month or less from regional providers. With all this in place, a modem is used to call into the SLIP/PPP server. Once the connection is made, you are on the Internet and can access any of its resources. The costs include the cost of the modem and software, plus the service provider's charges for access to the SLIP/PPP server. The speed of this Internet connection is limited to the slower of your computer's modem and the speed of the modem of the SLIP/PPP server to which you connect.

Connect via an On-Line Service

This approach requires nothing more than what is required to connect to any of the on-line information services—a modem, standard communications software, and an on-line information service account. There is normally a fixed monthly cost for basic services, including e-mail. The on-line information services provide a wide range of services, including e-mail and access to the World Wide Web. America Online, Microsoft Network, and Prodigy are examples of such services.

Internet Service Providers

An **Internet service provider (ISP)** is any company that provides individuals and organizations with access to the Internet. ISPs do not offer the extended informational services offered by commercial on-line services such as America Online or Prodigy. There are literally thousands of Internet service providers, ranging from universities making unused communications line capacity available to students and faculty to major communications giants such as AT&T and MCI. To use this type of connection, you must have an account with the service provider and software that allows a direct link via TCP/IP.

In choosing an Internet service provider, the important criteria are cost, reliability, security, the availability of enhanced features, and the service provider's general reputation. Reliability is critical because if your connection to the ISP fails, it interrupts your communications with customers and suppliers. Among the value-added services ISPs provide are electronic commerce, networks to connect employees, networks to connect with business partners, host computers to establish your own Web site, Web transaction processing, network security and administration, and integration services. Many corporate IS managers welcome the chance to turn to ISPs for this wide range of services because they do not have the

serial line Internet protocol (SLIP)
a communications protocol that transmits packets over telephone lines

point-to-point protocol (PPP)
a communications protocol that transmits packets over telephone lines

Internet service provider (ISP)
any company that provides individuals or organizations with access to the Internet

To use an ISP such as AT&T, you must have an account with the service provider and software that allows a direct link via TCP/IP.

TABLE 4.6

A Representative List of Internet
Service Providers

Internet Service Provider	Web Address
AT&T's WorldNet Service	www.att.com
BellSouth	www.bellsouth.com
Digex, Inc.	www.digex.net
Earthlink/Sprint	www.earthlink.net
GTE Internetworking	www.gte.net
IBM Internet Connection	www.ibm.net
WorldCom	www.wcom.com
Sprintlink	www.sprint.net
UUnet Technologies	www.uu.net

in-house expertise and cannot afford the time to develop such services from scratch. In addition, when organizations go with an ISP-hosted network, they can also tap the ISP's national infrastructure at minimum cost. That's important when a company has offices spread across the country.

In most cases, ISPs charge a monthly fee that can range from $15 to $30 for unlimited Internet connection. The fee normally includes use of e-mail. Some ISPs, however, are experimenting with no-fee Internet access.[14] But there are strings attached to the no-fee offers in most cases. Some free ISPs require that customers provide detailed demographic and personal information. In other cases, customers must put up with extra advertising banners on every Web site. Free net ISPs include NetZero, with approximately three million subscribers, and AltaVista. AltaVista's free net service obtained about 1.5 million customers in its first five months of operation. Table 4.6 identifies the major corporate Internet service providers.[15]

An increasing number of ISPs are offering Internet connection via satellite. Other options offered by some ISPs include DSL, ISDN, and cable modems.[16]

INTERNET AND TELECOMMUNICATION SERVICES

Telecommunications and networks are constantly applied to support information systems and organizational goals. For example, suppose a business needs to develop an accurate monthly production forecast. Doing so requires a manager to download data from customers' databases of sales forecasts. Telecommunications can provide a network link so that the manager can access the data needed for the production forecast report, which in turn supports the company's objective of better financial planning.

The consumer goods giant Procter & Gamble uses local area networks in all its plants to link office and plant workers to common software and shared databases and to provide e-mail services. The result is faster, more cost-effective, higher-quality product manufacturing. Other organizations transfer millions of important and strategic messages from one location to another every day. Telecommunications has become a critical component of information systems. In some industries it is almost a requirement for doing business; most companies could not survive without it. This section will look at some significant business applications of networks.

Voice Mail, Electronic Mail, and Instant Messaging

voice mail

technology that enables users to leave, receive, and store verbal messages for and from other people around the world

With **voice mail**, users can leave, receive, and store verbal messages for and from other people around the world. People can also send messages to others via electronic mail, also called e-mail. With the right hardware and software, a sender can connect his or her computer to a network, type in a message, and send it to

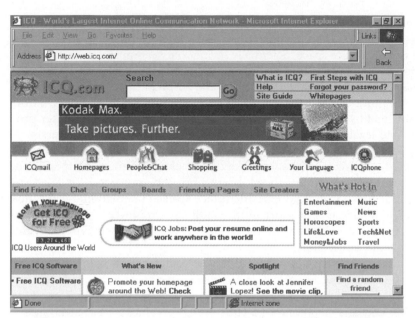

ICQ is a client program that informs you who's on-line and enables you to contact them and chat with them in real time.

instant messaging

a method that allows two or more individuals to communicate on-line using the Internet

telecommuting

a work arrangement in which employees work away from the office using personal computers and networks to communicate via e-mail with other workers and to pick up and deliver results

Many companies have adopted policies for telecommuting that enable employees to work away from the office using personal computers and networks. (Source: Stone/Ian Shaw.)

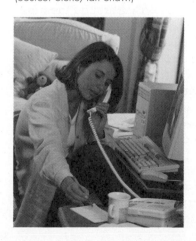

another person on the network. It is also possible to attach files (spreadsheets, word processing documents, etc.) to an e-mail message, and e-mail messages can be forwarded to other people. E-mail services are typically included with an Internet connection, through a company such as America Online or Prodigy. In addition, there are a number of free Internet providers with excellent features. Yahoo!, Hotmail, and Eudora, for example, offer free e-mail.[17] In some cases, you have to watch ads with these services to be able to use the e-mail features.

While e-mail has become an indispensable part of business, it also has potential problems. E-mail attachments can contain viruses that infect entire computer systems. E-mail can also be used to threaten people and organizations. From an employee's perspective, organizations can also use e-mails sent at work in lawsuits against employees. Some employees have been disciplined or fired because of inappropriate e-mail messages they sent from the office. In addition, some e-mails may contain confidential corporate information that should not be shared with everyone. E-mail can also be used to invade personal privacy. To combat some of these problems, software programs that provide security for e-mail transmissions have been developed. For example, a product from Qvtech, a Colorado Springs company, is developing a program that will make e-mail self-destruct after a certain length of time.[18] To protect privacy, public-key infrastructure (PKI) can be used to help keep e-mails, such as those used to send medical and sensitive corporate information, private and secure.[19]

Instant messaging is on-line, real-time communication between two or more people who are connected to the Internet.[20] With instant messaging, two or more screens open up. Each screen displays what one person is typing. Because the typing is displayed on the screen in real time, it is like talking to someone using the keyboard.

A number of companies offer instant messaging, including America Online, Yahoo!, and Microsoft Network. In addition to being able to type messages on a keyboard and have the information instantly displayed on the other person's screen, some instant messaging programs enable voice communication or connection to cell phones. One wireless service provider announced that it has developed a technology that can detect when a person's cell phone is turned on.[21] With this technology, it will be possible for someone on the Internet using instant messaging to communicate with someone on a cell phone anywhere in the world.

Telecommuting, Videoconferencing, and Internet Phone Service

More and more work is being done away from the traditional office setting. Many enterprises have adopted policies for **telecommuting** that enable employees to work away from the office using personal computers and networks. According to one study sponsored by AT&T, a company can save about $10,000 per year for each employee it allows to telecommute.[22]

There are several reasons why telecommuting is popular among workers. Single parents find that it helps in balancing family and work responsibilities because it eliminates the daily commute. It also enables qualified workers who may be unable to participate in the normal workforce (e.g., those who are physically challenged or who live in rural areas too far from the city office to commute on a regular basis) to become productive workers. Extensive use of telecommuting can lead to decreased need for office space, potentially saving a large company millions of dollars. Corporations are also being encouraged by public policy to try telecommuting as a means of reducing traffic congestion and air pollution.

Some types of jobs are better suited for telecommuting than others. These include jobs held by salespeople, secretaries, real estate agents, computer programmers, and legal assistants, to name a few. It also takes a special personality type to be effective while telecommuting. Telecommuters need to be strongly self-motivated, organized, and able to stay on track with minimal supervision, and must have a low need for social interaction. Jobs not good for telecommuting include those that require frequent face-to-face interaction, need much supervision, and have lots of short-term deadlines. Employees who choose to work at home must be able to work independently, manage their time well, and balance work and home life.

Videoconferencing enables people to have a conference by combining voice, video, and audio transmission. Not only are travel expenses and time reduced, but managerial effectiveness is increased through faster response to problems, access to more people, and less duplication of effort at geographically dispersed sites.[23] Almost all videoconferencing systems (Figure 4.12) combine video and phone call capabilities with data or document conferencing. You can see the other person's face, view the same documents, and swap notes and drawings. With some of the systems, callers can make changes to live documents in real time. Many businesses find that the document- and application-sharing features of the videoconference enhance group productivity and efficiency. Meeting over phone lines also fosters teamwork and can save corporate travel time and expense. Group videoconferencing is used daily in a variety of businesses as an easy way to connect work teams. Members of a team go to a specially prepared videoconference room equipped with sound-sensitive cameras that automatically focus on the person speaking, large TV-like monitors for viewing the participants at the remote location, and high-quality speakers and microphones. It costs around $60,000 to set up a typical group videoconferencing room. There are additional expenses associated with use of the telecommunications network to relay voice, video, data, and images.

The Royal Bank of Scotland uses videoconferencing to save its managers and executives the time and hassle of traveling between Scotland and England.[24] The system has been in place for more than ten years and is constantly being updated and improved. Today, the videoconferencing system has 13 videoconferencing studios for general use, 6 facilities in directors' offices, and another 30 desktop videoconferencing devices. "We do about 16 or 17 videoconferences a day between the studios," says George Clark, senior manager of telecommunications.

Internet phone service enables you to communicate with other Internet users around the world who have equipment and software compatible to yours.[25] This service is relatively inexpensive and can make sense for international calls. With some services, such as Net2Phone, it is now possible for someone using the Internet to make a call to someone using a standard phone.

Using **voice-over-IP (VOIP)** technology, network managers can route phone calls and fax transmissions over the same network they use for data—which means no more phone bills. Gateways installed at both ends of the communications link convert voice to IP packets and back. With the advent of widespread, low-cost Internet

---

videoconferencing

a telecommunication system that combines video and phone call capabilities with data or document conferencing

voice-over-IP (VOIP)

technology that enables network managers to route phone calls and fax transmissions over the same network they use for data

FIGURE 4.12

Videoconferencing

Videoconferencing allows participants to conduct long-distance meetings "face to face" while eliminating the need for costly travel.
(Source: Courtesy of Zydacron.)

telephone services, traditional long-distance providers are being pushed to either respond in kind or trim their own long-distance rates.

What is especially interesting about VOIP is the promise of new ways for merging voice with video and data communications over the Web or a company's data network. In the long run, it's not the cost savings that will boost the market, it's the multimedia capabilities it gives us and the smart call-management capabilities. Travel agents could use voice and video over the Internet to discuss travel plans; Web merchants could use it to show merchandise and take orders; customers could show suppliers problems with their products.

Electronic Data Interchange (EDI)

electronic data interchange (EDI)

an intercompany, application-to-application communication of data in standard format, permitting the recipient to perform the functions of a standard business transaction

Electronic data interchange (EDI) is an intercompany, application-to-application communication of data in standard format, permitting the recipient to perform the functions of a standard business transaction, such as processing purchase orders. EDI uses network systems and follows standards and procedures that allow output from one system to be processed directly as input to other systems, without human intervention. With EDI, the computers of customers, manufacturers, and suppliers can be linked (Figure 4.13). This technology eliminates the need for paper documents and substantially cuts down on costly errors. Customer orders and inquiries are transmitted from the customer's computer to the manufacturer's computer. The manufacturer's computer can then determine when new supplies are needed and can automatically place orders by connecting with the supplier's computer.

For some industries, EDI is a necessity. Some companies will do business only with suppliers and vendors using compatible EDI systems, regardless of the expense or effort involved. For small companies that may not be able to afford their own EDI system, there are vendors that provide EDI services, such

FIGURE 4.13

Two Approaches to Electronic Data Interchange

Many organizations now insist that their suppliers operate using EDI systems. Often the EDI connection is made directly between vendor and customer (a); alternatively, the link may be provided by a third-party clearinghouse, which provides data conversion and other services for the participants (b).

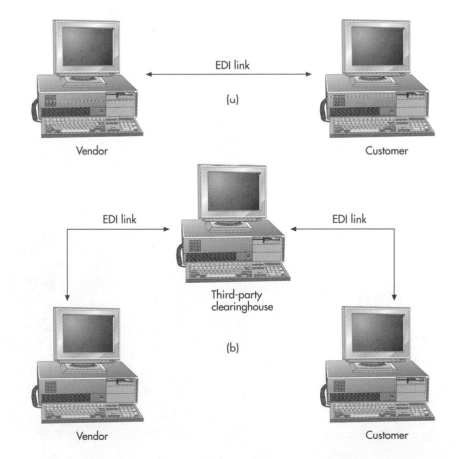

as St. Paul Software.[26] These companies allow smaller companies to provide products to larger corporations that require EDI capabilities.

Public Network and Specialized Services

public network services

systems that give personal computer users access to vast databases and other services, usually for an initial fee plus usage fees

Public network services give personal computer users access to vast databases and the Internet. They also allow them to send and receive e-mail, book airline reservations, check weather forecasts, get information on TV programs, analyze stock prices and investment information, communicate with others on the network, play games, and receive articles and government publications. Fees, based on the services used, can range from less than $15 to more than $500 per month (Figure 4.14). Providers of public network services include Microsoft, America Online, and Prodigy.

With millions of personal computers in businesses across the country, interest in specialized and regional information services is increasing. Specialized services, which can be expensive, include professional legal, patent, and technical information. For example, investment companies can use systems such as Quotron and Shark to get up-to-the-minute information on stocks, bonds, and other investments.

Regional services, also called metropolitan services, include local electronic bulletin boards and electronic mail facilities that offer information regarding local club, school, and government activities. An electronic bulletin board is a message center that displays messages in electronic form, much as a bulletin board displays paper messages in schools and offices. In addition to regional bulletin boards, national and international bulletin boards are available for people and groups with special interests or needs. These types of bulletin boards exist for many users, such as users of certain software packages and users with certain hobbies. Many public network services, including Prodigy and America Online, provide access to hundreds of different bulletin boards on a variety of topics and interest areas.

Global positioning systems (GPSs) are other types of specialized telecommunications services. They have long been used by the military to find locations of troops, equipment, and the enemy—within yards in some cases. Today, GPS is being used by companies to survey land and buildings and by individuals to

FIGURE 4.14

Public network services provide users with the latest information required to remain competitive. AOL, for example, enables subscribers to obtain up-to-the-minute stock quotes.

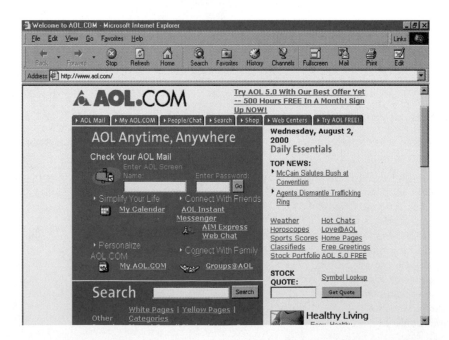

locate their positions while camping or exploring. The Garmin GPS III Plus, for example, is a nine-ounce system that displays your location on built-in maps.[27] The system can also determine speed and altitude. The Axiom Sports Tracker can be used on the slopes at Vail and Copper Mountain. After a day of skiing, you can download and print the runs you completed during the day. Some auto companies have placed GPS systems into their cars to assist travelers in need. The auto systems and some advanced GPSs combine basic GPS features with a cellular phone.

The use of pagers is also on the increase, with many features, including two-way paging. Airline companies have placed phone and Internet services on many of their aircraft, allowing people to stay in touch at 30,000 feet. With all these telecommunications systems and services, it is no wonder that managers and workers are able to conduct business from remote locations. Often called *virtual workers,* these managers and employees can conduct business at any time and at any place. Today, new telecommunications systems and services are being introduced every month. In the future, we can expect even more innovations in telecommunications that will dramatically alter how businesses and individuals stay connected and in touch.

Distance Learning

Telecommunications can be used to extend the classroom. General Motors, for example, has developed an interactive system to reach its 175,000 employees at more than 7,000 dealerships.[28] The new system allows employees to view a live course without leaving their dealership. Travel costs are being dramatically cut with the new system, and the quality of the training is expected to greatly improve. Often called **distance learning**, these electronic classes are likely to thrive in the future.

distance learning

the use of telecommunications to extend the classroom

With distance learning software and systems, instructors can easily create course home pages on the Internet. Students can access the course syllabus and instructor notes on the Web page. E-mail mailing lists can be established so students and the instructor can easily mail one another as a means of turning in homework assignments or commenting and asking questions about material presented in the course. It is also possible to form chat groups so that students can work together as a "virtual team" that meets electronically to complete a group project.

On-line Music, Radio, and Video

Music, radio, and video are now available on the Web. In fact, music is one of the hottest growth areas. Audio and video programs can be played on the Internet, or files can be downloaded. Using music players and music formats such as MP3, it is possible to download music from the Internet and listen to it anywhere using small, portable music players. A number of companies are now offering music over the Internet.[29] A few weeks after the announcement of the huge America Online (AOL)–Time Warner merger, the company also announced that the new merged company would acquire EMI Group for $20 billion to allow AOL–Time Warner to offer music over the Internet. According to AOL's CEO, Steve Case, "We have the opportunity to create a personal jukebox in the house and the car." Another key executive of AOL said, "This is the takeoff point for the music business." Music on the Internet, however, is not without controversy. One Internet site that allows people to share perfect digital copies of songs and music has been sued.[30] In Germany, a court ruled that a major Internet service provider may be liable for music copyright violations, and in China, imported music and videos have been banned.

It is also possible to listen to radio broadcasts over the Internet and download radio programs. Entire audio books can also be downloaded for later listening, using devices such as the Audible Mobile Player. This technology is similar to the popular books-on-tape media, except you don't need a cassette tape or a tape player. Worldstream Communications is now offering interactive talk shows on the Internet, with a format that resembles the popular TV talk shows. Topics range from politics to economics to news.[31] A typical program has a reporter interviewing a guest. On the Internet, you can see pictures of the guest and hear the live talk and audience reactions.

Some corporations have also started to use Internet video to broadcast corporate messages or to advertise on the Web.[32] Victoria's Secret, for example, used Internet video to advertise its lingerie line. The video was so popular that the 1.5 million viewers jammed the Internet site. In addition to advertising, some companies are now investigating the use of Internet video to broadcast stockholder meetings, statements to the public from top-level executives, and other messages. Doctors can also use Internet video to monitor and even control surgical operations that take place thousands of miles from them. Internet video is also being used successfully for teleconferencing, which can connect employees, managers, and corporate executives around the world in private conversations. Using Internet video, it is also possible to receive TV programs from an Internet site.

Telnet, FTP, and Content Streaming

Telnet
a terminal emulation protocol that enables users to log on to other computers on the Internet to gain access to public files

file transfer protocol (FTP)
a protocol that describes a file transfer process between a host and a remote computer and allows users to copy files from one computer to another

content streaming
a method for transferring multimedia files over the Internet so that the data stream of voice and pictures plays more or less continuously, without a break, or very few of them; enables users to browse large files in real time

chat room
a facility that enables two or more people to engage in interactive "conversations" over the Internet

Telnet is a terminal emulation protocol that enables you to log on to other computers on the Internet to gain access to their publicly available files. Telnet is particularly useful for perusing library card files and large databases. It is also called *remote logon*.

File transfer protocol (FTP) is a protocol that describes a file transfer process between a host and a remote computer. Using FTP, you can copy a file from another computer to your computer. FTP is often used to gain access to a wealth of free software on the Internet.

Content streaming is a method for transferring multimedia files over the Internet so that the data stream of voice and pictures plays more or less continuously, without a break, or very few of them. It also enables users to browse large files in real time. For example, rather than wait the half-hour it might take for an entire 5-MB video clip to download before they can play it, users can begin viewing a streamed video as it is being received.

Chat Rooms

A **chat room** is a facility that enables two or more people to engage in interactive "conversations" over the Internet. In fact, when you participate in a chat room, there may be dozens of participants from around the world. Multiperson chats are usually organized around specific topics, and participants often adopt nicknames to maintain anonymity. One form of chat room, Internet Relay Chat (IRC), requires participants to type their conversation rather than speak. Voice chat is also an option, but you must have a microphone, sound card and speakers, a fast modem, and voice-chat software compatible with that of the other participants. Do not use extreme words or repeat rumors (you could risk libel or defamation lawsuits). Protect yourself by not offering personal information such as home address, employer, or phone number.

As you can see, the variety of Internet services is broad. Organizations are increasingly turning to the Internet for communications, whether voice or data networking is needed. We now take a look at a special portion of the Internet that is in the news daily—the World Wide Web.

THE WORLD WIDE WEB

World Wide Web (WWW, or W3)

an Internet service comprising tens of thousands of independently owned computers that work together as one

The **World Wide Web** (the Web, WWW, or W3) is an Internet service comprising tens of thousands of independently owned computers that work together as one. These computers, called Web servers, are scattered all over the world and contain every imaginable type of data. Thanks to the high-speed Internet circuits connecting them and some clever cross-indexing software, users are able to jump from one Web computer to another effortlessly—creating the illusion of using one big computer. Because of its ability to handle multimedia objects, including linking multimedia objects distributed on Web servers around the world, the Web is emerging as the most popular means of information access on the Internet today.

The Web is a menu-based system that uses the client/server model. It organizes Internet resources throughout the world into a series of menu pages, or screens, that appear on your computer. Each Web server maintains pointers, or links, to data on the Internet and can retrieve that data. However, you need the right hardware and telecommunications connections, or the Web can be painfully slow.

home page

a cover page for a Web site that has titles, graphics, and text

hypermedia

tools that connect the data on Web pages, allowing users to access topics in whatever order they wish

Data can exist on the Web as ASCII characters, word processing files, audio files, graphic and video images, or any other sort of data that can be stored in a computer file. A Web site is like a magazine, with a cover page called a **home page** that has color graphics, titles, and text. All the type that is underlined is hypertext, which links the on-screen page to other documents or Web sites. **Hypermedia** connects the data on pages, allowing users to access topics in whatever order they wish. As opposed to a regular document that you read linearly, hypermedia documents are more flexible, letting you explore related documents at your own pace and navigate in any direction. For example, if a document mentions the Egyptian pharaohs, you can choose to see a picture of the pyramids, jump into a description of the building of the pyramids, and then jump back to the original document. Hypertext links are maintained using URLs. Table 4.7 lists some interesting Web sites. Many PC and business magazines also publish interesting and useful Web sites, and they are often evaluated and reviewed in print media and on-line.[33]

TABLE 4.7

Several Interesting Web Sites

Site	Description	URL
Monster	This is a job-hunting site. You can search for a job by type or company, list your résumé, and perform basic company research. One feature, Talent Market, allows people to put their skills up for bid.	www.monster.com
AskMe	This site offers expert advice on a variety of topics.	www.askme.com
ICQ	This is a chat facility that offers free chat services for two or more people.	web.icq.com
MyHelpDesk	This site offers information and help on over 1,500 products. The site includes vendor phone numbers, chat room addresses, and message boards.	www.myhelpdesk.com
MSN MoneyCentral	This Microsoft site offers a large range of financial and investment information.	moneycentral.msn.com
Britannica	This site provides the popular encyclopedia on-line.	www.britannica.com
DealTime	This site helps you find what you want on the Internet.	www.dealtime.com
eBay	This is a popular auction site on the Internet.	www.ebay.com
Amazon.com	This popular site sells books, videos, music, furniture, and much more.	www.amazon.com
Travelocity	This large site offers travel information and bargains.	www.travelocity.com
WebMD	This site provides medical information and advice.	www.webmd.com

hypertext markup language (HTML)

the standard page description language for Web pages

HTML tags

codes that let the Web browser know how to format text: as a heading, as a list, or as body text and whether images, sound, and other elements should be inserted

Extensible Markup Language (XML)

markup language for Web documents containing structured information, including words, pictures, and other elements

Web browser

software that creates a unique, hypermedia-based menu on your computer screen that provides a graphical interface to the Web

search engine

a Web search tool

FIGURE 4.15

Sample Hypertext Markup Language

Shown at the left on the screen is a document, and at the right are the corresponding HTML tags.

Hypertext markup language (HTML) is the standard page description language for Web pages. One way to think about HTML is as a set of high-lighter pens in different colors that you use to mark up plain text to make it a Web page—red for the headings, yellow for bold, and so on. The **HTML tags** let the browser know how to format the text: as a heading, as a list, or as body text. HTML also tells whether images, sound, and other elements should be inserted. Users mark up a page by placing HTML tags before and after a word or words. For example, to turn a sentence into a heading, you place the <H1> tag at the start of the sentence. At the end of the sentence, you place the clos-ing tag </H1>. When you view this page in your browser, the sentence will be displayed as a heading. This means that a Web page is made up of two things: text and tags. The text is your message, and the tags are codes that mark the way words will be displayed. All HTML tags are encased in a set of less than (<) and greater than (>) arrows, such as <H2>. The closing tag has a forward slash in it, such as </B> for closing bold. Figure 4.15 shows a simple document and its corresponding HTML tags.

A number of new Web standards are undergoing definition and early use. These include Extensible Markup Language (XML), cascading style sheets (CSS), and Dynamic HTML (DHTML). **Extensible Markup Language (XML)** is markup language for Web documents containing structured infor-mation, including words, pictures, and other elements. CSSs improve Web page presentation, and DHTML provides dynamic presentation of Web content. These standards will provide quicker access to and display of Web pages.

Web Browsers

A **Web browser** creates a unique, hypermedia-based menu on your computer screen that provides a graphical interface to the Web. The menu consists of graph-ics, titles, and text with hypertext links. The hypermedia menu links you to Internet resources, including text documents, graphics, sound files, and news-group servers. As you choose an item or resource, or move from one document to another, you may be jumping between computers on the Internet without knowing it, while the Web handles all the connections. The beauty of Web browsers and the Web is that they make surfing the Internet fun. Just clicking with a mouse on a highlighted word or a graphical button whisks you effortlessly to computers halfway around the world. Most browsers offer basic features such as support for backgrounds and tables, the ability to view a Web page's HTML source code, and a way to create hot lists of your favorite sites.

Search Engines

Looking for information on the Web is a little like browsing in a library—without the card catalog, it is extremely difficult to find infor-mation. Web search tools—called **search engines**—take the place of the card catalog. Most search engines, such as Yahoo.com, are free. They make money by charging advertis-ers to put ad banners on their sites.

The Web is a huge place, and it gets big-ger with each passing day, so even the largest search engines do not index all Internet pages. Even if you do find a search site that suits you, your query might still miss the mark. So when searching the

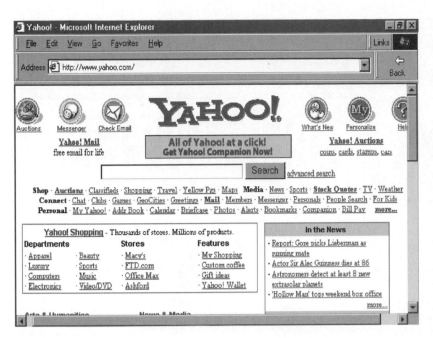

Yahoo! is one of the most popular search engines on the Web.

Java

an object-oriented programming language based on C++ that allows small programs (applets) to be embedded within an HTML document

push technology

automatic transmission of information over the Internet rather than making users search for it with their browsers

Web, you may wish to try more than one search engine to expand the total number of potential Web sites of interest. There are a number of Web search tools to choose from, as summarized in Table 4.8.

Java

Java is an object-oriented programming language from Sun Microsystems based on C++ that allows small programs called *applets* to be embedded within an HTML document. When the user clicks on the appropriate part of the HTML page to retrieve it from a Web server, the applet is downloaded onto the client workstation environment, where it begins executing (Figure 4.16).

Java lets software writers create compact "just-in-time" programs that can be dispatched across a network such as the Internet. On arrival, the applet automatically loads itself on a personal computer and runs—reducing the need for computer owners to install huge programs anytime they need a new function. And unlike other programs, Java software can run on any type of computer. So far, Java is used mainly by programmers to make Web pages come alive, adding splashy graphics, animation, and real-time updates. Java-enabled Web pages are more interesting than plain Web pages.

Push Technology

Push technology is used to send information automatically over the Internet rather than make users search for it with their browsers. Frequently, the information, or "content," is customized to match an individual's needs or profile. The use of push technology is also frequently referred to as "Webcasting." Most push systems rely on HTTP or Java technology to collect content from Web sites and deliver it to users' desktops. Before they can be "pushed," users must download and install software that acts like a TV antenna, capturing transmitted content. As with any new technology, the people paying for push have yet to venture beyond rudimentary applications. Most are focusing on improving communications with employees, customers, and business partners.

TABLE 4.8

Popular Search Engines

Search Engine	Web Address
AltaVista	http://www.altavista.com
Ask Jeeves	http://www.ask.com
Excite	http://www.excite.com
Galaxy	http://www2.galaxy.com
GO Network	http://www.go.com
Google	http://www.google.com
Lycos	http://www.lycos.com
WebCrawler	http://www.webcrawler.com
Yahoo!	http://www.yahoo.com

FIGURE 4.16

Web Page with a Java Applet

Free Java applets can be downloaded from this site for use on your own Web site.

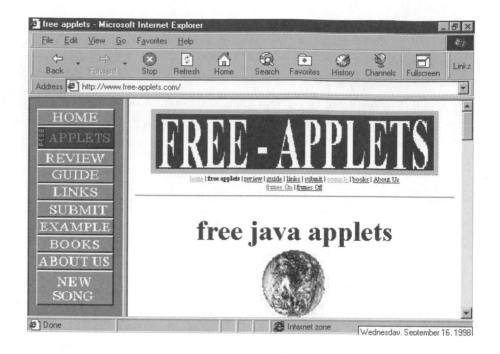

A number of companies are using push technology to deliver critical information over the Internet.[34] SAP, for example, uses push technology to deliver critical enterprise resource planning software over the Internet. Enterprise resource planning software is used by many large corporations to streamline operations. SAP uses software from BackWeb. Other companies, including oil service supplier Schlumberger and Carlson Wagonlit Travel, also use push technology to make priority deliveries of information over the Internet.

There are drawbacks to the use of push technology. One issue, of course, is information overload. Another is the volume of data being broadcast is so great that push technology can clog up the Internet communications links with traffic.

Business Uses of the Web

By linking buyers and sellers electronically on the Web, businesses are able to establish new and ongoing relationships with customers, allowing them access to information and products whenever it suits them. Businesses can use the Web as a tool for marketing, sales, and customer support. The Web can also serve as a low-cost alternative to fax, express mail, and other communications channels. It also can eliminate paperwork and drive down the cost per business transaction.

The Internet's business potential has just begun to be tapped. Read the "E-Commerce" box to learn how a home building company was able to use the Internet to reduce its inventory and related costs. As more and more people gain access to the World Wide Web, its functions are changing drastically. We discuss a couple of these applications—corporate intranets and extranets—next. Chapter 5 will cover e-commerce in more detail.

Drugstore.com is a Web shopping site that includes both a pharmacy and a retail store with health and beauty supplies.

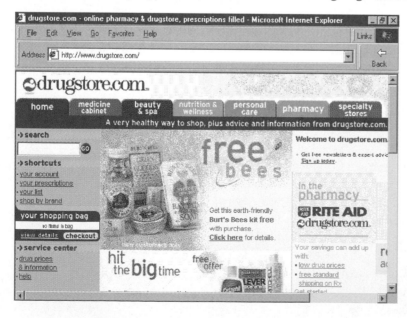

E-COMMERCE
Home Builders Build Sites on the Internet

The Internet has already transformed book sales, travel, and many other industries. Customers are flocking to the Internet for information and deals in record numbers. Today, the Internet is starting to have a dramatic impact even on older, more traditional businesses, such as home building.

Weather-Tek Design Center, a building company in Waukesha, Wisconsin, recently discovered the power of the Internet. The design center, located in a community close to Milwaukee and Lake Michigan, had inventory problems. The company had amassed too much inventory of custom door and window materials. Inventory that sits for months or years can be a huge expense. In addition to the money invested that can't be used for other purposes, excess inventory usually depreciates in value as new products are introduced, and there is always the threat that the inventory will be stolen or damaged. To solve its inventory problem, Weather-Tek turned to the Internet.

BuildersExpress.com is a Web site for builders looking for hard-to-find items. Builders can advertise what they want to unload, and other builders can search the BuildersExpress Web site to find what building materials they need. After stumbling on the Chicago-based BuildersExpress Web site, Weather-Tek decided to list all 140 inventory items it no longer needed, instead of letting the items collect dust. In 90 days, all 140 items were sold through the BuildersExpress Web site. According to Michael Gavin, founder of BuildersExpress, "This really wouldn't have been possible without the Internet."

Although there are a few big builders in the United States, most of the industry consists of tens of thousands of small builders. Many of them are traditional mom-and-pop operations. Because the industry is not organized or centralized, there are inefficiencies. Prices vary, materials can be scarce, and there can be a long wait for needed supplies. Building companies, such as Building Materials

Holding Corporation, are seeing dramatic results from putting lists of all their building materials on the Internet. Building Materials has also been able to use the Internet to receive payments faster from customers. "This is a way for us to add profits to a low-margin business," says Robert Mellor, CEO of Building Materials.

Other building companies are also starting to venture online. USBuild.com, for example, is developing software to help streamline pricing practices. The old system relied on subcontractors to bid on various projects, which was inefficient and not always cost effective. With USBuild.com, companies will be able to put jobs on the Internet for bid instead of relying on a handful of local subcontractors. The result could be lower total building costs. Equalfooting.com, another building Web site, offers price discounts to small builders. These discounts were typically available only to larger, high-volume builders. The Equalfooting.com Web site is backed by a number of influential individuals, including Steve Case, chairman of America Online, and Marc Andreessen, founder of Netscape Communications.

Discussion Questions

1. How was Weather-Tek able to use the Internet to its benefit?
2. Describe other building Internet sites.

Critical Thinking Questions

3. How was the Internet able to help small builders become more efficient and profitable?
4. What other industries could benefit from the Internet?

Sources: Adapted from Jim Carlton, "Home Builders Learn to Love the Internet," *The Wall Street Journal*, February 7, 2000, p. A2; and Brae Canlen, "Pro Dealers Seek E-Link to Builders," *National Home Center News*, January 24, 2000, p. 1.

INTRANETS AND EXTRANETS

intranet

an internal corporate network built using Internet and World Wide Web standards and products; used by the employees of the organization to access corporate information

An **intranet** is an internal corporate network built using Internet and World Wide Web standards and products. It is used by the employees of an organization to access corporate information. After getting their feet wet with public Web sites that promote company products and services, corporations are seizing the Web as a swift way to streamline–even transform—their organizations. These private networks use the infrastructure and standards of the Internet and the World Wide Web. A big advantage of using an intranet is that many people are already familiar with the Internet and Web, so they need little training to make effective use of their corporate intranet.

An intranet is an inexpensive yet powerful alternative to other forms of internal communications, including conventional computer setups. One of an intranet's most obvious virtues is its ability to slash the need for paper. Because Web browsers run on any type of computer, the same electronic information can be viewed by any employee. That means that all sorts of documents (such as internal phone books, procedure manuals, training manuals, and requisition forms) can be

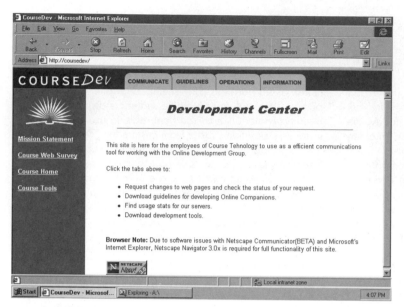

An intranet is an internal corporate network used by employees to access company information.

inexpensively converted to electronic form on the Web and be constantly updated. An intranet provides employees with an easy and intuitive approach to access information that was previously difficult to obtain. For example, it is an ideal solution to provide information to a mobile sales force that needs access to a lot of rapidly changing information. Intranets can also do something far more important. By presenting information in the same way to every computer, they can do what computer and software makers have frequently promised but never actually delivered: pull all the computers, software, and databases that dot the corporate landscape into a single system that enables employees to find information wherever it resides.

Universal reach is what made the Internet grow so rapidly. But Internet enthusiasts tended to focus on how to link far-flung people and businesses. When the Internet caught on, people were not considering it as a tool for running their business—but that is what is happening, with amazing speed. Just as the simple act of putting millions of computers around the world on speaking terms initiated the Internet revolution, so connecting all the islands of information in a corporation is sparking unprecedented collaboration. Corporate intranets are breaking down the walls within corporations.

Once a company sees the value of intranet access, it will want to move to the next stage of intranet usage—interactive transaction-based applications. At this stage, employees can query corporate databases to see the status of a customer order, a raw material shipment, or a manufacturing production run of a finished product.

More advanced use of the corporate intranet supports what has come to be known as workgroup computing. Workgroup computing involves many aspects, but basically it is an approach to supporting people working together in teams. One of the key aspects is the ability to store and share information in any form—text, video, sound, graphics, handwritten memos, or hand-drawn figures—which is often called a *knowledge base*. The key feature of workgroup computing is being able to organize and retrieve all this data simply. Group calendaring and scheduling allows an employee to check others' schedules and set up meetings. Another advantage of intranets is support for real-time meetings with people linked over networks, instead of making them travel to one place. Workgroup computing also supports work flow processes, tracking the status of documents—who has them, who is behind or ahead of schedule, and who gets them next.

A rapidly growing number of companies have advanced beyond the workgroup stage to offer limited network access to selected customers and suppliers. Such networks are referred to as extranets, which connect people who are external to the company. An **extranet** is a network that links selected resources of the intranet of a company with its customers, suppliers, or other business partners. Again, an extranet is built based on Web technologies.

extranet

a network that links selected resources of the intranet of a company with its customers, suppliers, or other business partners; based on Web technologies

Security and performance concerns are different for an extranet than for a Web site or network-based intranet. Authentication and privacy are critical on an extranet so that information is protected. Obviously, performance must be good to provide quick response to customers and suppliers. Table 4.9 summarizes the differences between users of the Internet, intranets, and extranets.

TABLE 4.9

Summary of Internet, Intranet, and Extranet Users

Type	Users	Need for User ID and Password
Internet	Anyone	No
Intranet	Employees	Yes
Extranet	Business partners	Yes

virtual private network (VPN)

a network that transfers information by encapsulating traffic in IP packets and sending the packets over the Internet

tunneling

the process by which VPNs transfer information by encapsulating traffic in IP packets over the Internet

firewall

a device that sits between your internal network and the outside Internet and limits access into and out of your network based on your organization's access policy

Secured intranet and extranet access applications usually require the use of a virtual private network (VPN). A **virtual private network (VPN)** is a secure connection between two points across the Internet.[35] VPNs transfer information by encapsulating traffic in IP packets and sending the packets over the Internet, a practice called **tunneling**. Most VPNs are built and run by Internet service providers. Companies that use a VPN from an Internet service provider have essentially outsourced their networks to save money on wide-area network equipment and personnel. An Internet **firewall**, a device that sits between your internal network and the outside Internet, limits access into and out of your network based on your organization's access policy. A firewall can be set up to allow access only from specific hosts and networks or to prevent access from specific hosts. In addition, you can give different levels of access to various hosts; a preferred host may have full access, whereas a secondary host may have access to only certain portions of your host's directory structure. In using a VPN, a user sends data from his or her personal computer to the company's firewall, which converts the data into a coded form that cannot be easily read by an intruder. The coded data is then sent via an access line to the company's Internet service provider. From here, the data is transmitted through tunnels across the Internet to the recipient's Internet service provider and then over an access line to the receiving company's firewall, where it is decoded and sent to the receiver's personal computer (Figure 4.17).

Companies and governmental agencies are big users of VPNs. NASA, for example, uses a VPN to transfer data between the space shuttles and earth.[36] The technology makes the shuttle appear as a node on the Internet. This allows NASA specialists on earth to control experiments being done in space. It also allows people to monitor shuttle operations using a standard Web browser.

NET ISSUES

With such widespread use of networks today, organizations are dealing with new issues almost daily. Control, access, hardware, and security issues affect all networks, so it is important to mention some of these management issues.

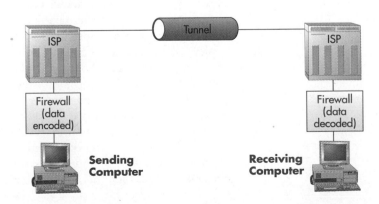

FIGURE 4.17

Virtual Private Network

Management Issues

Although the Internet is a huge, global network, it is managed at the local level; no centralized governing body controls the Internet. Although the U.S. federal government provided much of the early direction and funding for the Internet, the government does not own or manage it. The Internet Society and the Internet Activities Board (IAB) are the closest the Internet has to centralized governing bodies. These societies were formed to foster the continued growth of the Internet. The IAB oversees a number of task forces and committees that deal with Internet issues. One of the main functions of the IAB is to manage the network protocols used by the Internet, including TCP/IP. Some universities and government agencies are investigating how the Internet can be controlled to prevent sensitive information and pornographic material from being placed on the Internet.

Service Bottlenecks

Traffic on networks increases every day. The primary cause of service bottlenecks is simply the phenomenal growth. Traffic volume on company intranets is growing even faster than the Internet. Companies setting up an Internet or intranet Web site often underestimate the amount of computing power and communications capacity they need to service all the "hits" (requests for pages) they get from Web cruisers. Web server computers can be overwhelmed with thousands of hits per hour.

Slow modems and the copper-based telephone wire system that carries the signal into an office or home are the two current primary Internet bottlenecks. For most users, these two limit a user's maximum access speed to around 56 Kbps, which is still too slow for 16-bit stereo sound and smooth, full-screen video.

Connection agreements exist among the backbone companies to accept one another's traffic and provide a certain level of service. Some Internet providers do a good job and provide a high level of quality and service. Others do not do such a good job, creating wide variations in the quality of the Internet. At some interconnect points where major Internet operators hand off to one another, one operator may not be able to accept incoming traffic fast enough because its lines are overloaded with traffic. For example, linking with Pacific Rim nations is especially difficult. Most providers lease their lines from phone companies, and the cost of even a medium-speed line to connect California with Australia is more than $1.2 million per year—ten times the cost of a New York–to–San Francisco link, so some providers scrimp. This leads to inadequate capacity, which slows transmissions.

Routers, the specialized computers that send packets down the right network pathways, can also become bottlenecks. For each packet, every router along the way must scan a massive address book of about 40,000 area destinations (akin to Internet ZIP codes) to pick the right one. These routers can get overloaded and lose packets. The TCP/IP protocol compensates for this by detecting a missing packet and requesting the sending device to resend the packet. However, this leads to a vicious circle, as the network devices continually try to resend lost packets, further taxing the already overworked routers. This leads to long response times or loss of the connection to the network.

Several actions are being taken to open up the bottleneck. One solution involves the various backbone providers upgrading their backbone links. In some cases they are installing bigger, faster "pipes," and in others they are converting to newer transmission technologies that can send a message down the right path more quickly than standard packet-switching technology. A second solution to the bottleneck problem is provided by router manufacturers, who are working to develop improved models with increases in hardware capacity and more efficient software to provide quick access to addresses. Yet a third solution is to prioritize traffic. Today, all network traffic travels through the same big backbone

FIGURE 4.18

Cryptography is the process of converting a message into a secret code and changing the encoded message back into regular text.

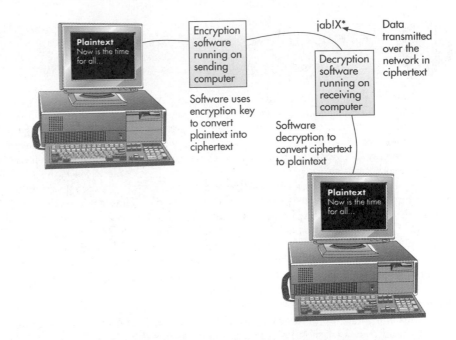

cookie

a text file that an Internet company can place on the hard disk of a computer system

cryptography

the process of converting a message into a secret code and changing the encoded message back to regular text

encryption

the conversion of a message into a secret code

The Federal Computer Incident Response Capability assists civilian government agencies with computer security incidents.

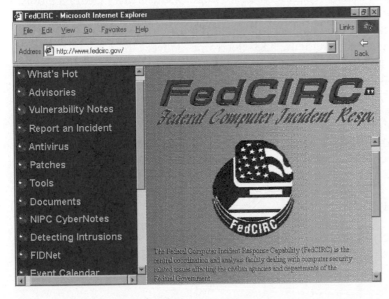

pipes. There is no way to make sure that your urgent message is not stalled behind someone downloading a magazine page. With prioritized service, customers could pay more for guaranteed delivery speed, much as an overnight package costs more than second-day delivery. If implemented, this solution could also affect the cost of network services that generate a lot of traffic, such as Internet phone and videoconference services. DSL and cable modems can also be used to speed Internet access.[37]

Privacy and Security

As use of the Internet grows, privacy and security issues become even more important. People and companies can be reluctant to embrace the Internet unless these issues are successfully addressed. From a consumer perspective, the protection of individual privacy is essential. Yet many people use the Internet without realizing their privacy may be in jeopardy. Many Internet sites use cookies to gather information about people who visit their sites. A **cookie** is a text file that an Internet company can place on the hard disk of a computer system. These text files keep track of visits to the site and the actions people take. To help prevent this potential problem, some companies are developing software to help prevent these files from being placed on computer systems. CookieCop, for example, allows Internet users to accept or reject cookies by Internet site.[38] Read the "Ethical and Societal Issues" box, which discusses privacy issues.

From a corporate strategy perspective, security of data is essential. Such approaches as cryptography can help. **Cryptography** is the process of converting a message into a secret code and changing the encoded message back to regular text. The original conversion is called **encryption**. The unencoded message is called *plaintext;* the encoded message is called *ciphertext.* Decryption converts ciphertext back into plaintext (see Figure 4.18). For much of the

ETHICAL AND SOCIETAL ISSUES
Outrage on the Web

Advertising on the Web has always been a way to generate revenues for companies. Instead of charging users when they visit a Web site, companies often place banners on their sites to advertise the products and services of other companies. Some Internet companies specialize in using Internet banner ads.

One of the top ad-server companies on the Internet has its banner ads on about 1,500 Web sites. Many of these ads are aimed at specific customer types by using "cookies." A cookie is a text file that collects information about people's behaviors and viewing patterns on the Web. Using cookies, some of these companies have collected an estimated 100 million profiles, each one containing information about a person's browsing and Internet viewing habits. Once collected, this information can be combined with other data. One company, for example, recently paid over $1 billion for a data-warehouse company that has detailed records of customer catalog purchases and related activities. By combining personal data from several sources, companies can put together comprehensive and detailed profiles on people. These profiles not only contain shopping information but can contain people's home addresses, their places of employment, phone numbers, and much more.

Marketing and advertising companies can greatly benefit from this detailed information about people. Some advertising companies, for example, are willing to pay 10 percent to 20 percent more for this targeted customer information. According to Susan Nathan, a senior vice president of an advertising company, "The Internet is the first medium that offers advertisers the ability to speak to your customers." While advertising companies are embracing the wealth of personal information that comes from combining databases from the Internet and other sources, many people are very concerned about their privacy. Will highly personal information be for sale to the highest bidder?

Privacy lawsuits are now being filed against some Internet companies. Harriet Judnick, for example, recently sued an Internet company for privacy invasion. After visiting a few Web sites, she started to get phone calls and e-mails from insurance companies, loan companies, and other companies trying to sell her something. She even found her medical insurance information on the Internet.

Discussion Questions

1. Describe banner ad companies discussed in this box.
2. What is their primary business?

Critical Thinking Questions

1. How can people protect their privacy when using the Internet?
2. What would you like to see in a corporate privacy statement to guarantee your privacy?

Sources: Adapted from Heather Green, "Privacy: Outrage on the Web," *Business Week*, February 14, 2000, p. 38; and Thomas Weber, "Can Your Complaints Build a Web Business," *The Wall Street Journal*, January 10, 2000, p. B1.

Cold War era, cryptography was the province of military and intelligence agencies; uncrackable codes were reserved for people with security clearance only.

Widespread deployment of cryptography requires additional hardware and software but is becoming increasingly necessary to support electronic commerce, copyright management, and electronic delivery of services. Without cryptography, people will not trust that electronic financial transactions, secret or private data, and valuable intellectual property will remain confidential across networks.

A cryptosystem is a software package that uses an algorithm, or mathematical formula, plus a key to encrypt and decrypt messages. The algorithm is calculated with the key and converts every character of the plaintext into other coded characters, thus creating the ciphertext. Only someone with the correct key should be able to decode the ciphertext. Good ciphertext appears to be nothing more than random characters. Encryption makes information useless to hackers and thieves.

U.S. banks and brokerage houses use the federal government's Data Encryption Standard (DES) algorithm to protect the integrity and confidentiality of fund transfers totaling some $2.3 trillion a day worldwide. Organizations encrypt the

INFORMATION SYSTEMS IN ACTION

The Post Office: Remaining Competitive in the Internet Age

Brian Kovar,
Kansas State University

In today's Internet age, the services provided by the U.S. Postal Service are often referred to as "snail mail," since e-mail can be exchanged in a matter of minutes or hours while the Post Office's first-class mail delivery takes one to three days. A greater number of Americans are turning to the Internet and e-mail as their preferred method of communication, especially since e-mail has no apparent cost to those who already have Internet access. According to Postal Service projections, first-class mail volume will decline at an average annual rate of 2.5 percent from 2003 to 2008, with losses attributed to e-commerce competition accounting for approximately $17 billion over the next few years. Even greeting cards and income tax returns, once thought to be reliable sources of income for the Postal Service, can now be delivered using the Internet.

The Postal Service, slow to enter the Internet game, has begun to take action and launch programs in order to remain competitive. It now has a vice president for e-commerce and a five-person e-braintrust. Plans are being made to provide e-mail service to the 120 million U.S. residential street addresses by linking a residential entry in the Postal Service's giant residential address database to a corresponding e-mail address. E-mail could be kept at an individual's Postal Service account, or it could be forwarded to another e-mail address. Individuals without Internet access could receive printouts of their e-mail delivered with their regular mail, at a cost of 41 cents for a two-page document, which is paid by the sender.

Other programs the Postal Service are using to remain competitive in the Internet Age include the following:

1. eBillPay, which lets individuals receive and pay bills electronically through the Postal Service Web site. Paper checks can be issued and sent through the regular mail system to companies that do not accept electronic payments.
2. Electronic merchandise return, in which participating online merchants can authorize customers to download a postage-paid label from the merchant's Web site in order to return merchandise.
3. PC Postage, which allows users to download computerized postage from a PC onto an envelope or label.
4. Delivery confirmation for parcels and priority mail for customers who log on at www.usps.gov.
5. PosteCS (Post Electronic Courier Service), which was developed in conjunction with the International Post Corporation to support secure electronic communications.
6. Electronic Postmark, which will provide an official time-and-date stamp when electronic messages are sent.
7. NetPost Mailing Online, in which participants electronically transmit documents and mailing lists to the Postal Service, which will print the documents, stuff envelopes, and then transmit them to the nearest post office for delivery.

Sources: Stephen Barr, "Sorting Out Mail's Place in the Internet Age," *The Washington Post,* January 24, 2000; "Postal Service Will Offer Online Bill-Paying Service," *The Wall Street Journal,* April 6, 2000; Julia Angwin, "E-Mail Goes Postal," *The Wall Street Journal,* July 31, 2000; and the United States Postal Service Web site, http://www.usps.gov/.

---

words and videos of their teleconferencing sessions. Individuals encode their electronic mail. And researchers use encryption to hide information about new discoveries from prying eyes.

Encryption is not just for keeping secrets. It can also be used to verify who sent a message and to tell whether the message was tampered with en route. A **digital signature** is a technique used to meet these critical needs for processing on-line financial transactions. Digital signatures involve a complicated technique that combines the public-key encryption method with a "hashing" algorithm that prevents reconstructing the original message. The hashing algorithm provides further encoding by using rules to convert one set of characters to another set (e.g., the letter *s* is converted to a *v*, 2 is converted to 7). Thus, encryption also can prevent electronic fraud by authenticating senders' identities with digital signatures.

Network management issues will take an increasing amount of time for organizations, but any user needs to be aware of the basics to function effectively in business. Communications, service, and daily work are all at stake.

digital signature

an encryption technique used to verify the identity of a message sender for the processing of on-line financial transactions

● SUMMARY

PRINCIPLE • The effective use of telecommunications and networks can turn a company into an agile, powerful, and creative organization, giving it a long-term competitive advantage.

Telecommunications has the potential to create profound changes in business because it lessens the barriers of time and distance.

The elements of a telecommunications system start with a sending unit that originates the message. The sending unit transmits a signal to a telecommunications device. The telecommunications device performs a number of functions, which can include converting the signal into a different form or from one type to another. The telecommunications device then sends the signal through a medium, which carries the electronic signal and interfaces between a sending device and a receiving device. The signal is received by another telecommunications device that is connected to the receiving computer.

• • •

Networks can be used to share hardware, programs, and databases across the organization. They can transmit and receive information to improve organizational effectiveness and efficiency. They enable geographically separated workgroups to share documents and opinions, which fosters teamwork, innovative ideas, and new business strategies.

PRINCIPLE • The Internet is like many other new technologies—it provides a wide range of services, some of which are effective and practical for use today, others of which are still evolving, and still others of which will fade away from lack of use.

The Internet is the world's largest computer network—actually a collection of interconnected networks, all freely exchanging information. The Internet transmits data from one computer (called a host) to another. The set of conventions used to pass packets from one host to another is known as the Internet protocol (IP). Transport control protocol (TCP) includes rules for computers to establish and break connections. Each computer on the Internet has an assigned address to identify it from other hosts.

An Internet service provider is any company that provides individuals or organizations with access to the Internet. To use this type of connection, you must have an account with the service provider and software that allows a direct link via the TCP/IP protocols.

• • •

Internet services include voice mail, e-mail, and instant messaging; telecommuting, videoconferencing, and Internet phone service; electronic data interchange; public network and specialized services; distance learning; on-line music, radio, and video; telnet, FTP, and content streaming; and chat rooms.

The Web is a collection of tens of thousands of independently owned computers that work together as one in an Internet service. High-speed Internet circuits connect these computers, and cross-indexing software is employed to enable users to jump from one Web computer to another effortlessly. Because of its ability to handle multimedia objects and hypertext links between distributed objects, the Web is emerging as the most popular means of information access on the Internet today.

A Web site is like a magazine, with a cover page called a home page that has graphics, titles, and text. Hypertext links are maintained using URLs (uniform resource locators), a standard way of coding the locations of the HTML (hypertext markup language) documents. Web pages are loosely analogous to chapters in a book.

The client communicates with the server according to a set of rules called HTTP (Hypertext Transfer Protocol), which retrieves the document and presents it to the users. HTML is the standard page description language for Web pages. A Web browser reads HTML and creates a unique, hypermedia-based menu on the user's computer screen that provides a graphical interface to the Web.

A rapidly growing number of companies are doing business on the Web, enabling shoppers to search for and buy products on-line. The travel, entertainment, gift, greetings, book and music businesses are experiencing the fastest growth on the Web.

• • •

An intranet is an internal corporate network built using Internet and World Wide Web standards and products. It is used by the employees of the organization to gain access to corporate information.

An extranet is a network that links selected resources of the intranet of a company with its customers, suppliers, or other business partners. It is built based on Web technologies. Authentication, privacy, and performance are critical on an extranet.

• • •

Management issues, service bottlenecks, privacy and security, and firewalls are issues that affect all networks.

● REVIEW QUESTIONS

1. What is a telecommunications medium? List three media in common use.
2. What advantages and disadvantages are associated with the use of client/server computing?
3. Describe a local area network and its associated components.
4. What is EDI? Why are companies using it?
5. What is the Internet? Who uses it and why?
6. What is the TCP/IP protocol? How is it used?
7. What are Internet service providers? What services do they provide?
8. What is a newsgroup? How would you use one?
9. What is videoconferencing and what business benefits does it provide?
10. What is the Web? Is it another network like the Internet, or a service that runs on the Internet?
11. What is a URL and how is it used?
12. What is a Web browser? How is it different from a Web search engine?
13. What is an extranet? How is it different from an intranet?
14. What is cryptography?
15. What are firewalls? How are they used?

● DISCUSSION QUESTIONS

1. What sort of issues would you expect to encounter in establishing an international network for a large, multinational company?
2. Consider an industry that you are familiar with through work experience, coursework, or a study of industry performance. How could electronic data interchange be used in this industry? What limitations would EDI have in this industry?
3. What is telecommuting? What are the advantages and disadvantages of telecommuting? Do you think that you will telecommute in your future career?
4. Discuss the pros and cons of conducting this course as a distance learning course.
5. Instant messaging is being used to a greater extent today. Describe how this technology could be used in a business setting. Are there any drawbacks or limitations in using instant messaging in business?
6. Briefly describe how Internet phone service operates. Discuss the potential for this service to affect traditional telephone services and carriers.
7. The U.S. government is against the export of strong cryptography software. Discuss why this may be so. What are some of the pros and cons of this policy?
8. Identify three companies with which you are familiar who are using the Web to conduct business. Describe their use of the Web.
9. Briefly summarize the differences in how the Internet, a company intranet, and an extranet are accessed and used.

● PROBLEM-SOLVING EXERCISES

1. You have been hired as a telecommunications consultant for a small but growing consumer electronics manufacturer. The company wishes to develop a telecommunications system to link with its suppliers. There are many options, including EDI and the Internet. The company has hired you to review its needs for the new system and to select a telecommunications solution. How would you proceed? What questions need to be asked? Use word processing software to prepare a list of at least ten questions that you need to answer to evaluate this project. Make some assumptions about the answers and write your opinion of this project. Embed a spreadsheet that details the approximate cost to set up the telecommunications connection between the company and its suppliers.

2. You work in the Los Angeles national headquarters of a large multinational company. After years of fighting the smog and traffic, you and your fellow workers have decided to do something about it. Your co-workers have elected you as spokesperson to develop a recommendation for management to implement a telecommuting program. Use PowerPoint or similar software to make a convincing presentation to management to adopt such a program. Your presentation must address such questions as what the benefits are to the company, what the costs of the hardware and software are, how individuals will be selected for participation, and what benefits employees will receive.

3. You are a manager in a Fortune 1000 company leading a project to develop an extranet linking your company to key suppliers. The extranet will provide suppliers with access to production planning and inventory data so that the suppliers can ship raw materials, packing materials, and supplies to your plant just-in-time. The goal is to reduce the level of inventory you must carry while not adversely affecting production. Develop a one-page project charter that defines the scope and purpose of the project, identifies the key members of the project team (by title), outlines the key technical and non-technical issues you will face, and defines the key project success criteria.

● TEAM ACTIVITIES

1. With a group of your classmates, visit one of the following: a cellular phone company, the college computing center, a phone or cable company, a police department, or another interesting organization that uses telecommunications. Prepare a report on how the organization is planning to use communications to enhance access to information—both yours and its. Find out what kind of telecommunications media and devices it uses currently and what changes the organization might make to improve data and information access.

2. Carefully analyze the different options to connect a PC to the Internet, including a standard phone line, DSL, cable, and satellite. Summarize the costs, advantages, and disadvantages of these options. Investigate what options are available in your area. Discuss your findings with the class. How would you evaluate the competitiveness of the local service providers in your area?

3. Identify a company that is making effective use of a company extranet. Find out all you can about its extranet. Try to speak with one or more of the customers or suppliers who use the extranet and ask what benefits it provides from their perspective.

● WEB EXERCISES

1. There are a number of on-line job-search companies, including Monster.com. Investigate one or more of these companies and research the positions available for a business major with an interest in information technology. You may be asked to summarize your findings for your class in a written or verbal report.

2. This chapter covers a number of powerful Internet tools, including Internet phones, search engines, browsers, e-mail, newsgroups, Java, intranets, and much more. Pick one of these topics and get more information from the Internet. You may be asked to develop a report or send an e-mail message to your instructor about what you found.

3. Shopping on the Internet is becoming more popular. Using a search engine, locate several Internet sites that sell new cars. Compare the costs and features of several cars you find interesting. Would you purchase a car on-line? What are the advantages and disadvantages of buying a car on the Internet?

● CASES

 Mirage on the Net

When it comes to building a top-notch network, Mirage decided not to gamble, especially with its newest facility, the Bellagio. This new 3,000-room luxury hotel and casino has original art, a spacious decor, and a complex water fountain that moves to music, ranging from classic to rock. "When the president of Bellagio looks me in the eye and says, 'Glenn, is this going to work?' I have to be able to guarantee success," says Glenn Bonner, chief information officer at Mirage Resorts. "This industry is littered with careers that chose new systems for opening day."

Bonner's conservative approach to developing a telecommunications system for Mirage appears to have succeeded. First, Bonner developed a regional area network, also called a metropolitan-area network, to tie the four casinos in Las Vegas together. The network uses 3Com CoreBuilder 7000 backbone switches in the four hotels and the data center in Las Vegas. The links between the hotels and the data center use a 155-Mbps mesh. "The mesh provides the redundancy and resiliency we needed," Bonner said.

According to some, however, the jewel in the crown for Mirage is the local area network (LAN) at the Bellagio. The ethernet LAN consists of servers and high-speed switches, which can provide a speed of 10 Mbps. According to Bonner, "Networking is not an exact science; it's an art. So you go and architect for scalability and capacity, and when you need it, you turn it on." Because a casino never closes, it is hard to shut down the network for maintenance or updates. Thus, it is critical to build the best network possible at the start. And because of the nature of the business, the network must be secure.

Discussion Questions

1. Briefly describe the network system used by Mirage.
2. What are the advantages and disadvantages of this approach?

Critical Thinking Questions

3. What special network issues would a casino company have to consider versus a more traditional company?
4. What other types of computer technology would be useful for a casino company?

Sources: Adapted from Jon Fontana, "Mirage Makes Broadband Bet," *Computerworld,* January 25, 1999, p. 23; and Richard Siklos, "Weaving Yet Another Web for Women," *Business Week,* January 17, 2000, p. 101.

 Kaiser Permanente Looks for a Cure on the Web

Many healthcare companies have been on the cutting edge of medicine but lag behind in the use of information systems and the Internet. Loading old steel filing cabinets with tons of paper medical and insurance records was typical. These massive paper documents were stored in huge storage rooms and shipped around the country by the truckload. Today, some healthcare providers are starting to take a more healthy approach to their record keeping by turning to technology and the Internet.

In a $2 billion project, Kaiser Permanente is launching a massive technology initiative. The objective is to move all its operations to the Internet. The healthcare provider plans to develop a digital medical record of all of its nine million members. The record will be linked to the company's 361 hospitals and 10,000 doctors, nurses, and other healthcare providers. The heart of the system will be an Internet site. The current plans are to give companies, such as General Motors and Wells Fargo, access to coverage rates and related information. Separate Internet sites will allow doctors and healthcare administrators the ability to order a wide array of supplies and equipment, from inexpensive bandages to costly CAT scan devices. Kaiser's CEO, David Lawrence, believes that the Internet "will be the central nervous system for tying together all of the elements needed to care for patients better—and it will do so in ways now unimaginable."

The new Internet strategy is aimed not only at improving patient care but also at improving the bottom line for Kaiser. In 1998, Kaiser had losses of $288 million on revenues of $15.5 billion. Some blame the disappointing loss on underestimating the amount of care the company needed to give patients. The new Internet system could help reduce costs. For example, prescription drug costs are rising 16 percent to 18 percent annually. Kaiser hopes that its investment in technology will help the company locate alternative drugs that are less expensive or less dangerous to patients. Kaiser also hopes that the vast Internet system will allow it to compare doctor's bills, which can vary substantially. The hope is to find standard treatments that can reduce costs. Kaiser also hopes that the new system will cut paperwork and record-keeping costs. According to Lawrence, "Companies have already squeezed what savings they can out of managed care, so HMOs like Kaiser must find new improvements to compete, or falter."

Discussion Questions

1. Describe Kaiser's move to the Internet.
2. Do you think Kaiser's Internet initiative was motivated primarily to improve patient care or reduce costs?

Critical Thinking Questions

3. In general, what healthcare systems and services would you recommend for the Internet?
4. Which company operations should be moved to the Internet? Are there operations that should not be moved to the Internet?

Sources: Adapted from Douglas Gantenbein et al., "Kaiser Takes the Cyber Cure," *Business Week,* February 7, 2000, p. 80; and Barbara Bowers, "Driven to the Top," *Best's Review Life/Health Edition,* April 1999, p. 32.

The Washington Post Tries the Web

Today, more people are using the Internet to get their local, regional, and national news. Newspapers across the country are placing content from their publications on the Internet. There are many benefits to consumers. Most of the sites do not charge a subscription fee; depending on the publication, some subscribers can save hundreds of dollars annually. When traveling, people can easily keep up with local news on-line. Many news sites also provide past issues, and most have powerful search engines. If you want to get news on a particular topic or if you want to search the classified ads for a used car, the search engine can quickly return the desired information. Some sites also allow customers to tailor their news. By filling out a short questionnaire, the on-line news service will only give you the news you really want. In addition, on-line newspapers also help reduce the need for paper—and trees to make paper.

The Washington Post has developed an Internet site that has been extremely successful. The well-respected on-line newspaper is read by 20 percent of residents in the Washington, D.C., area. According to Media Metrix, a Web-traffic tracking company, *The Washington Post* Internet site is rated 13th in the region. But although on-line newspapers like *The Washington Post* provide useful information for people, are they profitable?

On-line newspapers generate revenues from advertisements, but there are also costs associated with developing and running any Internet site. *The Washington Post,* for example, received about $17 million in additional advertising revenues from its Internet site. But it cost the newspaper approximately $65 million to develop and operate the site. Why would any company spend more on an Internet site than it returns in revenue? Many newspapers with well-established brand names are worried that other start-up, on-line-only news sites will eventually take away their core business of subscriptions, classified ad revenues, and general advertising revenues. A wrong move by these established news organizations could spell disaster. According to Patrick Heane, a senior analyst at Jupiter Communications, "It's the newspapers' game to lose." But how long will newspapers be willing to take losses on their Internet sites?

According to John Borse, chief financial officer for *The Washington Post,* "If we knew that, we'd all feel a lot better."

Discussion Questions

1. Why is it important for a company like *The Washington Post* to have a site on the Internet?
2. If you were the CEO of *The Washington Post,* would you continue to operate an Internet site?

Critical Thinking Questions

3. What types of companies need an Internet site? What types of companies do not?
4. How can an Internet site like The Washington Post be cost-justified?

Sources: Adapted from Erin White, "Washington Post Stays the Course with Web Operations," *The Wall Street Journal,* February 2, 2000, p. B4; and "The Emergence of Convergence," *American Journalism Review,* January 2000, p. 88.

● NOTES

Sources for the opening vignette on p. 133: Adapted from David Rynecki, "They've Got Mail," *Fortune,* February 7, 2000, p. 101; and David Rocks, "Going Nowhere Fast in Cyberspace," *Business Week,* January 31, 2000, p. 58.

1. Eric Brown, "Broadband Blues," *PC World,* February 2000, p. 57.
2. Matthew Hamblen, "Digital Subscriber Line," *Computerworld,* February 7, 2000, p. 69.
3. Kate Murphy, "Cruising the New-In Hyperdrive," *Business Week,* January 24, 2000, p. 170.
4. Scott Wooley, "Kiss That Duopoly Good-Bye," *Forbes,* February 21, 2000, p. 135.
5. Young Myung et al., "Design of Communication Networks with Survivability Constraints," *Management Science,* February 1999, p. 238.
6. UPS Web site at http://www.upsnet.com, accessed February 18, 2000.
7. Kerry Capell, "Europe's Top e.Business Leaders," *Business Week,* February 7, 2000, p. 58.
8. Jim Rohwer, "Japan Goes Web Crazy," *Forbes,* February 7, 2000, p. 115.
9. Pamela Druckerman, "Brazil Is Test Case for Free Web Access," *The Wall Street Journal,* February 7, 2000, p. A28.
10. John Gantz, "Internet2 Is on the Way," *Computerworld,* March 1, 1999, p. 34.
11. Kathleen Murphy, "To Challenge the Empire," *Internet World,* January 15, 2000.
12. Jane Weaver, "The Name Game," *PC Computing,* April 2000, p. 108.
13. Lee Copeland, "Automakers Race After Alleged Cybersquatters," *Computerworld,* February 28, 2000, p. 12.
14. Larry Armstrong, "No-Fee Net Access Is Making the Old Profit Models Obsolete," *Business Week,* February 7, 2000, p. 46.
15. Roberta Furger, "The Best ISPs," *PC World,* March 1999, p. 124.
16. Les Freed et al., The Faster Web," *PC Magazine,* April 20, 1999, p. 158.
17. Oliver Rist, "Eudora Now Free (If You Can Stand the Ads)," *PC World,* March 2000, p. 90.
18. Neil Gross, "E-mail That Won't Come Back to Haunt You," *Business Week,* November 15, 1999, p. 136.
19. Roberta Fusary, Digital Protection," *Computerworld,* March 1, 1999, p. 14.
20. Frank Deffler, "Instant Messaging to Work," *PC Magazine,* January 18, 2220, p. 82.
21. Stan Miastkowski, "Instant Messaging Heads for Cell Phones," *CNN Online,* December 13, 1999.
22. Sonya Donaldson, "Teleworker," *Home Office Computing,* February 2000, p. 101.

23. Devin Burden, "Companies Now Find Videoconferencing a More Viable Alternative to Meeting in Person," *Computerworld,* January 11, 1999, p. 91.

24. "Videoconferencing," *Management Consultancy,* February 4, 2000, p. 16.

25. Rebecca Blumenstein, "The Next Web Battle: Phone Calls," *The Wall Street Journal,* December 27, 1999, p. B1.

26. Richard Karpinski, "EDI Developer Extends E-Trading Boundaries," *Internet Week,* May 3, 1999, p. 10.

27. Gordon Bass, "On a Road to Nowhere?" *PC Computing,* December 1999, p. 308.

28. "Log On for Company Training," *Business Week,* January 10, 2000, p. 140.

29. Ronald Grover et al., "A Little Net Music," *Business Week,* February 7, 2000, p. 34.

30. Walter Mossberg, "Behind the Lawsuit: Napster Offers Model for Music Distribution," *The Wall Street Journal,* May 11, 2000, p. B1.

31. Nikhil Jutheesing, "Webcasting," *Forbes,* February 8, 1999, p. 104.

32. Adam Penenberg, "Informercial.com," *Forbes,* March 8, 1999, p. 116.

33. "Best of the Web," *Forbes,* May 22, 2000, p. 60; and Don Willmott, "The Top 100 Web Sites," *PC Magazine,* February 8, 2000, p. 144.

34. "SAP Opts for Push Technology," *Computing,* May 13, 1999, p. 8.

35. Frank Derfler, "Virtual Private Networks," *PC Magazine,* January 4, 2000, p. 146.

36. "Virtual Private Networks," *Network News,* November 24, 1999, p. 4.

37. Eric Brown, "Broadband, Narrow Choices," *PC World,* February 2000, p. 139.

38. Mark Sweeney, "Accept Cookies by Site," *PC Magazine,* February 22, 2000, p. 99.

PART III

Business Information Systems

CHAPTER 5

Electronic Commerce and Transaction Processing Systems

Principles	Learning Objectives
E-commerce is a new way of conducting business that presents both opportunities for improvement and potential problems.	• *Identify several advantages of e-commerce.* • *Identify two major challenges to society associated with e-commerce.* • *Identify and discuss the business impact of several e-commerce applications.*
E-commerce requires the careful planning and integration of a number of technology components.	• *Outline the key hardware and software components that must be in place for e-commerce to succeed.* • *Identify the electronic payments systems used to support e-commerce.*
Organizations must define and execute a strategy to be successful in e-commerce.	• *Outline the key components of an effective e-commerce strategy.*
An organization's transaction processing systems (TPSs) must support the routine day-to-day activities that occur in the normal course of business and help a company add value to its products and services.	• *Identify the basic components and processing activities common to all transaction processing systems.* • *Identify and discuss two basic methods of transaction processing.* • *Identify the transaction processing systems associated with order processing.*
Implementation of an enterprise resource planning (ERP) system enables a company to achieve numerous business benefits through the creation of a highly integrated set of systems.	• *Define the term* enterprise resource planning system. • *Discuss the advantages and disadvantages associated with the implementation of such a system.*

Weirton Steel

E-Commerce Pioneer

Weirton Steel Corporation is the nation's largest producer of tin-coated steel. Surprising as it may seem, this old-line company is recognized as an e-commerce pioneer.

Metals companies have increasingly found themselves squeezed by aggressive price competition, particularly from imports, substitution with other materials, and increasing customer demands for higher quality and faster, more reliable delivery times. Also, a large portion of the metals industry is a commodity market, which means customers can buy the same or equivalent products from multiple sources at about the same price. So metals companies must somehow differentiate themselves in this crowded market. In such a tough economic climate, Weirton had to learn new ways to remain profitable to survive.

Weirton formed an independent company called MetalSite to offer a secure Web-based marketplace for on-line purchases of metals products from many U.S. suppliers. The new venture was launched in phases beginning in the fall of 1998. Weirton spent more than $3 million and two years of research and validation of the site's concept, and it spent an additional $2 million to launch the site. Currently, 15 suppliers, including steel producers and service centers, offer a wide range of products through the MetalSite marketplace.

For metals buyers, MetalSite is a one-stop, up-to-date resource for comprehensive information about its industry and the products manufacturers are offering. For sellers, it provides an opportunity to expand their customer base, increase efficiencies in the selling process, turn inventory more quickly, and free up sales staff from nonproductive tasks. Sellers provide MetalSite with daily product availability and pricing information independently of each other. In effect, MetalSite serves as an independent clearinghouse for metals industry information, product availability, and on-line purchases. The individual sellers, and not MetalSite, are responsible for pricing, order fulfillment, and payment processing through traditional channels.

Buyers can access and use the site for free. Sellers are assessed a transaction fee for each on-line sale and an annual service fee for consulting and development of an individual market center, which basically is a mini Web site for each seller within MetalSite. MetalSite also offers users numerous ways to communicate with their peers throughout the world—much as they do at industry trade shows. The on-line tools they can use include bulletin boards, presentations, and forums.

The site features Internet security technology that is the most widely used by businesses throughout the world. Arthur Andersen LLP conducts ongoing security audits, as well as periodic audits of MetalSite's information technology, processes, and procedures.

As you read this chapter, consider the following:

- How are organizations using e-commerce to provide improved customer service, become more productive, and remain competitive?

- What are the key issues that must be addressed to ensure effective implementation of e-commerce?

- How do e-commerce systems integrate with the everyday transaction processing systems used to capture data and update records of business transactions?

Weirton Steel is just one example of how businesses and individuals use e-commerce to reduce transaction costs, speed the flow of goods and information, improve the level of customer service, and enable close coordination of actions among manufacturers, suppliers, and customers. E-commerce also enables consumers and companies to gain access to worldwide markets, and consumers must understand e-commerce so that they can use it to their advantage. Business managers must also understand and employ e-commerce so that their companies can remain competitive. In addition, e-commerce can capture detailed data and pass it on to transaction processing systems to update an organization's databases of fundamental business operations. So it is essential that you understand the role of a firm's transaction processing systems and the ways they interrelate with its e-commerce operations.

AN INTRODUCTION TO ELECTRONIC COMMERCE

business-to-business (B2B) e-commerce

a form of e-commerce in which the participants are organizations

business-to-consumer (B2C) e-commerce

a form of e-commerce in which customers deal directly with the organization, avoiding any intermediaries

E-commerce involves conducting business activities using electronic data transmission with computers, telecommunications networks, and streamlined work processes. Weirton Steel in the opening vignette provides an example of **business-to-business (B2B) e-commerce**, in which the participants are organizations. There is also **business-to-consumer (B2C) e-commerce**, in which customers deal directly with the organization and avoid any intermediaries. Worldwide business-to-business transactions over the Internet are projected to grow eighteenfold, from $403 billion in 2000 to more than $7,300 billion in 2004[1]—roughly 15 times the dollar volume of business-to-consumer e-commerce. Even though on-line purchases accounted for less than 1 percent of total sales dollar volume as recently as 1999, traditional retailers view e-commerce as a potent threat to brick-and-mortar stores because of its rapid growth throughout the world. As a result, they are making e-commerce part of their brick-and-mortar operations. For example, Borders Books & Music offers in-store kiosks, allowing shoppers access to its immense on-line inventory.

Business processes that are strong candidates for conversion to e-commerce are those that are paper based and time consuming and those that can make business more convenient for customers. As a result, it is no surprise that the first business processes that companies converted to an e-commerce model were those related to buying and selling. For example, Cisco Systems, the maker of Internet routers and other telecommunications equipment, saved $350 million in paperwork and transaction costs in the first year of its e-commerce operation.[2]

Some companies, such as those in the automotive and aerospace industries, have been conducting e-commerce for decades through the use of electronic data interchange (EDI), which transfers business data (invoices, purchase orders, etc.) between companies using a standard data format. Many companies have now gone beyond EDI to launch e-commerce initiatives with suppliers, customers, and employees to address business needs in new areas.

Value Chains in E-Commerce

supply chain management

a key value chain composed of demand planning, supply planning, and demand fulfillment

All business organizations contain a number of value-added processes. The supply chain management process is a key value chain that, for most companies, offers tremendous business opportunities if converted to e-commerce. **Supply chain management** is composed of three subprocesses: demand planning to anticipate market demand, supply planning to allocate the right amount of enterprise resources to meet demand, and demand fulfillment to fill

Demand Planning

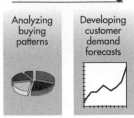

Analyzing buying patterns

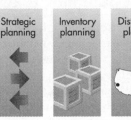

Developing customer demand forecasts

Supply Planning

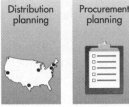

Strategic planning

Inventory planning

Distribution planning

Procurement planning

Transportation planning

Supply allocation

Demand Fulfillment

Order fulfillment

Backlog management

Order promising

Customer verification

Order capture

FIGURE 5.1

Supply Chain Management

customer orders quickly and efficiently (Figure 5.1). The objective of demand planning is to understand customers' buying patterns and develop overall long-term, intermediate-term, and short-term forecasts of customer demand. Supply planning includes inventory planning, distribution planning, procurement planning, transportation planning, and supply allocation. The goal of demand fulfillment is to provide fast, accurate, and reliable delivery of customer orders. Demand fulfillment includes order capturing, customer verification, order promising, backlog management, and order fulfillment.

Conversion to e-commerce supply chain management provides businesses an opportunity to (1) increase revenues or decrease costs by eliminating time-consuming and labor-intensive steps throughout the order and delivery process, (2) improve customer satisfaction by enabling customers to view detailed information about delivery dates and order status, and (3) reduce inventory including raw materials, safety stocks, and finished goods. Achieving these goals requires integrating all subprocesses that exchange information and move goods between suppliers and customers, including manufacturers, distributors, retailers, and any other enterprise within the extended supply chain.

Business to Business (B2B)

Business-to-business e-commerce offers enormous opportunities. It is considerably larger and expected to grow much more rapidly than business-to-consumer (B2C). It allows manufacturers to buy at a low cost worldwide, and it offers enterprises the chance to sell to a global market right from the start. In addition, e-commerce offers great promise for developing countries, helping them enter the prosperous global marketplace, and so helping reduce the gap between rich and poor countries.

The rapid devolpment of e-commerce presents great challenges to society, however. Even though e-commerce is creating new job opportunities, it could also cause a loss of employment in some traditional jobs such as order processing or customer service areas. And as we are already seeing, many companies may fail in the intense competitive environment of e-commerce. Because of these threats, it is vital that the opportunities and implications of e-commerce be understood.

Business to Consumer (B2C)

Even though it has attracted a lot of media attention, e-commerce for consumers is still in its early stages. Most shoppers are not yet convinced that it is worthwhile to connect to the Internet, search for shopping sites, wait for the images to download, try to figure out the ordering process, and then worry about whether their credit card numbers will be stolen by a hacker. But attitudes are changing, and an increasing number of shoppers are beginning to appreciate the importance of e-commerce. For time-strapped households, consumers wonder, why waste time fighting crowds in shopping malls when from the comfort of home, you can shop on-line anytime and have the goods delivered directly? These shoppers have found that many goods and services are cheaper when purchased via the Web—for example, stocks, books, newspapers, airline tickets, and hotel rooms. They can also get information about automobiles, cruises, and homes to cut better deals. Internet shoppers can unleash shopping bots or access sites such as www.jango.com to browse the Internet and obtain lists of items, prices, and merchants. More than a new way to place orders, the Internet is emerging as a paradise for comparison shoppers.

electronic retailing (e-tailing)

the direct sale from business to consumer through electronic storefronts, typically designed around an electronic catalog and shopping cart model

E-COMMERCE APPLICATIONS

E-commerce is being applied to retail and wholesale, manufacturing, marketing, investment and finance, and auctions. Here are some current uses in those areas.

Retail and Wholesale

cybermall

a single Web site that offers many products and services at one Internet location

---

AutoNation.com is the largest auto dealership network, with 376 dealerships and 37 used-vehicle megastores.

There are numerous examples of e-commerce in retail and wholesale. **Electronic retailing**, sometimes called e-tailing, is the direct sale from business to consumer through electronic storefronts typically designed around the familiar electronic catalog and shopping cart model. Companies such as Office Depot, Wal-Mart, and many others have used the same model to sell wholesale to employees of corporations. There are tens of thousands of electronic retail Web sites—selling literally everything from soup to nuts. Another means to support retail shopping is the **cybermall**, a single Web site that offers many products and services at one Internet location. The cybermall employs the Internet's ubiquity to pull together multiple buyers and sellers into one virtual place, easily reachable through a Web browser.[3]

Auto retailing provides an excellent example of how e-commerce is transforming retail selling. AutoNation, with $17 billion in sales, 376 dealerships, and 37 used-vehicle megastores, is the largest auto dealership network. Its goal is to make every step of the purchase process available on-line—everything but the test drive and handing over the keys. The e-commerce unit has only 22 employees, but it expects to generate three-quarters of a billion dollars in on-line sales using Compass, AutoNation's dealer extranet that delivers and manages on-line sales leads. Now sales reps, instead of waiting to pounce on the next customer who walks in, use the computer to e-mail interested customers with information or to set up appointments.[4]

Spending on manufacturing, repair, and operations (MRO) goods and services—from simple office supplies to mission-critical equipment, such as the motors, pumps, compressors, and instruments that keep manufacturing facilities up and running smoothly—is an example of e-commerce applied to wholesale selling. Spending on MRO goods and services often approaches 40 percent of a manufacturing company's total revenues. Despite its importance, spending can be haphazard without automated controls for purchasing materials. In addition, companies face significant internal costs resulting from ineffective and cumbersome MRO management processes. Estimates show that a high percentage of manufacturing downtime is often caused by not having the right part at the right time in the right place. E-commerce software for plant operations enables users to identify duplicate and functionally equivalent items, which provides opportunities for cost savings. Comparing various suppliers and consolidating spending with fewer suppliers leads to decreased costs. In addition, automating the work process reduces time delays in getting necessary MRO goods and services.

Manufacturing

One approach many manufacturers take to raise profitability and improve customer service is to move their supply chain operations onto the Internet. Here they can form an **electronic exchange** to join with competitors and suppliers alike using computers and Web sites to buy and sell goods, trade market information, and run back-office operations, such as inventory control, as shown in Figure 5.2.

electronic exchange

an electronic forum where manufacturers, suppliers, and competitors buy and sell goods, trade market information, and run back-office operations

FIGURE 5.2

Model of an Electronic Exchange

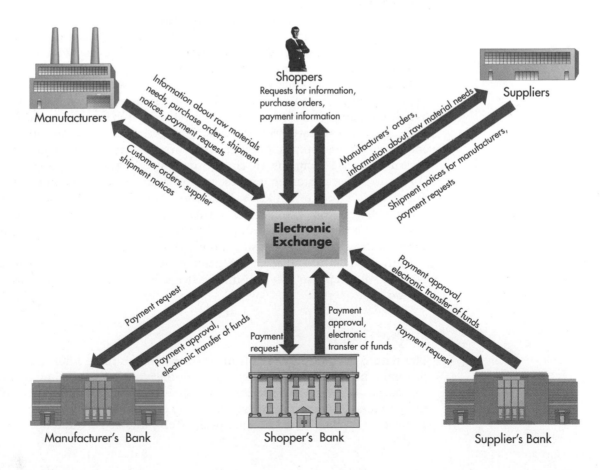

With such an exchange, the business center is not a physical building but a network-based location where business interactions occur. This approach was taken by Weirton Steel, discussed in the opening vignette, and has greatly speeded the movement of raw materials and finished products among all members of the business community, thus reducing the amount of inventory that must be maintained. It has also led to a much more competitive marketplace and lower prices.

By mid-2000, nearly 1,000 on-line marketplaces in 70 industries had been announced by Internet and brick-and-mortar companies, including the following: the Worldwide Retail Exchange led by Safeway and Kmart; a giant marketplace for high-tech firms overseen by Hewlett-Packard and Compaq Computer; a global exchange for the auto industry run by automakers Ford, General Motors, and DaimlerChrysler; Chemdex in the chemical industry; and eSteel in the steel industry. It is predicted that more than 10,000 on-line exchanges will sweep the business world between 2000 and 2003; however, Arthur Andersen estimates barely one in four will survive.[5] To date, only a few exchanges are believed to be profitable.

Several strategic and competitive issues could limit the use of exchanges. Many companies distrust their corporate rivals and fear they may lose trade secrets through participation in such exchanges. Suppliers worry that the on-line marketplaces and their auctions will drive down the prices of goods and favor buyers. Suppliers also can spend a great deal of money in setting up to participate in multiple exchanges. For example, more than a dozen new exchanges have appeared in the oil industry, and the printing industry has more than 20 on-line marketplaces. Until a clear winner emerges in particular industries, suppliers are more or less forced to sign on to several or all of them.[6] Yet another issue is potential government scrutiny of exchange participants—any time competitors get together to share information, it raises questions of collusion or antitrust behavior.

In addition to formal exchanges, manufacturers are using e-commerce to improve the efficiency of the selling process. For example, with annual revenue of $1.7 billion, Timken is the world's largest manufacturer of tapered roller bearings, a product used in everything from automobile transmissions to paper mills. Timken sells primarily to distributors and has used electronic data interchange (EDI) networks to exchange purchase orders and invoices with its business partners for years. But globalization, consolidation, an emphasis on one-stop shopping, and comprehensive contracts increased pressure on the company to change its ways. Timken developed Timken Direct to move customer queries on-line and improve customer relationships. With links to Timken's back-end inventory databases, the secure Timken Direct Web site handles the two customer concerns that occupied most of the service representatives' time: product availability and price. Distributors can go the Web site and get this information for themselves. As a result, Timken has freed up 15 percent of its service reps and converted them to sales reps, so they can make sales calls instead of fielding availability and cost queries.[7]

Marketing

The nature of the Web allows firms to gather much more information about customer behavior and preferences than they could with other marketing approaches. Marketing organizations measure browsing activities as customers and potential customers gather information and make their purchases. Analysis of this data is complicated because of the Web's interactivity and because each

DoubleClick is a leading global Internet advertising company that leverages technology and media expertise to help advertisers use the power of the Web to build relationships with customers.

market segmentation

the identification of specific markets and targeting them with advertising messages

visitor voluntarily provides or refuses to provide personal data such as name, address, e-mail address, telephone number, and demographic data. Internet advertisers use the data they gather to identify specific portions of their markets and target them with tailored advertising messages. This practice, called **market segmentation**, divides the pool of potential customers into segments, which are usually defined in terms of demographic characteristics such as age, sex, marital status, income level, and geographic location.

The idea of technology-enabled relationship management has become possible when promoting and selling on the Web. *Technology-enabled relationship management* occurs when a firm obtains detailed information about a customer's behavior, preferences, needs, and buying patterns and uses that information to set prices, negotiate terms, tailor promotions, add product features, and otherwise customize its entire relationship with that customer.

DoubleClick is a leading global Internet advertising company that uses its technology and media expertise to help advertisers harness the power of the Web and build relationships with customers. The DoubleClick Network is its flagship product, a collection of high-traffic and well-recognized sites on the Web (AltaVista, Dilbert, US News, Macromedia, and more than 1,500 others). This network of sites is coupled with DoubleClick's proprietary DART targeting technology, which allows advertisers to target their best prospects with precise profiling criteria. DoubleClick then places a company's ad in front of those prospects. Comprehensive on-line reporting lets advertisers know how their campaign is performing and what type of users are seeing and clicking on their ads. The system is also designed to track and summarize these user transactions in the form of reports, and to compute DoubleClick Network member fees.

Investment and Finance

The Internet has revolutionized the world of investment and finance. Perhaps the changes have been so great because this industry had so many inefficiencies and so much opportunity for improvement.

On-Line Stock Trading

Before the World Wide Web, if you wanted to invest in stocks, you called your broker and asked what looked promising. He'd tell you about two or three companies and then would try to sell you shares of a stock or perhaps a mutual fund. The sales commission was well over $100 for the stock (depending on the price of the stock and the number of shares purchased) or as much as an 8 percent sales charge on the mutual fund. If you wanted information about the company before you invested, you would have to wait two or three days for a one-page Standard and Poor's stock report providing summary information and a chart of the stock price for the past two years. Once you purchased or

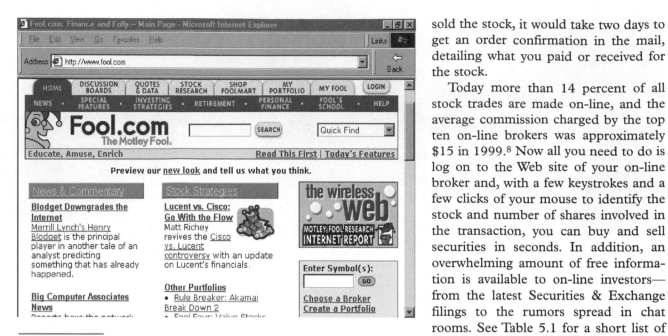

Fool.com is a Web site that allows users to track the performance of their stock portfolios.

sold the stock, it would take two days to get an order confirmation in the mail, detailing what you paid or received for the stock.

Today more than 14 percent of all stock trades are made on-line, and the average commission charged by the top ten on-line brokers was approximately $15 in 1999.[8] Now all you need to do is log on to the Web site of your on-line broker and, with a few keystrokes and a few clicks of your mouse to identify the stock and number of shares involved in the transaction, you can buy and sell securities in seconds. In addition, an overwhelming amount of free information is available to on-line investors— from the latest Securities & Exchange filings to the rumors spread in chat rooms. See Table 5.1 for a short list of the more valuable sites.

One indispensable tool of the on-line investor is a portfolio tracker that allows you to enter information about the securities you own—ticker symbol, number of shares, price paid, and date purchased—at a tracker Web site. You can then access the tracker site to see how your stocks are doing. In addition to reporting the current value of your portfolio, most sites provide access to news, charts, company profiles, and analyst ratings on each of your stocks. You can also program many of the trackers to watch for certain events (e.g., stock price change of more than +/- 3 percent in a single day). When one of the events you specified occurs, an "alert" symbol is posted next to the affected stock. Table 5.2 lists a number of the more popular tracker Web sites.

On-Line Banking
On-line banking customers can check balances of their savings, checking, and loan accounts; transfer money among accounts; and pay their bills. On-line banking customers like the convenience of having current knowledge of account balances, of eliminating the need to write checks in longhand, and of reducing the amount they spend on envelopes and stamps. All of the nation's

TABLE 5.1

Web Sites Useful to Investors

Name of Site	URL	Description
411Stocks	www.411stocks.com	One-stop location to get lots of information about a stock—price data, news, discussion groups, charts, basic data, financial statements, and delayed quotes
MarketReporter	www.marketreporter.com	Provides financial news, recommendations, upgrades, downgrades, message boards, stock market simulation game
Thomson Investors Network	www.thomsoninvest.com	Financial commentary from a number of stock market publications, including *First Watch* and *Stocks to Watch*
Elite Trader	www.elitetrader.com	Virtual gathering place for day traders with bulletin boards and chat rooms
Dayinvestor.com	www.dayinvestor.com	News and stock alerts with frequent briefs on market activity and rumors

Priceline.com is a patented Internet bidding system that enables consumers to save money by naming their own price for goods and services.

electronic bill presentment

a method of billing in which the biller posts an image of your statement on the Internet and alerts you by e-mail that your bill has arrived

major banks and many of the smaller banks enable their customers to pay bills on-line. Nearly 7.5 million customers banked on-line by the end of 2000, and that number may double to 15.7 million by 2003.[9]

Here's how electronic bill payment works. You first set up a list of frequent payees, along with their addresses and a code describing the payment, such as "home mortgage." When you go on-line to pay your bills, you simply enter the code or name assigned to the check recipient, the amount of the check, and the date you want it paid. In many cases, the bank still prints and mails a check, so you have to time your on-line transactions to allow for bank processing and mail delays. But most bill-paying programs allow you to schedule recurring payments for every week, month, or quarter, which you might want to do for your auto loan or health insurance bill.

The next advance in on-line bill paying is **electronic bill presentment**, which eliminates all paper, including the bill itself. Under this process, the biller posts an image of your statement on the Internet and alerts you by e-mail that your bill has arrived. You then direct your bank to pay it. As of May 2000, CheckFree (http://www.checkfree.com) offers such a service from 76 companies, including major utility companies, Amoco, Phillips 66, and Sears. Many major banks, including Key, Norwest, and BancOne, offer a similar service run by Microsoft and First Data Corp. directly at the banks' Web sites.

Auctions

The Internet has created many new options for electronic auctions, where geographically dispersed buyers and sellers can come together. Priceline.com is the patented Internet bidding system that enables consumers to achieve significant savings by naming their own price for goods and services. Priceline.com takes these consumer offers and then presents them to sellers, who can fill as much of that guaranteed demand as they wish at price points determined by buyers.

Now that we've examined some of the applications of e-commerce, let's look at some technical issues related to information systems and technology that make it possible.

TABLE 5.2

Popular Stock Tracker Web Sites

Name of Web Stock Tracker Site	URL
MSN MoneyCentral	moneycentral.msn.com/investor
Quicken.com	www.quicken.com
The Motley Fool	www.fool.com
Yahoo!	quote.yahoo.com
Morningstar	www.morningstar.com

E-COMMERCE TECHNOLOGY COMPONENTS

For e-commerce to succeed, a complete set of hardware, software, and network components must be chosen carefully and integrated to support a large volume of transactions with customers, suppliers, and other business partners worldwide. On-line consumers frequently complain that poor Web site performance (e.g., slow response time and "lost" orders) drives them to abandon some e-commerce sites in favor of those with better, more reliable performance. This section provides a brief overview of the key technology infrastructure components (Figure 5.3).

Hardware

A Web server complete with the appropriate software is key to successful e-commerce. The amount of storage capacity and computing power required of the Web server depends primarily on two things—the software that must run on the server and the volume of e-commerce transactions that must be processed. Although business managers and information systems staff can define the software to be used, it is difficult for them to estimate how much traffic the site will generate. As a result, the most successful e-commerce solutions are designed to be highly scalable so that they can be upgraded to meet unexpected user traffic.

Many companies decide that a third-party Web service provider is the best way to meet their initial e-commerce needs. A Web service rents out space on its computer system and provides a high-speed connection to the Internet, which minimizes the initial setup costs for e-commerce. The service provider can also provide personnel trained to operate, troubleshoot, and manage the Web server. Other companies decide to take full responsibility for acquiring, operating, and supporting their own Web server hardware and software, but this approach requires considerable up-front capital and a set of skilled and trained individuals. Whichever approach is taken, there must be adequate hardware backup to avoid a major business disruption in case of a failure of the primary Web server.

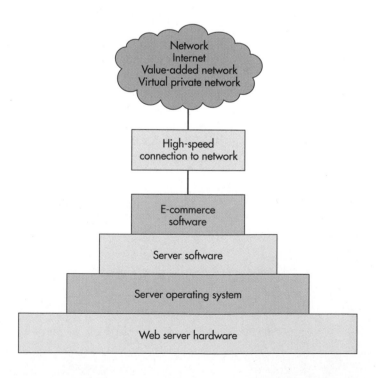

FIGURE 5.3

Key E-Commerce Technical Components

Web Server Software

Each e-commerce Web server must have software to perform a number of fundamental services, including security and identification authentication, retrieval and sending of Web pages, and Web page construction. The two most popular Web server software packages are Apache HTTP Server and Microsoft Internet Information Server.

Web page construction software

software that uses Web editors to produce both static and dynamic Web pages

Web page construction software is used to produce Web pages—either static or dynamic. *Static Web pages* always contain the same information—for example, a page that provides text about the history of the company or a photo of corporate headquarters. *Dynamic Web pages* contain variable information and are built in response to a specific Web visitor's request. For example, if a Web site visitor inquires about the availability of a certain product by entering a product identification number, the Web server will search the product inventory database and generate a dynamic Web page based on the current product information it found. This same request by another visitor later in the day may yield different results due to changes in product inventory.

E-Commerce Software

Once an organization has located or built a host server, including the hardware, operating system and Web server software, it must investigate and install e-commerce software. **E-commerce software** supports the following tasks: catalog management, product configuration, shopping cart facilities, e-commerce transaction processing, and Web traffic data analysis.

e-commerce software

software that supports catalog management, product configuration, shopping cart facilities, and e-commerce transaction processing

Catalog Management

Catalog management software combines different product data formats into a standard format for uniform viewing, aggregating, and integrating catalog data into a central repository for easy access, retrieval, and updating of pricing and availability changes.

catalog management software

software that automates the process of creating a real-time interactive catalog and delivering customized content to a user's screen

Product Configuration

Customers need help when an item they are purchasing has many components and options. Product configuration software tools were originally developed in the 1980s to help B2B salespeople match their company's products to customer needs. Buyers use the new Web-based **product configuration software** to build the product they need on-line with little or no help from salespeople. For example, Dell customers use product configuration software to build the computer of their dreams. Use of such software can expand into the service arena as well, with consumer loans and financial services to help people decide what sort of loan or insurance is best for them.

product configuration software

software used by buyers to build the product they need on-line

Shopping Cart

Today many e-commerce sites use an **electronic shopping cart** to track the items selected for purchase, allowing shoppers to view what is in their cart, add new items to it, or remove items from it, as shown in Figure 5.4. To order an item, the shopper simply clicks on that item. All the details about it, including its price, product number, and other identifying information, are stored automatically. If the shopper later decides to remove one or more items from the cart, he or she can do so by viewing the cart's contents and removing any unwanted items. When the shopper is ready to pay for the items, he or she clicks a button (usually labeled "proceed to checkout") and begins a purchase transaction. Clicking the "Checkout" button displays another screen that usually asks the shopper to fill out billing, shipping, and payment method information and to confirm the order.

electronic shopping cart

a model used by many e-commerce sites to track the items selected for purchase, allowing shoppers to view what is in their cart, add new items to it, and remove items from it

FIGURE 5.4

Electronic Shopping Cart

An electronic shopping cart (or bag) allows on-line shoppers to view their selections and add or remove items.

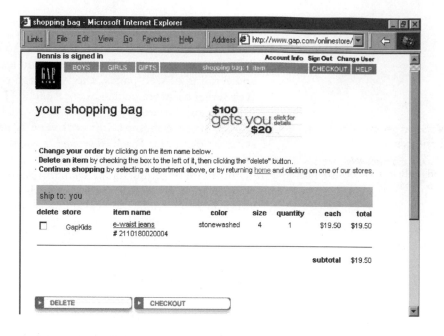

E-Commerce Transaction Processing

e-commerce transaction processing software

software that provides the basic connection between participants in the e-commerce economy, enabling communications between trading partners regardless of their technical infrastructure

E-commerce transaction processing software takes data from the shopping cart and calculates volume discounts, sales tax, and shipping costs to arrive at the total cost. In some cases, the software determines shipping costs by connecting directly to shipping companies such as UPS, FedEx, and Airborne. In other cases, shipping cost may be a predetermined amount for each item ordered.

More and more companies are outsourcing the actual inventory management and order fulfillment process to a third party. In this situation, the e-commerce transaction processing software routes order information to the third party to ship from inventory under their management. For example, Hewlett-Packard has an arrangement with FedEx to manage inventory and the shipment of orders for HP printers.

Web Site Traffic Data Analysis

Web site traffic data analysis software

software that processes and analyzes data from the Web log file to provide useful information to improve Web site performance

Web site traffic data analysis software captures visitor information, including who is visiting the Web site (the visitor's URL), what search engine and key words they used to find the site, how long their Web browser viewed the site, the date and time of each visit, and which pages were displayed. These data are placed into a *Web log file* for future analysis. The software makes sense of all the data captured in the Web log file—turning it into useful information to improve Web site performance.

Network and Packet Switching

Obviously, for e-commerce to take place, some sort of network is needed for secure transmission of data, whether over the Internet, corporate extranets, value-added networks (VANs), or virtual private networks (VPNs). All these approaches rely on basic packet switching technology and the use of routers to help each packet arrive at its destination quickly and economically. In choosing among the options, the considerations are cost, availability, reliability, security, and redundancy. Cost includes the initial development cost as well as the ongoing operational and support costs. Availability concerns the hours the network is *scheduled* to be available for normal use. Reliability is the percentage of available time the network is actually fully operational, typically 99 percent or more.

A security system assures customers that information they provide to retailers, such as credit card numbers, cannot be viewed by anyone else on the Web.

digital certificate

an attachment to an e-mail message or data embedded in a Web page that verifies the identity of a sender or a Web site

electronic cash

an amount of money that is computerized, stored, and used as cash for e-commerce transactions

electronic wallet

a computerized stored value that holds credit card information, electronic cash, owner identification, and address information

Security is the ability to keep messages from being intercepted. Redundancy is the ability of the network to keep operating if key elements fail. All are important factors that must be considered.

Electronic Payment Systems

Electronic payment systems are a key component of the e-commerce infrastructure. Electronic payments make up only about 1 percent of all consumer settlements now, but they may reach 5 percent by 2005—a growth from $43 billion to $325 billion in just a few short years.[10] E-commerce technology relies on user identification and encryption to safeguard business transactions. Actual payments are made in a variety of ways, including electronic cash; electronic wallets; and smart, credit, charge, and debit cards.

Authentication technologies are used by organizations to confirm the identity of a user requesting access to information or assets. A **digital certificate** is an attachment to an E-mail message or data embedded in a Web site that verifies the identity of a sender or Web site.

One tip to the security of a transaction is visible on screen. Look at the bottom left corner of your browser before sending your credit card number to an e-commerce vendor. If you use Netscape Navigator, make sure you see a solid key in a small blue rectangle. If you use Microsoft Internet Explorer, the words "Secure Web site" appear near a little gold lock. If you're worried about how secure a secure connection is, visit the Netcraft.com site. At this site you can type in any Web site address and determine the equipment being used for secure transactions. One more tip: To ensure security, you should always use the newest browser available. The newer the browser, the better the security.

Electronic cash is an amount of money that is computerized, stored, and used as cash for e-commerce transactions. A consumer must open an account with a bank to obtain electronic cash. Whenever the consumer wants to withdraw electronic cash to make a purchase, he or she accesses the bank via the Internet and presents proof of identity—typically a digital certificate. After the bank verifies the consumer's identity, it issues the consumer the requested amount of electronic cash and deducts the same amount from the consumer's account. The electronic cash is stored in the consumer's electronic wallet on his or her computer's hard drive, or on a smart card (both are discussed later).

Consumers can spend their electronic cash when they locate e-commerce sites that accept electronic cash for payment. The consumer sends electronic cash to the merchant for the specified cost of the goods or services. The merchant validates the electronic cash to be certain it is not forged and belongs to the customer. Once the goods or services are shipped to the consumer, the merchant presents the electronic cash to the issuing bank for deposit. The bank then credits the merchant's account for the transaction amount, minus a small service charge.

On-line shoppers quickly tire of repeatedly entering their shipment and payment information each time they make a purchase. An **electronic wallet** holds credit card information, electronic cash, owner identification, and address information. It provides this information at an e-commerce site's checkout counter. When consumers click on items to purchase, they can then click on their electronic wallet to order the item, thus making on-line shopping much faster and easier. Household International, a leading provider of consumer finance, credit card, auto finance, and credit insurance products, in partnership with General Motors Corporation and CyberCash, a provider of e-commerce technologies and services for merchants, introduced the GM Card easyPay electronic wallet marketed to a large portion of GM card members. The GM Card easyPay Wallet stores a shopper's name, credit card information, shipping details, and other pertinent facts that can be called up to make an on-line purchase with a single click of a computer mouse. The wallet is available to GM card members and interested consumers at www.GMCard.com.[11]

Smart, Credit, Charge, and Debit Cards

On-line shoppers use credit and charge cards for the majority of their Internet purchases. A *credit card,* such as Visa or MasterCard, has a preset spending limit based on the user's credit limit, and each month the user can pay off a portion of the amount owed or the entire credit card balance. Interest is charged on the unpaid amount. A *charge card,* such as American Express, carries no preset spending limit, and the entire amount charged to the card is due at the end of the billing period. Charge cards do not involve lines of credit and do not accumulate interest charges. *Debit cards* operate like cash or a personal check. While a credit card is a way to "buy now, pay later," a debit card is a way to "buy now, pay now." When you use a debit card, your money is quickly deducted from your checking or savings account. Credit, charge, and debit cards currently store limited information about you on a magnetic stripe. This information is read each time the card is swiped to make a purchase. Credit card customers are protected by law from paying more than $50 for fraudulent transactions.

smart card

a credit card–sized device with an embedded microchip to provide electronic memory and processing capability

The **smart card** is a credit card–sized device with an embedded microchip to provide electronic memory and processing capability. Smart cards can be used for a variety of purposes, including storing a user's financial facts, health insurance data, credit card numbers, and network identification codes and passwords. They can also store monetary values for spending. American Express launched its Blue smart card in 1999. The card comes with a "reader" that attaches to your PC monitor with sticky tape. You also must visit the American Express Web site to get an electronic wallet to store your credit card information and shipping address. When you want to buy something on-line, you go to the checkout screen of a Web merchant, swipe your Blue card through the reader, type in a password, and you're done. The digital wallet automatically tells the vendor your credit card number, its expiration date, and your shipping information.[12]

The "E-Commerce" box outlines why companies are preparing their Web sites to handle customers from outside the United States and what steps they must take to do so.

E-COMMERCE
Get Ready for Global E-Commerce

E-commerce sites need to shift their focus from North American consumers as the use of the Internet spreads rapidly in markets throughout Europe, Asia, and Latin America. The majority of Internet users will live outside the United States by 2003, and the U.S. share of all e-commerce revenues is projected to shrink from 69 percent in 2000 to 59 percent by 2003. On-line retail sales in Europe will grow from 2.9 billion euros in 1999 to 175 billion euros in 2005. Companies that want to succeed on the Web cannot ignore the global shift, and developing a sound strategy is critical to ensure that Web sites are relevant to the consumers and businesses the company wants to reach, whether those customers are in Cleveland, Singapore, or Frankfurt.

The first step in developing a global e-commerce strategy is to determine which global markets make the most sense for selling products or services on-line. One approach is to target regions and countries in which a company already has on-line customers. Companies can track the country from which current users are visiting, and established global companies can look to their overseas offices to help determine the languages and countries to target for their Web sites.

Once the company decides which global markets it wants to reach with its Web site, it must adapt the existing U.S.-centric site to another language and culture—a process called *localization*. Localization requires companies to have a deep understanding of the country, its people, and the market, which means either building a physical presence in the country or forming partnerships so that detailed knowledge can be gathered. Companies must take painstaking steps to ensure that e-commerce customers have a local experience even though they're shopping at the Web site of an American company.

Some of the steps involved in localization require recognizing and conforming to the nuances, subtleties, and tastes of local cultures, as well as supporting basic trade laws and technological capabilities such as each country's currency, local connection speeds, payment preferences, laws, taxes, and tariffs. For example, when Dell Computer launched an e-commerce site to sell PCs to consumers in Japan, it made the mistake of surrounding most of the site with black borders, a negative sign in Japanese culture. Japanese Web shoppers took one look at the site and fled. Support for Asian languages is difficult because Asian alphabets are more complex and not all Web development tools are capable of handling them. As a result, many companies choose to tackle Asian markets last. In addition, great care must be taken to choose icons that are relevant to a country. For example, the use of mailboxes and shopping carts may not be familiar to global consumers. Europeans don't take their mail from large, tubular receptacles, nor do many of them shop in stores large enough for wheeled carts.

One of the most important and most difficult decisions in a company's global Web strategy is whether Web content should be generated and updated centrally or locally. Companies that expand through international partnerships may be tempted to hand control to the new international entities to tap the expertise of employees in the new markets. But turning over too much control can lead to a muddle of country-specific sites with no consistency and a scattered corporate message. A mixed model of control may be best. Centralizing decisions about corporate identity, brand representation, and Web technology can minimize Web development and support effort as well as present a consistent corporate and brand message. But a local authority can decide on content and services best tailored for given markets.

Companies must also be aware that consumers outside the United States will access sites with different devices, so they must modify their site design accordingly. In Europe, for example, closed-system iDTVs (interactive digital televisions) are becoming a popular way to access on-line content, with iDTVs projected to reach 80 million European households by 2005. Such devices have better resolution and more screen space than the PC monitors U.S. consumers use to access the Internet. So users of iDTVs expect more ambitious graphics.

A new group of software and service vendors has emerged to address Web globalization issues. The group includes companies such as Idiom, GlobalSight, and Uniscape.com. Their software can integrate with popular e-commerce and Web content management software from vendors such as Vignette, BroadVision, and Interwoven. The multilingual Web site management software can work especially well for global sites with central management.

E-COMMERCE
Get Ready for Global E-Commerce (continued)

The only way to compete with global companies is to be a global company. Successful firms operate with storefronts designed for each target market, with shared sourcing and infrastructure to support the network of stores, and with local marketing and business development teams to take advantage of local opportunities. Service providers continue to emerge to solve the cross-border logistics, payments, and customer service needs of these pan-European retailers.

Discussion Questions

1. Outline the major steps in taking a U.S. Web site global.
2. What is meant by central control versus local control of the Web site content? Which approach is better? Why?

Critical Thinking Questions

3. Which approach is better to gain a deep understanding of a country, its people, and the market—form a partnership with a company in the country, or

hire a software vendor familiar with Web globalization issues. Why?

4. Your company has just completed globalization of its Web site to address the needs of customers in seven countries in Latin America. How would you evaluate the success of this effort?

Sources: Adapted from Matt Hicks and Anne Chen, "Dress for Global Success," *PC Week*, April 3, 2000, pp. 65–73; Anne Chen and Matt Hicks, "Going Global? Avoid Cultural Clashes," *PC Week*, April 3, 2000, p. 6; Idiom Web, "E-Business Globalization" page for global consulting, http://www.idiominc.com, accessed May 20, 2000; "European Online Retail Will Soar to 175 Billion Euros by 2005," press release, http://www.forrester.com/ER/Press/Release/0,1769,266,FF.html, accessed March 28, 2000; and Clay Shirky, "Go Global or Bust," *Business 2.0*, March 2000, http://www.business2.com/articles/2000/03/content/break_3.html, accessed May 20, 2000.

STRATEGIES FOR SUCCESSFUL E-COMMERCE

With all of the constraints to e-commerce just covered, it is important for a company to develop an effective Web site—one that is easy to use, accomplishes the goals of the company, yet is affordable to set up and maintain. We cover several issues for a successful e-commerce site here.

Developing an Effective Web Presence

According to International Data Corporation, as of June 2000, 60 percent of U.S. companies have in-house Web sites but only about 20 percent of them actually sell anything over the Web.[13] When building a Web site, the first thing to decide is which tasks the Web site must accomplish. Most people agree that an effective Web site is one that creates an attractive presence and that meets the needs of its visitors, including one or more of the following:

- Obtain general information about the organization
- Obtain financial information for making an investment decision
- Learn the organization's position on social issues
- Learn about the products or services that the organization sells
- Buy the products or services that the company offers
- Check the status of an order
- Get advice or help on effective use of the products
- Register a complaint about the organization's products
- Register a complaint concerning the organization's position on social issues
- Provide a product testimonial or idea for a product improvement or new product
- Obtain information about warranties or service and repair policies for products
- Obtain contact information for a person or department in the organization

Once a company determines which objectives its site should accomplish, it can proceed to the details of actually putting up a site.

Putting Up a Web Site

Companies large and small can establish Web sites. Previously, companies had to develop their sites in-house or find and hire contractors to develop them. But you no longer have to learn the intricacies of HTML or Java, master Web design software, or hire someone to build your site. Web site hosting services and the use of storefront brokers are two options to designing, building, operating, and maintaining a Web site. Both options offer the advantage of getting the Web site up and running faster and cheaper than doing it yourself, especially for a firm with few or no experienced Web developers.

Web Site Hosting Services

Web site hosting companies such as Bigstep.com, eCongo.com, and freemerchant.com have made it possible to set up a Web page and conduct e-commerce within a matter of days and with little up-front cost. Such companies have packaged all the basic development tools and made them available for free. Using only a browser, you can choose a Web-site template suited to your specific type of business, whatever it may be—clothing, collectibles, or sports equipment. Using the tools provided, you can write the descriptive text and add images of the products you want to sell. Within minutes, your site is up on the Web and accessible to millions.

Web site hosting companies

companies that provide the tools and services required to set up a Web page and conduct e-commerce within a matter of days and with little up-front cost

Storefront Brokers

Another model for setting up a Web site is the use of a **storefront broker** to serve as an intermediary between your Web site and on-line merchants that have the actual products and retail expertise. At sites such as esaler.com (www.eSaler.com) or Vstore (www.Vstore.com), you pick and choose what to sell to match the themes of your site and the products you think might interest your visitors. The products are displayed on your own Web pages, creating, in effect, a virtual storefront stocked with merchandise that is actually handled by another on-line merchant.

storefront brokers

companies that act as intermediaries between your Web site and on-line merchants that have the products and retail expertise

The storefront broker deals with the details of the transactions, including who gets paid for what, and is responsible for bringing together merchants and reseller sites. The storefront broker is similar to a distributor in standard retail operations, but in this case no product moves—only electronic data flows back and forth. Products are ordered by a customer at your site, transacted through a user interface provided by the storefront broker, and shipped by the merchant.

Storefront brokers make their money by taking a commission from the merchant—anywhere from 5 percent to 25 percent—or by collecting a finder's fee per customer from the merchant. The storefront operator usually takes a commission, again in the range of 5 to 25 percent. Although these brokered sales represent a loss of margin for the original merchant, the increased volume can often make up for it. By multiplying the number of outlets for the products—sometimes by the thousands—merchants stand to make more money.[14]

Consider the case of Fruit of the Loom. The underwear manufacturer supplies plain white T-shirts to distributors who, in turn, sell the shirts to designers who add logos touting colleges and other organizations. Fruit of the Loom's e-commerce system automatically ships T-shirts to distributors at the negotiated price whenever stocks run low.

Building Traffic to Your Web Site

The Internet includes hundreds of thousands of e-commerce Web sites. With all of those potential competitors, a company must take strong measures to ensure that the customers it wants to attract can find its Web site. The first step is to obtain and register a domain name. It helps if your domain name says something about your business. For instance, an unusual or interest-grabbing name might seem to be a good catchall, but it might not describe the nature of the business to casual browsers. If you want to sell soccer uniforms and equipment, then you'd try to get a domain name like www.soccerstuff.com, www.soccerworld.com, or www.soccervillage.com. The more specific the Web address, the better.

The next step to attracting customers is to make your site search-engine-friendly by including a meta tag in your store's home page. A **meta tag** is a special HTML tag, not visible on the displayed Web page, that contains keywords representing your site's content, which search engines use to build indexes pointing to your Web site. Again, the selection of keywords is critical to attracting customers, so they should be chosen carefully.

meta tag

a special HTML tag, not visible on the displayed Web page, that contains keywords representing your site's content, which search engines use to build indexes pointing to your Web site

You can also use Web site traffic data analysis software, mentioned earlier, to tell you the URLs from which your site is being accessed, the search engines and keywords that find your site, and other useful information. Using this data can help you identify search engines to which you need to market your Web site, allowing you to submit your Web pages to them for inclusion in the search engine's index.

These tips and suggestions are only a few ideas that can help a company set up and maintain an effective e-commerce site. With technology and competition changing continually, managers should read articles in print and on the Web to keep up to date on ever-evolving issues.

As we pointed out earlier, e-commerce applications often provide customers the ability to order and pay for products and request service and information. As such, these applications form the basis for a class of information systems called transaction processing systems, which we discuss next.

AN OVERVIEW OF TRANSACTION PROCESSING SYSTEMS

Every organization has manual and automated transaction processing systems (TPSs), which process the detailed data necessary to update records about the fundamental business operations of the organization. These systems include order entry, inventory control, payroll, accounts payable, accounts receivable, and general ledger, to name just a few. The input to these systems includes basic business transactions such as customer orders, purchase orders, receipts, time cards, invoices, and payroll checks. The result of processing business transactions is that the organization's records are updated to reflect the status of the operation at the time of the last processed transaction. Automated TPSs consist of databases, telecommunications, people, procedures, software, and hardware devices used to process transactions. The processing activities include data collection, data edit, data correction, data manipulation, data storage, and document production.

UPS adds value to its service by providing timely and accurate data on-line on the exact location of a package.

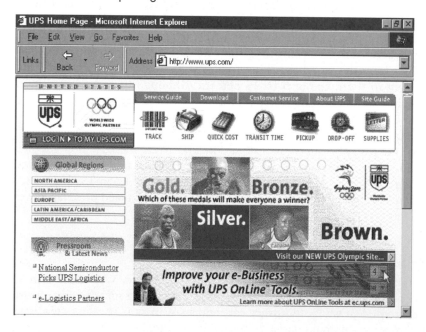

Because TPSs often perform activities related to customer contacts—such as order processing and invoicing—these information systems play a critical role in providing value to the customer. For example, by capturing and tracking the movement of each package, United Parcel Service (UPS) can provide timely and accurate data on the exact location of a package. Shippers and receivers can access an on-line database and, by providing the airbill number of a package, find the package's current location. Such a system provides the basis for added value through improved customer service.

Without transaction processing information systems, recording and processing business transactions would consume huge amounts of an organization's resources. The transaction processing system (TPS) also provides employees involved in other business processes—the management information system/decision support system (MIS/DSS) and the artificial intelligence/expert systems (AI/ES)—with data to help them achieve their goals. A transaction processing system serves as the foundation for the other systems (Figure 5.5). Transaction processing systems perform routine operations such as sales ordering and billing, often performing the same operations daily or weekly. The amount of support for decision making that a TPS directly provides managers and workers is low.

These systems require a large amount of input data and produce a large amount of output without requiring sophisticated or complex processing. As we move from transaction processing to management information/decision support, and artificial intelligence/expert systems, we see less routine, more decision support, less input and output, and more sophisticated and complex processing and analysis. But the increase in sophistication and complexity in moving from transaction processing does not mean that it is less important to a business. In most cases, all these systems start as a result of one or more business transactions.

Traditional Transaction Processing Methods and Objectives

batch processing system

method of computerized processing in which business transactions are accumulated over a period of time and prepared for processing as a single unit or batch

With **batch processing systems**, business transactions are accumulated over a period of time and prepared for processing as a single unit or *batch* (Figure 5.6a). The time period during which transactions are accumulated is whatever length of time is needed to meet the needs of the users of that system. For example, it may be important to process invoices and customer payments for the accounts receivable system daily. On the other hand, the payroll system may receive time cards and process them biweekly to create checks and update employee earnings records as well as to distribute labor costs. The essential characteristic of a batch processing system is that there is some delay between the occurrence of the event and the eventual processing of the related transaction to update the organization's records.

on-line transaction processing (OLTP)

computerized processing in which each transaction is processed immediately, without the delay of accumulating transactions into a batch

With **on-line transaction processing (OLTP)**, each transaction is processed immediately and the affected records are updated, without the delay of accumulating transactions into a batch (Figure 5.6b). Consequently, at any time, the data in an on-line system always reflects the current status. When you make an airline reservation, for instance, the transaction is processed and all databases, such as seat occupancy and accounts receivable, are updated immediately. This type of processing is absolutely essential for businesses that require data quickly and update it often, such as airlines, ticket agencies, and stock

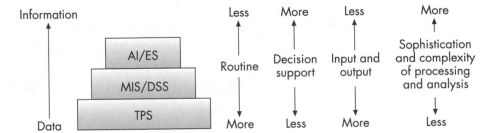

FIGURE 5.5

TPS, MIS/DSS, and AI/ES in Perspective

FIGURE 5.6

Batch versus On-Line
Transaction Processing

Batch processing (a) inputs
and processes data in groups.
In on-line processing (b),
transactions are completed as
they occur.

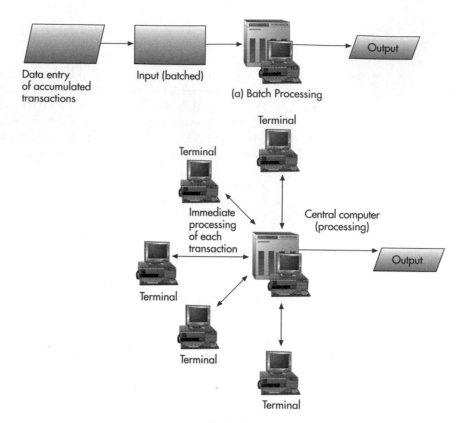

investment firms. Many companies have found that OLTP helps them provide faster, more efficient service—one way to add value to their activities in the eyes of the customer. Increasingly, companies are using the Internet to perform many OLTP functions.[15]

Even though the technology exists to run TPS applications using on-line processing, it is not done for all applications. For many applications, batch processing is more appropriate and cost-effective. Payroll transactions and billing are typically done via batch processing. Specific goals of the organization define the method of transaction processing best suited for the various applications of the company. Figure 5.7 (on page 194) shows the total integration of a firm's transaction processing systems.

One objective of any TPS is error-free data input and processing. An editing program, for example, should have the ability to determine that an entry that should read "40 hours" is not entered as "400 hours" or "4000 hours" because of a data entry error. As seen in the "Ethical and Societal Issues" box, preventing transaction processing fraud is also an important part of maintaining a high degree of accuracy.

When a TPS is developed or modified, the personnel involved should carefully consider how the new or modified system might provide a significant and long-term benefit. Some of the ways that companies can use transaction processing systems to achieve competitive advantage are summarized in Table 5.3.

Transaction Processing Activities

All transaction processing systems perform a common set of basic data processing activities. TPSs capture and process data that describes fundamental business transactions. This data is used to update databases and to produce a variety of reports for use by people both within and outside the enterprise

ETHICAL AND SOCIETAL ISSUES
Credit Card Fraud and Transaction Processing

Some believe that companies that do not ride the wave of the Internet era will not survive in the long term. Companies, from retail firms to job-search services, are going on-line in record numbers. For most companies, this means putting most or all of their transaction processing systems, including sales, order processing, inventory control, and billing, on the Internet. Although the specific products and Internet approaches vary tremendously from one firm to the next, one transaction processing activity seems to be similar for all Internet companies: The billing application of most Internet companies involves collecting fees and funds from credit cards. For most of these companies, there is a fear of credit card fraud.

Although customers and banks feel the impact of credit card fraud, the companies operating on the Internet often are dealt the biggest blow. First, fraud is one of the biggest reasons for consumers and merchants alike to avoid e-commerce. According to the technology officer of one company, "There were days when we had more fraud than legitimate sales." One fraud scam ripped off hundreds of thousands of dollars in merchandise from Amazon.com, Cyberian Outpost, and other companies. The merchandise included computers, software, books, and other items.

Law enforcement agencies are striving to protect companies and consumers from credit card fraud, but many companies think that these efforts are too little, too late. As a result, companies are increasingly looking to other approaches to stop credit card fraud, including fraud-busting software. With names like SecurePay and Prism CardAlert, these software products are helping companies and customers gain more confidence in buying and selling on-line.

Discussion Questions

1. Why is bill paying such an important part of a company's transaction processing system?
2. What other transaction processing applications are critical for successfully completing transactions on the Internet?

Critical Thinking Questions

3. If you were starting to sell products on the Internet, what steps would you take to help reduce credit card fraud?
4. What new laws, if any, would you recommend to help curb credit card fraud?

Sources: Adapted from Margaret Mannix, "High Tech Card Fraud Goes on Right Behind You," *U.S. News & World Report,* February 14, 2000, p. 54; Cynthia Morgan, "Ripped Off," *Computerworld,* March 8, 1999, p. 1; and Lucy Dixon, "Unibanco Offers Virtual Credit Card," *Precision Marketing,* September 20, 1999, p. 10.

---

transaction processing cycle

the process of data collection, data editing, data correction, data manipulation, data storage, and document production

data collection

the process of capturing and gathering all data necessary to complete transactions

TABLE 5.3

Examples of Transaction Processing Systems for Competitive Advantage

(see Figure 5.8 on page 195). The business data goes through a **transaction processing cycle** that includes data collection, data editing, data correction, data manipulation, data storage, and document production (see Figure 5.9 on page 196).

Data Collection

The process of capturing and gathering all data necessary to complete transactions is called **data collection**. In some cases this can be done manually, such as by collecting handwritten sales orders or changes to inventory. In other cases, data collection is automated via special input devices such as scanners, point-of-sale devices, and terminals.

Competitive Advantage	Example
Customer loyalty increased	Use of customer interaction system to monitor and track each customer interaction with the company
Superior service provided to customers	Use of tracking systems that are accessible by customers to determine shipping status
Better relationship with suppliers	Use of an Internet marketplace to allow the company to purchase products from suppliers at discounted prices
Superior information gathering	Use of order configuration system to ensure that products ordered will meet customer's objectives
Costs dramatically reduced	Use of warehouse management system employing scanners and bar-coded product to reduce labor hours and improve inventory accuracy

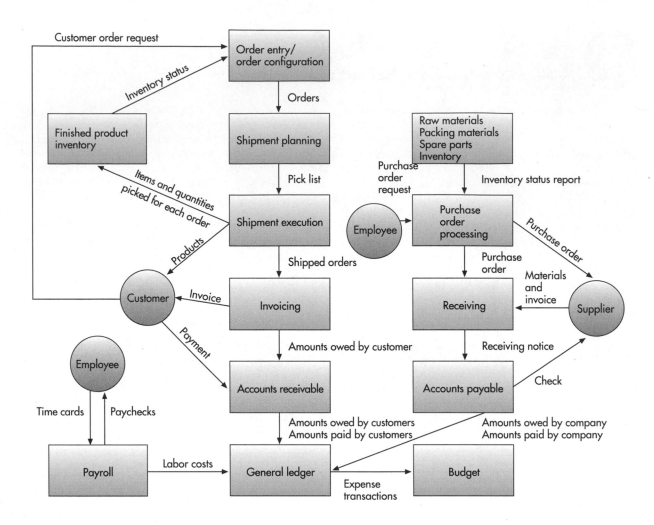

FIGURE 5.7

Integration of a Firm's TPSs

source data automation

the process of capturing data at its source to record it accurately, in a timely fashion, with minimum manual effort, and in a form that can be directly entered into the computer rather than keying the data from a separate document

Data should be captured at its source, and it should be recorded accurately, in a timely fashion, with minimal manual effort, and in a form that can be directly entered to the computer rather than keying the data from some type of document. This approach is called **source data automation**. An example of source data automation is the use of scanning devices at the grocery checkout to read the Universal Product Code (UPC) automatically. Reading the UPC bar codes is quicker and more accurate than having a cash register clerk enter codes manually. The scanner reads the bar code for each item and looks up its price in the item database. The point-of-sale transaction processing system uses the price data to determine the customer's bill. The number of units of this item purchased, the date, the time, and the price are also used to update the store's inventory database, as well as its database of detailed purchases. The inventory database is used to generate a management report notifying the store manager to reorder items whose sales have reduced the stock below the reorder quantity. The detailed purchases database can be used by the store (or sold to market research firms or manufacturers) for detailed analysis of sales (see Figure 5.10 on page 197).

Data Editing

data editing

the process of checking data for validity and completeness

An important step in processing transaction data is to perform **data editing** for validity and completeness to detect any problems with the data. For example, quantity and cost data must be numeric and names must be alphabetic; otherwise, the data is not valid. Often the codes associated with an individual transaction are

FIGURE 5.8

A Simplified Overview of a
Transaction Processing System

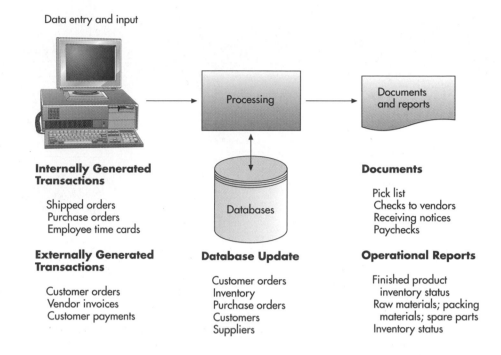

Data entry and input

Processing

Documents and reports

Databases

Internally Generated Transactions

Shipped orders
Purchase orders
Employee time cards

Externally Generated Transactions

Customer orders
Vendor invoices
Customer payments

Database Update

Customer orders
Inventory
Purchase orders
Customers
Suppliers

Documents

Pick list
Checks to vendors
Receiving notices
Paychecks

Operational Reports

Finished product
   inventory status
Raw materials; packing
   materials; spare parts
Inventory status

edited against a database containing valid codes. If any code entered (or scanned) is not present in the database, the transaction is rejected.

Data Correction

It is not enough to reject invalid data. The system should provide error messages that alert those responsible for the data edit function. These error messages must specify what problem is occurring so that corrections can be made. **Data correction** involves reentering miskeyed or misscanned data that was found during data editing. For example, a UPC that is scanned must be in a master table of valid UPCs. If the code is misread or does not exist in the table, the checkout clerk is given an instruction to rescan the item or key in the information manually.

Data Manipulation

Another major activity of a TPS is **data manipulation**, the process of performing calculations and other data transformations related to business transactions. Data manipulation can include classifying data, sorting data into categories, performing calculations, summarizing results, and storing data in the organization's database for further processing. In a payroll TPS, for example, data manipulation includes multiplying an employee's hours worked by the hourly pay rate. Overtime calculations, federal and state tax withholdings, and deductions are also performed.

Data Storage

Data storage involves updating one or more databases with new transactions. Once the update process is complete, this data can be further processed and manipulated by other systems so that it is available for management decision making. Thus, although transaction databases can be considered a by-product of transaction processing, they affect nearly all other information systems and decision-making processes in an organization.

data correction

the process of reentering miskeyed or misscanned data that was found during data editing

data manipulation

the process of performing calculations and other data transformations related to business transactions

data storage

the process of updating one or more databases with new transactions

FIGURE 5.9

Data Processing Activities
Common to Transaction
Processing Systems

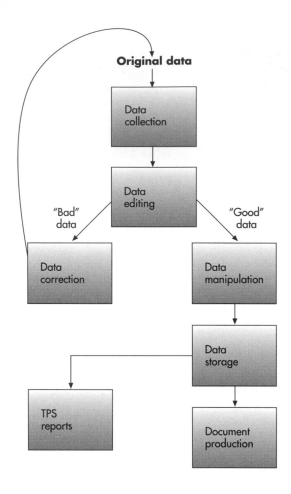

Original data

Data collection

Data editing

"Bad" data → Data correction

"Good" data → Data manipulation

Data storage

TPS reports

Document production

---

business resumption planning

the process of anticipating and providing for disasters

Companies like Iron Mountain provide a secure, off-site environment for records storage. In the event of a disaster, vital data can be recovered.
(Source: © 2000 Photodisc.)

Document Production and Reports

TPSs produce important business documents that may be paper reports or displays on computer screens. Paychecks, for example, are hard-copy documents produced by a payroll TPS, while an outstanding balance report for invoices might be displayed by an accounts receivable TPS. Often, results from one TPS are passed downstream as input to other systems (as shown in Figure 5.8), where the results of updating the inventory database are used to create the stock exception report (a type of management report) of items whose inventory level is less than the reorder point.

In addition to major documents such as checks and invoices, most transaction processing systems provide useful reports that help managers and employees perform various activities. These reports can be printed or displayed on a computer screen. A report showing current inventory is one example; another might be a document listing items ordered from a supplier to help a receiving clerk check the order for completeness when it arrives. A TPS can also produce reports required by local, state, and federal agencies, such as statements of tax withholding and quarterly income statements.

Most organizations would grind to a screeching halt if their transaction processing systems failed. **Business resumption planning** is the process of anticipating and minimizing the effects of disasters. Disasters can be natural

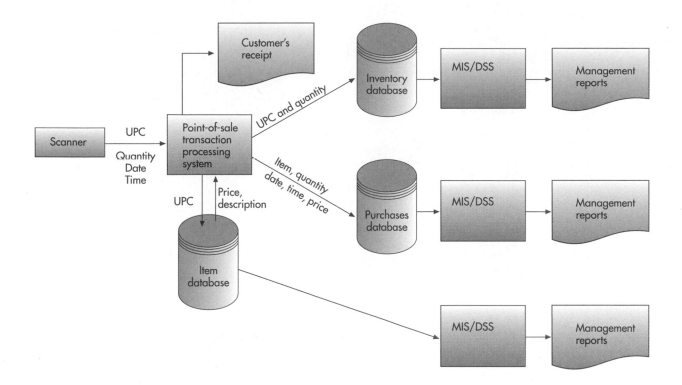

FIGURE 5.10

Point-of-Sale Transaction
Processing System

Scanning items at the checkout
stand results in updating a
store's inventory database and
its database of purchases.

emergencies such as a flood, a fire, or an earthquake, or interruptions in business processes due to such causes as labor unrest or erasure of an important file. Business resumption planning focuses primarily on two issues: maintaining the integrity of corporate information and keeping key information systems running until normal operations can be resumed.

One of the first steps of business resumption planning is to identify potential threats or problems, such as natural disasters, employee misuse of personal computers, and poor internal control procedures. Business resumption planning also involves disaster preparedness. The primary tools used in disaster planning and recovery are backups for hardware, software and databases, telecommunications, and personnel. A common backup for hardware is a similar or compatible computer system owned by another company or a specialized backup system provided by an organization from which a written hardware backup agreement is obtained. Software and databases can be backed up by making duplicate copies of all programs and data. Some business recovery plans call for the backup of vital telecommunications, with the most critical nodes on the network backed up by duplicate components. Business and IS managers should occasionally hold an unannounced "disaster test"—similar to a fire drill—to ensure that the disaster plan is effective.

Order Processing Systems

Since transaction processing systems were first built to handle the give and take between customers and product suppliers, we can gain a better understanding of how they work by examining several common transaction processing systems that support order processing (Table 5.4).

Order processing systems include order entry, sales configuration, shipment planning, shipment execution, inventory control, invoicing, customer interaction, and routing and scheduling. The business processes supported by these systems are so critical to a firm's operation that collectively they are sometimes referred to as the "lifeblood of the organization." Figure 5.11 is a system-level

order processing systems

systems that process order entry,
sales configuration, shipment
planning, shipment execution,
inventory control, invoicing,
customer interaction, and routing
and scheduling

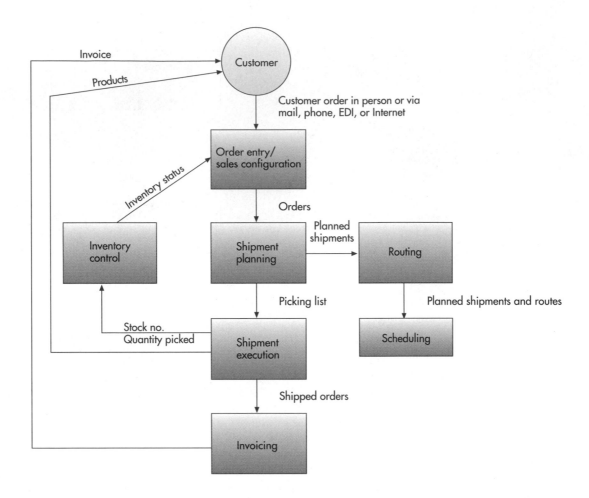

FIGURE 5.11

Order Processing Systems

TABLE 5.4

The Systems That Support
Order Processing

flowchart that shows the various systems and the information that flows between them. A rectangle represents a system, a line represents the flow of information from one system to another, and a circle represents any entity outside the system—in this case, the customer. What is key to note here is how these transaction processing systems work together as an integrated whole to support major business processes.

Order Processing System	Purpose
Order entry	Captures the basic data needed to process a customer order
Sales configuration	Ensures that the products and services ordered are sufficient to accomplish the customer's objectives and will work well together
Shipment planning	Determines which open orders will be filled and from which location they will be shipped
Shipment execution	Coordinates the outflow of all products from the organization, with the objective of delivering quality products on time to customers
Inventory control (finished product)	Updates computerized inventory records to reflect the exact quantity on hand of each stock-keeping unit
Invoicing and billing	Generates customer invoices based on records received from the shipment execution transaction processing system
Customer interaction	Monitors and tracks each customer contact with the company
Routing and scheduling	Determines the best way to get products from one location to another

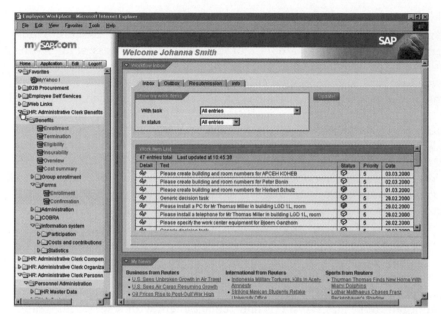

ENTERPRISE RESOURCE PLANNING

Flexibility and quick response are hallmarks of business competitiveness. Access to information at the earliest possible time can help businesses serve customers better, raise quality standards, and assess market conditions. Enterprise resource planning (ERP) is a key factor in instant access.[16] Although some think that ERP systems are only for extremely large companies, this is not the case. Medium-sized companies can also benefit from the ERP approach. A few leading vendors of ERP systems are listed in Table 5.5.

An Overview of Enterprise Resource Planning

The key to ERP is real-time monitoring of business functions, which permits timely analysis of key issues such as quality, availability, customer satisfaction, performance, and profitability. Financial and planning systems receive "triggered" information from manufacturing and distribution. When something happens on the manufacturing line that affects a business situation—for example, packing material inventory drops to a certain level, delaying delivery of a customer's order—a message is triggered for the appropriate person in purchasing. In addition to manufacturing and finance, ERP systems can also support human resources, sales, and distribution. This sort of integration is breaking through traditional corporate boundaries.

ERP systems accommodate the different ways each company runs its business by either providing vastly more functions than one business could ever need or including customization tools that allow firms to fine-tune what should already be a close match. SAP R/3 is the undisputed king of the first approach.[17] R/3 is easily the broadest and most feature-rich ERP system on the market. Thus, rather than compete on size, most rivals focus on customizability. ERP systems have the ability to configure and reconfigure all aspects of the IS environment to support whatever way your company runs its business.

Advantages and Disadvantages of ERP

Increased global competition, new needs of executives for control over the total cost and product flow through their enterprises, and ever-more-numerous customer interactions are driving the demand for enterprisewide access to real-time

Software Vendor	Name of Software
Oracle	Oracle Manufacturing
SAP America	SAP R/3
Baan	Triton
PeopleSoft	PeopleSoft
J. D. Edwards	World

TABLE 5.5

Some ERP Software Vendors

information. ERP offers integrated software from a single vendor to help meet those needs. The primary benefits of implementing ERP include elimination of inefficient systems, easing adoption of improved work processes, improving access to data for operational decision making, and technology standardization. ERP vendors have also developed specialized systems for specific applications and market segments. SAP, for example, has developed a Customer Relationship Management (CRM) package for its ERP system.[18] Osram Sylvania, a lighting manufacturer, plans on using CRM to allow lighting buyers to place orders directly over the Internet. Of course, developing custom packages for every market need and segment would be a huge undertaking for ERP vendors.[19] As a result, major ERP vendors are increasingly seeking help from other software vendors to develop specialized programs to tie directly into their ERP systems. Even with the benefits of ERP, most companies have found it surprisingly difficult to justify implementation of an ERP system based strictly on cost savings.

Elimination of Costly, Inflexible Legacy Systems

Adoption of an ERP system enables an organization to eliminate dozens or even hundreds of separate systems and replace them with a single integrated set of applications for the entire enterprise. In many cases, these systems are decades old, the original developers are long gone, and the systems are poorly documented. As a result, they are extremely difficult to fix when they break, and adapting them to meet new business needs takes too long. They become an anchor on the organization that keeps it from moving ahead and remaining competitive. An ERP system helps match the capabilities of an organization's information systems to its business needs—even as these needs evolve.

Improvement of Work Processes

Competition requires companies to structure their business processes to be as effective and customer-oriented as possible. ERP vendors do research to define the best business processes. They gather requirements of leading companies within the same industry and combine them with findings from research institutions and consultants. The individual application modules included in the ERP system are then designed to support these best practices, the most efficient and effective ways to complete a business process. As a result, implementation of an ERP system ensures good work processes based on **best practices**. For example, for managing customer payments, the ERP system's finance module can be configured to reflect the most efficient practices of leading companies in an industry. This increased efficiency ensures that everyday business operations follow the optimal chain of activities, with all users supplied the information and tools they need to complete each step.

best practices

the most efficient and effective ways to complete a business process

Increase in Access to Data for Operational Decision Making

ERP systems operate via an integrated database and use essentially one set of data to support all business functions. So, decisions on optimal sourcing or cost accounting, for instance, can be run across the enterprise from the start, rather than looking at separate operating units and then trying to coordinate that information manually or reconciling data with another application. The result is an organization that looks seamless, not only to the outside world but also to the decision makers who are deploying resources within the organization.

The data is integrated to provide excellent support for operational decision making and allows companies to provide greater customer service and support, strengthen customer and supplier relationships, and generate new business opportunities. For example, once a salesperson makes a new sale, the business data captured during the sale is available to the financial, sales, distribution and manufacturing business functions in other departments.

Upgrade of Technology Infrastructure

An ERP system provides an organization with the opportunity to upgrade and simplify the information technology it employs. In implementing ERP, a company must determine which hardware, operating systems, and databases it wants to use. Centralizing and formalizing these decisions enables the organization to eliminate the hodgepodge of multiple hardware platforms, operating systems, and databases it is currently using—most likely from a variety of vendors. Standardization on fewer technologies and vendors reduces ongoing maintenance and support costs as well as the training load for those who must support the infrastructure.

Expense and Time in Implementation

Getting the full benefits of ERP is not simple or automatic. Although ERP offers many strategic advantages by streamlining a company's transaction processing system, ERP is time consuming, difficult, and expensive to implement. Some companies have spent years and tens of millions of dollars implementing ERP systems. And when there are problems with an ERP implementation, it can be expensive. A $350 million maker of home textile furnishings, for example, blamed a 24 percent drop in its sales in one quarter on problems with its ERP system.[20]

Difficulty Implementing Change

In some cases, a company has to make radical changes in how it operates to conform with the work processes (best practices) supported by the ERP. These changes can be so drastic to longtime employees that they retire or quit rather than go through the change. This exodus can leave a firm short of experienced workers.

Difficulty Integrating with Other Systems

Most companies have other systems that must be integrated with the ERP. These systems can include financial analysis programs, Internet operations, and other applications. Many companies have experienced difficulties making these other systems operate with their ERP.

Risks in Using One Vendor

The high cost to switch to another vendor's ERP system makes it extremely unlikely that a firm will do so. Thus, the initial ERP vendor knows it has a "captive audience" and there is less incentive to listen to and respond to customer issues. The high cost to switch also creates a high level of risk if the ERP vendor allows its product to become outdated or goes out of business. Picking an ERP system involves not just choosing the best software product but also the right long-term business partner.

Even with the high cost, long installation times, and complexity, there is no indication that the enthusiasm for this powerful software is slowing. Companies are continuing to look to ERP systems for increased business competitiveness.

INFORMATION SYSTEMS IN ACTION
The Future of Transaction Processing in Financial Services

Kathleen S. Hartzel, *Duquesne University*
Demetrios D. Mahramas, *Deloitte Consulting*

In no other industry are the rules surrounding the exchange of information changing as quickly as they are in financial services. Expectations of instant access to information, increasing stock market activity, growing complexity in managing risk and liquidity, mounting profit pressure, and the pervasive influence of the Internet are among the forces demanding a new model for transaction processing. In response to these market pressures, the industry is embracing the idea of *Straight Through Processing (STP)*. STP is the real-time, automated flow of information among systems both within and outside an organization.

The adoption of STP architecture is often not possible given the current information technology infrastructure of most financial service industry members. Traditionally, their systems are built in separate product-based systems, decentralized processes are used, and batch processing is common. These infrastructure choices limit an organization's ability to provide accurate and meaningful data in a timely manner. An STP environment is based on key principles that include automating and simplifying processes, reducing redundancy, and increasing the availability of essential data through a centrally shared, real-time database. In an effort to establish an STP environment, an organization must redefine the way it processes transactions by implementing an enterprisewide model for efficient trade processing. This precursor to STP is what Deloitte Consulting calls the Strategic Processing Environment.

The real-time automated flow of information in the Strategic Processing Environment incorporates rules-based processes that replace most human intervention and thus shortens transaction processing times and reduces human error. External information is exchanged between companies via an electronic trade-processing service. This electronic intermediary standardizes and validates organizations' transactions. Real-time matching and confirmation of transactions occurs, and the status of trades is available continuously. Thus, operating costs and the number of unsettled claims are reduced. In addition, centralizing trade processing applications reduces data redundancy and allows critical information to be used across product offerings, which in turn can lead to increased revenues. Furthermore, internal communication is improved because of the standardized information required by a single data source.

The benefits of a Strategic Processing Environment are significant. Information is more timely, accurate, and comprehensive. In addition, the organization can provide better customer service; costs and risks are reduced; and its information technology infrastructure is more flexible and scalable. Overall, by establishing a Strategic Processing Environment, a securities organization has built an e-business infrastructure to better differentiate itself and excel within the financial services industry.

● SUMMARY

PRINCIPLE • E-commerce is a new way of conducting business that presents both opportunities for improvement and potential problems.

Businesses and individuals use e-commerce to reduce transaction costs, speed the flow of goods and information, improve the level of customer service, and enable the close coordination of actions among manufacturers, suppliers, and customers. E-commerce also enables consumers and companies to gain access to worldwide markets. The nature of the Web allows firms to gather much more information about customer behavior and preferences than they could with other marketing approaches. Detailed information about a customer's behavior, preferences, needs, and buying patterns allow companies to set prices, negotiate terms, tailor promotions, add product features, and otherwise customize a relationship with a customer.

• • •

The rapid development of e-commerce presents great challenges to society, however. Even though e-commerce is creating new job opportunities, it could also cause a loss of employment in some traditional jobs such as order processing or customer service areas. And as we are already seeing, many companies may fail in the intense competitive environment of e-commerce.

• • •

In business-to-business (B2B) e-commerce, the participants are organizations. Manufacturers are joining electronic exchanges in which they can join with competitors and suppliers to use computers and Web sites to buy and sell goods, trade market information, and run back-office operations such as inventory control. They are also using e-commerce to improve the efficiency of the selling process by moving customer queries about product availability and prices online. In business-to-consumer (B2C) e-commerce, customers deal directly with an organization and avoid any intermediaries. Electronic retailing (e-tailing) is the direct sale from business to consumer through electronic storefronts designed around an electronic catalog and shopping cart model. The Internet has revolutionized the world of investment and finance, especially on-line stock trading and on-line banking. The Internet has also created many options for electronic auctions, in which geographically dispersed buyers and sellers can come together.

PRINCIPLE • E-commerce requires the careful planning and integration of a number of technology components.

A number of technology components must be chosen and integrated to support a large volume of transactions with customers, suppliers, and other business partners worldwide. These components include hardware, Web server software, e-commerce software, and network and packet switching.

• • •

A number of electronic payment systems are used to support e-commerce, including electronic cash; electronic wallets; and smart, credit, charge, and debit cards.

PRINCIPLE • Organizations must define and execute a strategy to be successful in e-commerce.

Most people agree that an effective Web site is one that creates an attractive presence and meets the many and varied needs of its visitors. E-commerce start-ups must decide whether they will build and operate the Web site themselves or outsource this function. Web site hosting services and storefront brokers provide alternatives to building your own Web site. It is also critical to build traffic to your Web site by registering a domain name that is relevant to your business, making your site search-engine-friendly by including a meta tag in your home page, and using Web site traffic data analysis software to attract additional customers.

PRINCIPLE • An organization's transaction processing system (TPSs) must support the routine, day-to-day activities that occur in the normal course of business and help a company add value to its products and services.

TPSs consist of all the components of a CBIS, including databases, telecommunications, people, procedures, software, and hardware devices to process transactions. All TPSs perform the following basic activities: Data collection involves the capture of source data needed to complete a set of transactions; data editing checks for data validity and completeness; data correction involves providing feedback of a potential problem and enabling users to change the data; data manipulation is the performance of calculations, sorting, categorizing, summarizing, and storing for further processing; data storage involves placing transaction data into one or more databases; and document production involves outputting records and reports.

• • •

The methods of transaction processing systems include batch and on-line. Batch processing involves the collection of transactions into batches, which are entered into the system at regular intervals as a group. On-line transaction processing (OLTP) allows transactions to be entered as they occur.

• • •

The order processing systems include order entry, sales configuration, shipment planning, shipment execution, inventory control, invoicing, customer interaction, and routing and scheduling.

PRINCIPLE • **Implementation of an enterprise resource planning (ERP) system enables a company to achieve numerous business benefits through the creation of a highly integrated set of systems.**

Enterprise resource planning (ERP) software is a set of integrated programs that manage a company's vital business operations for an entire multisite, global organization. It must be able to support multiple legal entities, multiple languages, and multiple currencies. Although the scope of an ERP system may vary from vendor to vendor, most ERP systems provide integrated software to support manufacturing and finance. In addition to these core business processes, some ERP systems are capable of supporting additional business functions such as human resources, sales, and distribution.

• • •

Implementation of an ERP system can provide many advantages, including elimination of costly, inflexible legacy systems; providing improved work processes; providing access to data for operational decision making; and creating the opportunity to upgrade technology infrastructure. Some of the disadvantages associated with ERP systems are that they are time-consuming, difficult, and expensive to implement.

● REVIEW QUESTIONS

1. Define *e-commerce*. What is the expected size of the worldwide e-commerce market?
2. What sort of business processes are good candidates for conversion to e-commerce?
3. What is e-commerce supply chain management, and what is its goal?
4. What are some of the business opportunities presented by business-to-business e-commerce? What are some of the key issues?
5. What is an electronic exchange? How does it work? What are some of the issues associated with the use of electronic exchanges?
6. What are some of the advantages of using a service like Bigstep.com or eCongo.com to set up a Web site? What are some of the disadvantages?
7. Outline the steps a firm needs to take to adapt an existing U.S.-centric Web site to another language and culture.

8. What role do digital certificates play in e-commerce?
9. Briefly explain the differences among smart, credit, charge, and debit cards.
10. What is technology-enabled relationship management?
11. Describe the basic activities common to all transaction processing systems (TPSs).
12. List several characteristics of transaction processing systems.
13. What is an enterprise resource planning system?
14. List and briefly discuss the business objectives common to all TPSs.
15. What is the difference between batch processing and on-line processing systems?
16. What is business resumption planning?
17. What systems are included in the order processing family of systems?

● DISCUSSION QUESTIONS

1. Why is it important for effective e-commerce applications such as order taking to be tightly integrated to applications such as inventory control and production planning?
2. Why hasn't on-line banking caught on more quickly than it has?
3. Wal-Mart, the world's number one retail chain, has turned down several invitations to join

exchanges in the retail and consumer goods industries. Is this good or bad for the overall U.S. economy? Why?
4. How would you decide whether to use a Web service or a storefront broker to set up a Web site for a small business you want to run out of your home?

5. Briefly discuss actions e-commerce companies have taken to ensure the legitimacy of their customers. Which of these approaches do you think is best? Why?

6. Discuss the pros and cons of e-commerce companies capturing data about you as you visit their sites.

7. Assume that you are the owner of a small business. Describe the day-to-day transaction processing activities that you would encounter.

8. Your company is a medium-sized firm with sales of $500 million per year. It has been decided that the organization will implement an ERP system to support all operations at headquarters, three plants, and four distribution centers. What are some of the key questions that must be answered to further define the scope of this effort?

9. What is the advantage of implementing ERP as an integrated solution to link multiple business processes? What are some of the issues and potential problems?

10. You are building your firm's first-ever customer interaction system. Discuss the features you would design into the system. How might you include suggestions from your customers into your design?

11. You are in charge of a complete overhaul of your firm's order processing systems. How would you define the requirements for this collection of systems? What features would you want to include? What steps would you use to plan for a potential disaster?

● PROBLEM-SOLVING EXERCISES

 1. Imagine that you work for a company like DoubleClick that captures information about e-commerce shoppers for use in marketing. Develop a list of the data items that represents the data you would ideally like to capture for each e-commerce shopper. Develop a second list of the data you would like to capture about each visit of an e-commerce shopper to a specific Web site.

 2. Do research to get current data about the growth of B2B or B2C e-commerce—either in the United States or worldwide. Use a graphics software package to create a line graph representing this growth. Extend the growth line ten years beyond the available data using two different modeling tools available with the software package. Write a paragraph discussing the accuracy of your ten-year projection and the likelihood that e-commerce will achieve this forecast.

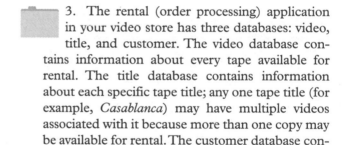 3. The rental (order processing) application in your video store has three databases: video, title, and customer. The video database contains information about every tape available for rental. The title database contains information about each specific tape title; any one tape title (for example, *Casablanca*) may have multiple videos associated with it because more than one copy may be available for rental. The customer database contains each customer's ID number and address.

Specific fields in each database are listed in the accompanying table. Key fields are indicated with an asterisk.

Title	Video	Customer
TapeNumber*	VideoNumber*	CustomerNumber*
Title	TapeNumber	CustomerName
Category	Status	Address
Rating		Phone Number
Rental Rate		Rental Amount Y-T-D
Rental Amount Y-T-D		

Build a simple transaction processing system to support the video store operation. Enter the complete database definitions into your database management software and create a data entry screen to allow store personnel to efficiently enter customer rentals and returns. This screen must include basic information such as tape name and number, customer name and number, date out, date returned, and charge for rental per day. The data from the screen updates all appropriate data in each database.

a. Enter several of your favorite movies to create at least ten entries for the title database.

b. For each title entry, create one to five entries in the video database.

c. Make up at least ten entries for the customer database.

d. Enter the data necessary to handle the checkout of at least six specific videos by different customers. Can your simple transaction processing system handle a situation in which one customer checks out more than one video at a time? What happens if a customer wants to check out a video but there are no copies remaining?

e. Check to see whether the Y-T-D fields in the title and customer databases are updated correctly.

● TEAM ACTIVITIES

1. As a team, choose an idea for a Web site—products or services you would provide. Develop an implementation plan that outlines the steps you need to take and the decisions you must make to set up the Web site and make it operational.

2. Assume that your team owns a T-shirt company. Assign each team member to a transaction processing activity. Describe the detailed activities that would be needed for your T-shirt company.

3. As a team, interview a business owner about the company's transaction processing system. Develop a report that describes this company's transaction processing system.

● WEB EXERCISES

1. Some of the most private areas of your life are those surrounding the healthcare of you and your family. Details such as medical history, prescriptions, insurance—even your basic health and beauty aid purchases—are no one's business but your own. Access the Web sites of several on-line drugstores, such as Drug Emporium, Walgreens, and CVS, and read their privacy policies and perform an assessment of their security. How do you rate these sites?

2. CyberCash provides e-commerce technologies and services, enabling commerce across the entire spectrum from electronic retailing to the Internet. CyberCash provides a complete line of software products and services allowing merchants, financial institutions, and consumers to conduct secure transactions and other e-commerce functions using a broad array of popular payment forms. Visit the CyberCash Web site at http://www.cybercash.com and become familiar with its products and services. Find the Web site of a CyberCash competitor and compare and contrast the products and services of the two companies.

3. A number of companies, including SAP, sell enterprise resource planning (ERP) software to coordinate a company's transaction processing system. Search the Internet to get more information about ERP or one of the companies that makes and sells this powerful software. You may be asked to develop a report or send an e-mail message to your instructor about what you found.

● CASES

1 Starting a Procurement Business

Companies are starting partnerships to form Internet exchanges. Although some exchanges are general purpose, most are oriented toward the procurement function of transaction processing. In supporting these exchanges, businesses hope to save significant time and money in purchasing the parts, supplies, and services needed to manufacture their products—ranging from cars to agricultural products. In addition to companies forming these strategic alliances, others are developing procurement businesses and Internet exchanges for the general market. One example of this move is the alliance of Chase Manhattan Bank and Deloitte Consulting.

In early 2000, Chase Manhattan Bank and Deloitte Consulting decided to form a new company. The overall objective of the new firm is to offer Internet-based procurement services to larger, Fortune 1000 companies. Chase and Deloitte hope to make a profit from the new company and also save money for participating companies. On average, Chase and Deloitte forecast that each company participating in the new Internet procurement system should save from $200 million to $350 million annually. The cost savings are expected to come from combining the purchasing needs of all participating companies to obtain better volume discounts.

The new Internet procurement firm, however, is still in the planning stage. The name of the new firm, the technology to be used, and who will head the new company have not been determined.

Discussion Questions

1. Will this new Internet procurement firm be attractive to the average Fortune 1000 company?
2. Do you think that other companies will also try to start Internet procurement companies?

Critical Thinking Questions

3. Assume that you are the manager of a Fortune 1000 manufacturing company. Discuss the advantages and disadvantages of this approach to procurement. Would you be willing to sign a long-term contract to be a part of this firm?
4. Assume that several Fortune 1000 manufacturing companies are considering forming a strategic partnership to form an Internet exchange. Compare and contrast this approach with the type of Internet procurement firm discussed in this case.

Sources: Adapted from Craig Stedman, "Chase, Deloitte Start Procurement Firm," *Computerworld*, February 14, 2000, p. 20; "Local Government Equals Big Business," *Computing*, March 2, 2000, p. 4; and Clinton Wilder et al., "Sabre to Open Internet Marketplace," *Information Week*, March 6, 2000.

2 ERP in Mergers and Acquisitions

Enterprise resource planning can offer a company many benefits over traditional transaction processing approaches. By integrating a firm's TPS functions into a unified system, companies can process basic transactions faster, more efficiently, and at a lower cost, in most cases. In addition, because many ERP packages include best practices approaches to running a business, the ERP approach can help a company improve its overall approach to doing business. ERP implementations, however, can be costly and time-consuming.

Lyondell Chemical is a Houston-based company with 9,000 employees and annual revenues of $8 billion. From 1995 to 1997, Lyondell implemented SAP R/3. After the ERP system was operational in 1997, some believed that only maintenance of the system would be required in the future. But this turned out not to be the case. Like other companies, Lyondell is

making acquisitions in its effort to grow. Because of the increased scope of the company, the result was two additional SAP implementations in 1999. According to Robert Tolbert, Lyondell's chief information officer, "If you're merging supply-chain organizations to be efficient, you need to be on the same system to share data." Lyondell decided to implement SAP R/3 in each of the merged companies. As a result, the costs and time required to implement an ERP system had to be incurred several times, and it is likely that Lyondell will continue to have ERP implementation costs as it continues to acquire other companies. Fortunately, Lyondell was able to learn how to implement the ERP systems more easily and efficiently. "After the first couple of times, you have some in-house experience. So the third time you don't need as many consultants," says CIO Tolbert.

Lyondell is not alone. Other companies have also experienced multiple ERP implementations. Celestia, for example, used one ERP program in

Canada and another ERP program in its Asian operations. The company is an IBM spin-off based in Toronto with annual revenues of $3.8 billion. Celestia is now considering the possibility of standardizing on a single ERP system. As with Lyondell, this would require another ERP implementation in either Canada or Asia, with all the accompanying costs and time requirements.

Discussion Questions

1. What are the advantages of implementing ERP in a merger or acquisition?
2. What are some of the disadvantages?

Critical Thinking Questions

3. How would you minimize the cost of implementing multiple ERP systems over time?
4. Assume that you are the chief information officer of Celestia. What factors would you consider in deciding which ERP implementation to use in both Canada and Asia? How would you decide which system to use worldwide?

Sources: Adapted from Erik Sherman, "Early ERP Implementations Have Proved to be Costly and Time-Consuming," *Computerworld*, February 14, 2000, p. 53; "Projects Hot Sheet," *Chemical Week*, January 19, 2000, p. 27; and "SAP Branches Out into Applications Hosting Market," *Network News*, March 1, 2000, p. 4.

Reapplication of General Electric Web Site

General Electric (GE) is a diversified services, technology, and manufacturing company that operates in more than 100 countries and employs nearly 340,000 people worldwide, including 197,000 in the United States. GE Aircraft Engines (GEAE) is a division of GE, with annual sales in excess of $10 billion and 33,000 employees worldwide. It is the world's largest producer of large and small jet engines for commercial and military aircraft. GEAE also supplies engines for boats and provides aviation services. Throughout the 1990s, more than 50 percent of the world's large commercial jet engine orders were awarded to GE or CFM International, a joint company of GE and Snecma of France. W. James McNerney is the president and CEO of GEAE, which is headquartered in Cincinnati.

Under the leadership of its highly respected chairman and CEO, John F. Welch, GE has earned a number of awards as the World's Most Admired Company (by *Fortune* magazine in 1998 and 1999) and the World's Most Respected Company (by the *Financial Times* in 1998 and 1999). However, in May 1999, it became clear that a key e-commerce project was floundering. To fix the situation, McNerney recruited a former Green Beret, John Rosenfeld, to take on part of the failed project, a customer Web center for the complex spare-parts business, and build it at lightning speed. The mandate was to be operational within seven months. This challenging goal was met, and today GEAE's Customer Web Center has the following functions:

- *Spare-parts order management.* Checks parts availability, finds alternative parts, places orders, builds automated parts lists (for parts usually ordered together), checks status of orders, and tracks orders via UPS and FedEx.
- *Component repair and engine overhaul.* Checks status of repair and overhaul work in shops around the world and accesses initial findings reports and cost estimates, including corresponding high-resolution digital photos of damaged parts.
- *Spare-parts warranty.* Submits warranty claims, reviews status of claims, and generates reports.
- *Technical publications.* Finds information from illustrated parts catalogs, engine service manuals, standard practice manuals, fleet highlights, and service bulletins.
- *On-line wizards.* Determines return on investment for engine upgrades (e.g., part X will save Y dollars in maintenance over time frame Z).
- *On-line video training.* Will help users brush up on how to remove, install, or service a part.

Discussion Questions

Assume you are the manager of a critical e-commerce project in the Consumer Appliances division of GE that must be completed as soon as possible. You have traveled to Cincinnati to meet with Jim McNerney and John Rosenfeld to see if you can reapply the GEAE Web site for Consumer Appliances.

1. What questions will you ask about the technical infrastructure and capabilities of the Web site to assess whether it will meet the needs of the Consumer Appliances division?

2. To what degree would customer involvement be necessary to assess the performance of the GEAE Web site?

Critical Thinking Questions

3. What similarities are there between the spare-parts business for aircraft engines and consumer appliances? What are some of the differences?

4. How might you assess whether the GEAE Web site has the functionality needed to support the Consumer Appliances division?

Sources: Adapted from Marcia Stepanek, "How to Jump-Start Your E-Strategy," *Business Week,* June 5, 2000, pp. 96–100; "The $11 Billion Web Start-up," *Computerworld,* May 1, 2000, pp. 56–60; and "GE Fact Sheet," General Electric Web site, http://www.ge.com/factsheet.html, accessed May 26, 2000.

● NOTES

Sources for the opening vignette on p. 173: Adapted from "Investor Relations" and "About Us," Weirton Steel Corporation Web site, http://www.weirton.com, accessed May 4, 2000; MetalSite Web page, press releases, http://www.metalsite.com, accessed May 4, 2000; and Clinton Wilder, "E-Transformation," *Information Week,* September 13, 1999, pp. 44–62.

1. Edward Iwata, "Despite the Hype, B2B Marketplaces Struggle," *USA Today,* May 10, 2000, pp. 1B–2B.
2. David Judson, "Net Links Cut Costs for Firms," *The Cincinnati Enquirer,* June 4, 2000, p. F4.
3. Robert Hof, "E-Malls for Business," *Business Week,* March 3, 2000, pp. 32–34.
4. Clinton Wilder, "AutoNation: A Different Style of Sales Pitch," *Information Week,* September 13, 1999, p. 48.
5. Edward Iwata, "Despite the Hype, B2B Marketplaces Struggle," *USA Today,* May 10, 2000, pp. 1B–2B.
6. Julia King and Jaikumar Vijayan, "It's 'Supplier Beware' Online," *Computerworld,* March 27, 2000, p. 6.
7. Clinton Wilder, "Timken: A Big Step for an Old-Line Industry," *Information Week,* September 13, 1999, p. 50.
8. William Glassgall, "The Investor Revolution," *Business Week Online,* February 22, 1999, accessed at http://www.businessweek.com/search.htm.
9. Ellen Stark, "Banking from Home," Money.com Guide to Online Investing, http://www.money.com/money/onlineinvesting/banking, accessed May 30, 2000.
10. Bethany McLean, "The Fortune-50 Index, More Than Just Dot-Coms," *Fortune,* December 6, 1999, pp. 130–138.
11. "The GM Card to Drive E-Commerce with Introduction of the GM Card easyPay Electronic Wallet," News Releases section of the Cybercash Web site, http://www.cybercash.com/cybercash/company/, accessed May 13, 2000.
12. Geoffrey Smith, "New Money for the Net," *Business Week,* June 5, 2000, p. EB 14.
13. Marcia Stepanek, "How to Jump-Start Your E-Strategy," *Business Week,* June 5, 2000, pp. 96–100.
14. Neil Randall, "Setting Up Shop Online," *PC Magazine,* November 16, 1999, pp. 137–154.
15. Jaikumar Vijayan, "Manufacturing Group Launches B-to-B Hub," *Computerworld,* March 20, 2000, p. 16.
16. Alorie Gilbert, " ERP Vendors Look for Rebound after Slowdown," *Information Week,* February 14, 2000, p. 156.
17. Justin Fox, "Lumbering toward B2B," *Fortune,* June 12, 2000, p. 257.
18. Craig Stedman, "Meshing CRM, Business Poses Challenges," *Computerworld,* March 27, 2000, p. 50.
19. Jaikumar Vijayan, "ERP Vendors Admit They Can't Do It All," *Computerworld,* April 3, 2000, p. 2.
20. Alorie Gilbert, "A Question of Convenience," *Information Week,* February 21, 2000, p. 34.

Information and Decision Support Systems

*O*ur job was to give the
Emergency Preparedness
Division and the governor's
office the information they
needed to make decisions.

— Mark Hunter, assistant state
maintenance engineer for the
South Carolina Department of
Transportation, describing the
Hurricane Evacuation Decision
Support System

Principles

Learning Objectives

The management information system (MIS) must provide the right information to the right person in the right fashion at the right time.

- *Outline and briefly describe the stages of a problem-solving process.*
- *Define the term MIS and clearly distinguish between a TPS and an MIS.*
- *Discuss information systems in the functional areas of business organizations.*

Decision support systems (DSSs) are used when the problems are more unstructured.

- *List and discuss important characteristics of DSSs that give them the potential to be effective management support tools.*
- *Identify and describe the basic components of a DSS.*

Specialized support systems, such as group decision support systems (GDSSs) and executive support systems (ESSs), use the overall approach of a DSS in situations such as group and executive decision making.

- *State the goal of a GDSS and identify the characteristics that distinguish it from a DSS.*
- *Identify fundamental uses of an ESS and list the characteristics of such a system.*

Reuters Group

Providing Information and Decision Support on the Internet

If information is power, then Reuters Group is one of the chief sources of that power. It is also one of the oldest information sources. Founded almost 150 years ago, the company originally was a news agency, feeding stories and information to newspapers. About 20 years ago, Reuters saw a need to provide financial information to banks, brokerage companies, and wealthy individuals. Companies and people needed information for their investment and banking decisions, and Reuters decided to fill the need. Today, the company provides decision support for a half million users at about 60,000 companies around the world. Providing information and decision support on a wholesale basis to other companies and individuals generates revenues of $5 billion annually for the company. The average user pays $1,500 per month to get access to the vast financial information contained in Reuters's databases. In addition to financial information, Reuters still operates a news service business for newspapers, magazines, broadcasters, and Internet portals.

Unfortunately, the Internet has become the main competition for Reuters. Although the information on Internet sites may not be as complete and the decision support may not be as comprehensive as Reuters's databases, many Internet sites are free to users. Internet users have been happy to get free information, but Reuters stockholders have not. The stock price for Reuters has languished at about $50 per share, while other stock prices have generally soared. Until recently, Reuters did not have a coherent strategy for the future or a solution to its sagging stock price. The confusion ended, however, with Peter Job, the new chief executive of Reuters.

Peter Job decided that Reuters had to get on the Internet bandwagon. He decided to invest more than $800 million over four years to develop a comprehensive Internet site to sell and distribute information directly to individuals in addition to its traditional corporate customers. "Why does the individual have to get a different deal than the institution?" asked Job. The initiative has made individual customers and stockholders happy. The stock price of Reuters recently reached $137 a share after the announcement. Although most stockholders are happy with their new wealth, some are concerned. Will the move to the Internet mean that the company can't generate the hefty fees from its corporate clients?

As you read this chapter, consider the following:

- How valuable is the information and decision support provided by Reuters to its corporate clients? Do you think the information is worth the average $1,500 monthly cost to most users?

- What are the disadvantages of providing similar information and decision support to individuals over the Internet?

Information is the lifeblood of today's organizations. Thanks to information systems, managers and employees can obtain useful information in real time. As we saw in the preceding chapter, the TPS captures a wealth of data. When this data is filtered and manipulated, it can become powerful information and decision support for managers and employees. The ultimate goal of management information and decision support systems is to help managers and executives make better decisions and solve important problems. The results can be increased revenues, reduced costs, and the realization of corporate goals. We begin by investigating decision making and problem solving.

DECISION MAKING AND PROBLEM SOLVING

Every organization needs effective decision making to reach its objectives and goals. In most cases, strategic planning and the overall goals of the organization set the stage for value-added processes and the decision making required to make them work. Often, information systems assist with strategic planning and problem solving.

Decision Making as a Component of Problem Solving

In business, one of the highest compliments you can get is to be recognized by your colleagues and peers as a "real problem solver." Problem solving is a critical activity for any organization because decisions are made daily on issues ranging from routine tasks to strategic decisions. We examine the problem-solving process here so that you can be better prepared for your future career—in whatever field you choose. Once a problem has been identified, the problem-solving process begins with decision making. A well-known model developed by Herbert Simon divides the **decision-making phase** of the problem-solving process into three stages: intelligence, design, and choice. This model was later incorporated into an expanded model of the entire problem-solving process (Figure 6.1).

The first stage in the problem-solving process is the **intelligence stage**. During this stage, potential problems or opportunities are identified and defined. Information is gathered that relates to the cause and scope of the problem.

decision-making phase

the first part of problem solving, including three stages: intelligence, design, and choice

intelligence stage

The first stage of decision making, during which potential problems and opportunities are identified and defined

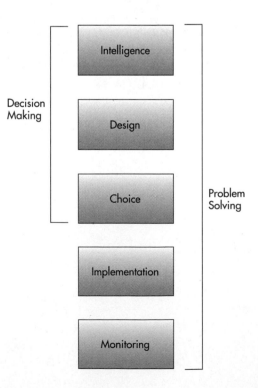

FIGURE 6.1

How Decision Making Relates to Problem Solving

The three stages of decision making—intelligence, design, and choice—are augmented by implementation and monitoring to result in problem solving.

During the intelligence stage, resource and environmental constraints are investigated. For example, exploring the possibilities of shipping tropical fruit from a farm in Hawaii to stores in Michigan would be done during the intelligence stage. The perishability of the fruit and the maximum price consumers in Michigan are willing to pay for the fruit are problem constraints. Aspects of the problem that must be considered in this case include federal and state regulations regarding shipment of food products.

In the **design stage**, alternative solutions to the problem are developed. In addition, the feasibility of these alternatives is evaluated. In our tropical fruit example, the alternative methods of shipment, including the transportation times and costs associated with each, would be considered. During this stage the problem solver might determine that shipment by freighter to California and then by truck to Michigan is not feasible because the fruit would spoil.

The last stage of the decision-making phase, the **choice stage**, requires selecting a course of action. In our tropical fruit example, the Hawaiian farm might select the method of shipping by air to Michigan as its solution. The choice stage would then conclude with selection of the actual air carrier. As we will see later, various factors influence choice; the apparently easy act of choosing is not as simple as it might first appear.

Problem solving includes and goes beyond decision making. It also includes the **implementation stage**, when the solution is put into effect. For example, if the Hawaiian farmer's decision is to ship the tropical fruit to Michigan as air freight on a specific carrier, implementation involves informing the farming staff of the new shipping method, getting the fruit to the airport, and actually shipping the product to Michigan.

The final stage of the problem-solving process is the **monitoring stage**. In this stage, decision makers evaluate the implementation to determine whether the anticipated results were achieved and to modify the process if not. Monitoring can involve both feedback and adjustment. For example, after the first fruit shipment, the Hawaiian farmer might learn that the air freight route often makes a stopover in Phoenix, Arizona, where the plane sits exposed on the runway for a number of hours while loading additional cargo. If this fluctuation in temperature and humidity makes the fruit spoil, the farmer might have to readjust his solution to include a new air freight firm that does not make a stopover, or perhaps he would consider a change in fruit packaging.

Programmed versus Nonprogrammed Decisions

In the choice stage, many factors influence the decision maker's selection of a solution. One such factor is whether the decision can be programmed. **Programmed decisions** are made using a rule, procedure, or quantitative method. For example, ordering inventory when inventory levels drop to 100 units is a rule. Programmed decisions are easy to computerize using traditional information systems. It is simple, for example, to program a computer to order more inventory when inventory levels for a certain item reach 100 units or fewer. Most of the processes automated through transaction processing systems share this characteristic: the relationships between system elements are fixed by rules, procedures, or numerical relationships. Management information systems are also used to solve programmed decisions by providing reports on problems that are routine and well defined (structured problems).

---

design stage

the second stage of decision making, during which alternative solutions to the problem are developed

choice stage

the third stage of decision making, which requires selecting a course of action

problem solving

a process that goes beyond decision making to include the implementation stage

implementation stage

the stage of problem solving during which a solution is put into effect

monitoring stage

the final stage of the problem-solving process, during which decision makers evaluate the implementation

programmed decisions

decisions made using a rule, procedure, or quantitative method

---

Ordering more inventory when inventory levels drop to specified levels is an example of a programmed decision.
(Source: Stone/Mitch Kezar.)

nonprogrammed decisions

decisions that deal with unusual or exceptional situations

Nonprogrammed decisions, however, deal with unusual or exceptional situations. In many cases, these decisions are difficult to quantify. Determining the appropriate training program for a new employee, deciding whether to start a new product line, and weighing the benefits and drawbacks of installing a new pollution control system are examples. Each of these decisions contains unique characteristics to which rules or procedures may not obviously apply. Today, decision support systems are used to solve a variety of nonprogrammed decisions, in which the problem is not routine and rules and relationships are not well defined (unstructured or ill-structured problems).

Optimization, Satisficing, and Heuristic Approaches

optimization model

a process to find the best solution, usually the one that will best help the organization meet its goals

In general, computerized decision support systems can either optimize or satisfice. An **optimization model** finds the best solution, usually the one that will best help the organization meet its goals. For example, an optimization model can find the appropriate number of products an organization should produce to meet a profit goal, given certain conditions and assumptions. Optimization models use problem constraints. A limit on the number of available work hours in a manufacturing facility is an example of a problem constraint. Some spreadsheet programs, such as Excel, have optimizing features (Figure 6.2).

satisficing model

a model that will find a good—but not necessarily the best—problem solution

A **satisficing model** is one that finds a good—but not necessarily the best—problem solution. Satisficing is usually used because modeling the problem precisely enough to get an optimal decision would be too difficult, complex, or costly. Satisficing normally does not look at all possible solutions but only at those likely to give good results. Consider a decision to select a location for a new plant. To find the optimal (best) location, you would have to consider all cities in the United States or the world. A satisficing approach would be to consider only five or ten cities that might satisfy the company's requirements. Limiting the options may not result in the best decision, but it will likely result in a good decision, without spending the time and effort to investigate all cities. Satisficing is a good alternative modeling method because it is sometimes too expensive to analyze every alternative to get the best solution.

heuristics

commonly accepted guidelines or procedures that usually find a good solution

Heuristics, often referred to as "rules of thumb"—commonly accepted guidelines or procedures that usually find a good solution—are very often used in decision making. An example of a heuristic rule in e-commerce is "the first company to market will gain the greatest name recognition and ultimate

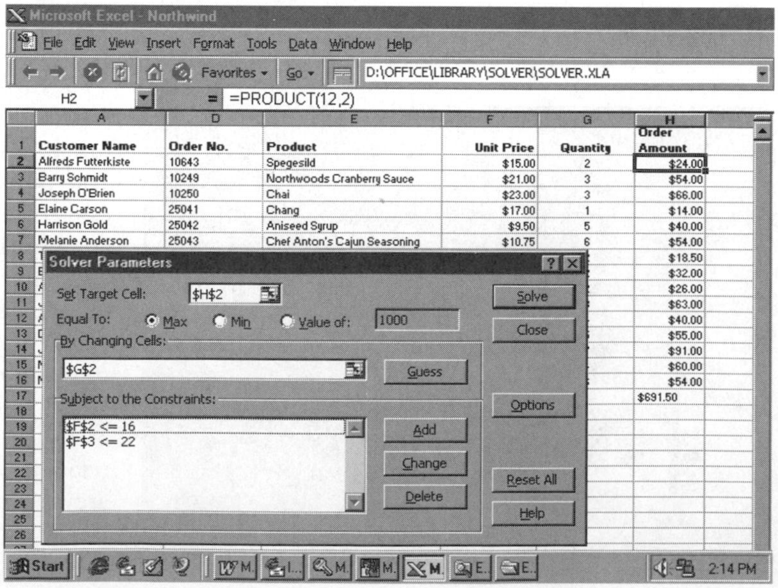

FIGURE 6.2

Some spreadsheet programs, such as Excel, have optimizing routines. This figure shows Solver, which can find an optimal solution given certain constraints.

profitability." A heuristic that baseball team managers use is to place batters most likely to get on base at the top of the lineup, followed by the power hitters who'll drive them in to score. An example of a heuristic used in business is to order four months' supply of inventory for a particular item when the inventory level drops to 20 units or fewer; even though this heuristic may not minimize total inventory costs, it may be a very good rule of thumb to avoid stockouts without too much excess inventory. One way to achieve better decision making and problem solving is through a management information system.

AN OVERVIEW OF MANAGEMENT INFORMATION SYSTEMS

Management information systems (MISs) can often give companies a competitive advantage by providing the right information to the right people in the right format and at the right time. In many cases, companies and individuals are willing to pay companies like Reuters for this type of information.

Management Information Systems in Perspective

The primary purpose of an MIS is to help an organization achieve its goals by providing managers with insight into the regular operations of the organization so that they can control, organize, and plan more effectively and efficiently. In short, an MIS provides managers with information and support for effective decision making and provides feedback on daily operations. An MIS adds value to processes within an organization. A manufacturing MIS, for example, is a set of integrated systems that can help managers monitor a manufacturing process to maximize the value of raw materials as they are assembled into finished products. For the most part, this monitoring is accomplished through summary reports produced by the MIS. These reports can be obtained by analyzing and summarizing the detailed data contained in transaction processing databases and presenting meaningful results to managers. These reports support managers by providing them with data and information for decision making. Figure 6.3 shows the role of MISs within the flow of an organization's information. Note that business transactions can enter the organization through traditional methods or via the Internet or an extranet connecting customers and suppliers to the firm's transaction processing systems.

As Figure 6.3 shows, the summary reports from the MIS are just one of many sources of information available to managers, and management information systems are used at all levels within an organization.

Each MIS is an integrated collection of subsystems, which are typically organized along functional lines within an organization. Thus, a financial MIS includes subsystems that address financial reporting, profit and loss analysis, cost analysis, and management of funds. Many functional subsystems share hardware resources, data, and often even personnel. But some subsystems are self-contained within one functional area and are specialized for its purposes. The overall efficiency of the MIS can be improved by integrating subsystems. For instance, two distinct functional departments may collect and maintain similar data (e.g., customer lists maintained by both sales and accounting). Or perhaps hardware resources are only partially used by one functional area and might be shared by another. Like other corporate resources, the investment in the MIS should be maximized by decreasing waste and underutilization. But the key goal of providing the right information to the right person in the right fashion at the right time should never be forgotten in trying to increase efficiencies, or the purpose of the MIS will be lost.

Inputs to a Management Information System

Data that enters an MIS originates from both internal and external sources (Figure 6.3). The most significant internal source of data for an MIS is the organization's various TPSs. One of the major activities of the TPS is to capture and store the data from ongoing business transactions. With every business transaction, TPS applications change and update the organization's databases. For example, the billing application helps keep the accounts receivable database up to date so that managers know who owes the company money. These updated databases are a primary internal source of data for the management information system. In companies that have implemented an ERP system, the collection of databases associated with this system are an important source of internal data for the MIS. Other internal data comes from specific functional areas throughout the firm. External sources of data can include customers, suppliers, competitors, and stockholders whose data is not already captured by the TPS, as well as other sources, such as the Internet. In addition, many companies have implemented extranets to link them to external organizations to exchange data and information.

The MIS uses the data obtained from these sources and processes it into information more usable to managers, primarily in the form of routine reports. For example, rather than simply obtaining a chronological list of sales activity over the past week, a national sales manager might obtain her organization's weekly sales data in a format that allows her to see sales activity by region, by local sales representative, by product, and even in comparison with last year's sales.

FIGURE 6.3

Sources of Managerial Information

The MIS is just one of many sources of managerial information. Decision support systems, executive support systems, and expert systems also assist in decision making.

Outputs of a Management Information System

The output of most management information systems is a collection of reports that are distributed to managers. These reports include scheduled reports, key-indicator reports, demand reports, exception reports, and drill down reports (Figure 6.4).

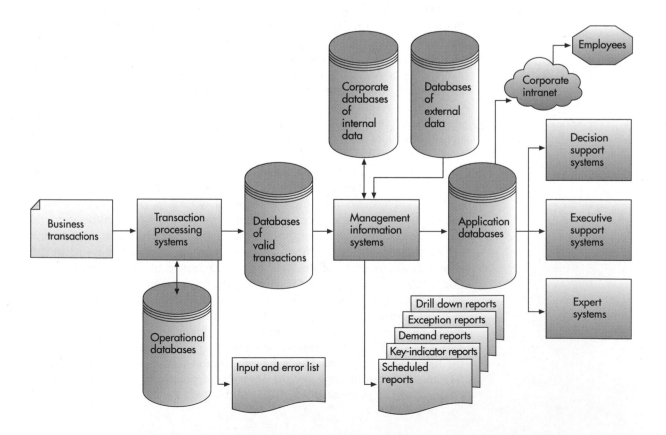

FIGURE 6.4

Reports Generated by an MIS

The five types of reports are
(a) scheduled, (b) key-indicator,
(c) demand, (d) exception, and
(e–h) drill down.
(Source: George W. Reynolds,
Information Systems for Managers,
3rd ed., St. Paul, MN: West
Publishing Co., 1995.)

(a) Scheduled Report

Daily Sales Detail Report

Prepared: 08/10/XX

Order #	Customer ID	Salesperson ID	Planned Ship Date	Quantity	Item #	Amount
P12453	C89321	CAR	08/12/96	144	P1234	$3,214
P12453	C89321	CAR	08/12/96	288	P3214	$5,660
P12454	C03214	GWA	08/13/96	12	P4902	$1,224
P12455	C52313	SAK	08/12/96	24	P4012	$2,448
P12456	C34123	JMW	08/13/96	144	P3214	$ 720
....	....	....	....	....	....	....

(b) Key-Indicator Report

Daily Sales Key Indicator Report

	This Month	Last Month	Last Year
Total Orders Month to Date	$1,808	$1,694	$1,914
Forecasted Sales for the Month	$2,406	$2,224	$2,608

(c) Demand Report

Daily Sales by Salesperson Summary Report

Prepared: 08/10/XX

Salesperson ID	Amount
CAR	$42,345
GWA	$38,950
SAK	$22,100
JWN	$12,350
....	....
....	....

(d) Exception Report

Daily Sales Exception Report—Orders Over $10,000

Prepared: 08/10/XX

Order #	Customer ID	Salesperson ID	Planned Ship Date	Quantity	Item #	Amount
P12345	C89321	GWA	08/12/96	576	P1234	$12,856
P22153	C00453	CAR	08/12/96	288	P2314	$28,800
P23023	C32832	JMN	08/11/96	144	P2323	$14,400
....	....	....	....	....	....	....
....	....	....	....	....	....	....

continued

Scheduled Reports

Scheduled reports are produced periodically, or on a schedule, such as daily, weekly, or monthly. For example, a production manager could use a weekly summary report that lists total payroll costs to monitor and control labor and job costs. A manufacturing report produced once a day to monitor the production of a new product is another example of a scheduled report. Other scheduled reports can help managers control customer credit, the performance of sales representatives, inventory levels, and more.

A **key-indicator report** summarizes the previous day's critical activities and is typically available at the beginning of each workday. These reports can summarize inventory levels, production activity, sales volume, and the like. Key-indicator reports are used by managers and executives to take quick, corrective action on significant aspects of the business.

scheduled reports

reports produced periodically, or on a schedule, such as daily, weekly, or monthly

key-indicator report

summary of the previous day's critical activities; typically available at the beginning of each workday

FIGURE 6.4 (continued)

Reports Generated by an MIS

(e) First-Level Drill Down Report

Earnings by Quarter (Millions)			
	Actual	Forecast	Variance
2nd Qtr. 1999	$12.6	$11.8	6.8%
1st Qtr. 1999	$10.8	$10.7	0.9%
4th Qtr. 1998	$14.3	$14.5	-1.4%
3rd Qtr. 1998	$12.8	$13.3	-3.8%

(f) Second-Level Drill Down Report

Sales and Expenses (Millions)			
Qtr: 2nd Qtr. 1999	Actual	Forecast	Variance
Gross Sales	$110.9	$108.3	2.4%
Expenses	$ 98.3	$ 96.5	1.9%
Profit	$ 12.6	$ 11.8	6.8%

(g) Third-Level Drill Down Report

Sales by Division (Millions)			
Qtr: 2nd Qtr. 1999	Actual	Forecast	Variance
Beauty Care	$ 34.5	$ 33.9	1.8%
Health Care	$ 30.0	$ 28.0	7.1%
Soap	$ 22.8	$ 23.0	-0.9%
Snacks	$ 12.1	$ 12.5	-3.2%
Electronics	$ 11.5	$ 10.9	5.5%
Total	$110.9	$108.3	2.4%

(h) Fourth-Level Drill Down Report

Sales by Product Category (Millions)			
Qtr: 2nd Qtr. 1999 Division: Health Care	Actual	Forecast	Variance
Toothpaste	$12.4	$10.5	18.1%
Mouthwash	$ 8.6	$ 8.8	-2.3%
Over-the-Counter Drugs	$ 5.8	$ 5.3	9.4%
Skin Care Products	$ 3.2	$ 3.4	-5.9%
Total	$30.0	$28.0	7.1%

Demand Reports

demand reports

reports developed to give certain information at a manager's request

Demand reports are developed to give certain information at a manager's request. In other words, these reports are produced on demand. For example, an executive may want to know the production of a particular item—a demand report can be generated to give the requested information. Other examples of demand reports include reports requested by executives to show the hours worked by a particular employee, total sales to date for a product, and so on.

Exception Reports

exception reports

reports automatically produced when a situation is unusual or requires management action

Exception reports are produced automatically when a situation is unusual or requires immediate action. For example, a manager might set a trigger point that generates a report of all inventory items with fewer than five days of sales on hand. This unusual situation requires prompt action to avoid running out of stock on the item. The exception report generated would contain only items with fewer than five days of sales in inventory. As with key-indicator reports, exception reports are most often used to monitor aspects important to an organization's success. Typically, when an exception report is produced, a manager or executive takes action. Trigger points for an exception report should be set carefully—if they are set too low, they may result in an abundance of exception reports; if they are set too high, critical problems could be overlooked. For example, if a manager wants a report that contains all projects over budget by $100 or more, he may find that almost every company project exceeds its budget by that amount. The $100 trigger point is probably too low. A trigger point of $10,000 might be more appropriate.

Drill Down Reports

drill down reports

reports providing increasingly detailed data about a situation

Drill down reports provide increasingly detailed data about a situation. Through the use of drill down reports, analysts can see data at a high level first (similar to a bag of cookies), then at a more detailed level (say, an Oreo), and then a very detailed level (an Oreo cookie's double-filling components).

Developing Effective Reports

Management information system reports can help managers develop better plans, make better decisions, and obtain greater control over the operations of the firm. But report types can overlap. For example, a manager can demand an exception report or set trigger points for items contained in a key-indicator report. Certain guidelines should be followed in designing and developing reports to yield the best results. Table 6.1 explains these guidelines.

FUNCTIONAL ASPECTS OF THE MIS

Most organizations are structured along functional lines or areas. This functional structure can usually be seen in an organization chart, which shows who reports to whom throughout the organization. Some of the traditional functional areas are accounting, finance, marketing, personnel, research and development (R&D), legal services, operations/production management, and information systems. As we mentioned previously, an MIS can be divided along those functional lines to produce reports tailored to individual functions (Figure 6.5).

financial MIS

an information system that provides financial information to all financial managers within an organization

A Financial Management Information System

A **financial MIS** provides financial information not only for executives but also for other people who need to make better daily decisions. Finding opportunities and quickly identifying problems can mean the difference between a business's success and failure.[1] Deutsche Bank and Chase Manhattan, for example, started a joint trading and reporting system for currency.[2] Some banks use a reporting

TABLE 6.1

Guidelines for Developing MIS Reports

Guidelines	Reason
Tailor each report to user needs.	The unique needs of the manager or executive should be considered, requiring user involvement and input.
Spend time and effort producing only reports that are useful.	Once instituted, many reports continue to be generated even though no one uses them anymore.
Pay attention to report content and layout.	Prominently display the information that is most desired. Do not clutter the report with unnecessary data. Use commonly accepted words and phrases. Managers can work more efficiently if they can easily find desired information.
Use management by exception reporting.	Some reports should be produced only when there is a problem to be solved or an action that should be taken.
Set parameters carefully.	Low parameters may result in too many reports; high parameters mean valuable information could be overlooked.
Produce all reports in a timely fashion.	Outdated reports are of little or no value.
Periodically review reports.	Review reports at least once a year to make sure all reports are still needed. Review report content and layout. Determine whether additional reports are needed.

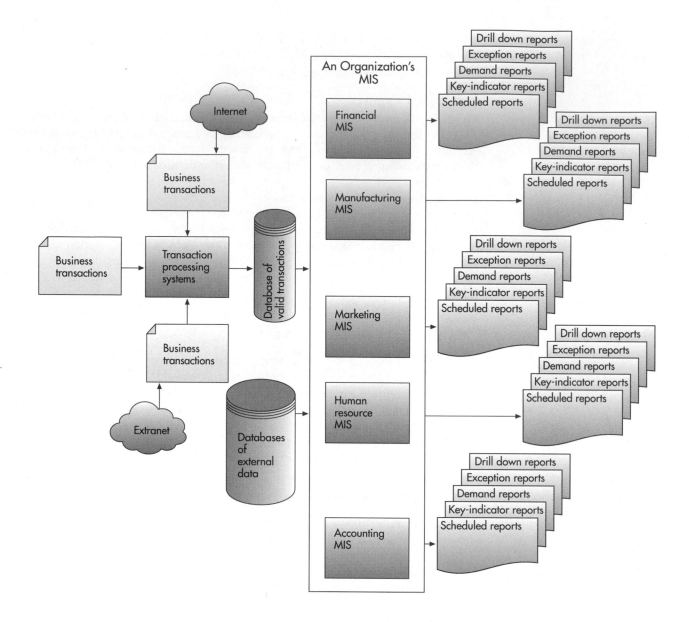

FIGURE 6.5

The MIS is an integrated
collection of functional
information systems, each
supporting particular
functional areas.

system that categorizes or grades customers, with excellent customers having fees forgiven and red tape cut. First Union Bank has a Web-based computer system that can pull up customer records and rankings in about 15 seconds.[3]

Specifically, the financial MIS performs the following functions:

- Integrates financial and operational information from multiple sources, including the Internet, into a single MIS
- Provides easy access to data for both financial and nonfinancial users, often through use of the corporate intranet to access corporate Web pages of financial data and information
- Makes financial data available on a timely basis to shorten analysis turn-around time
- Enables analysis of financial data along multiple dimensions—time, geography, product, plant, customer
- Analyzes historical and current financial activity
- Monitors and controls the use of funds over time

Figure 6.6 shows typical inputs, function-specific subsystems, and outputs of a financial MIS.

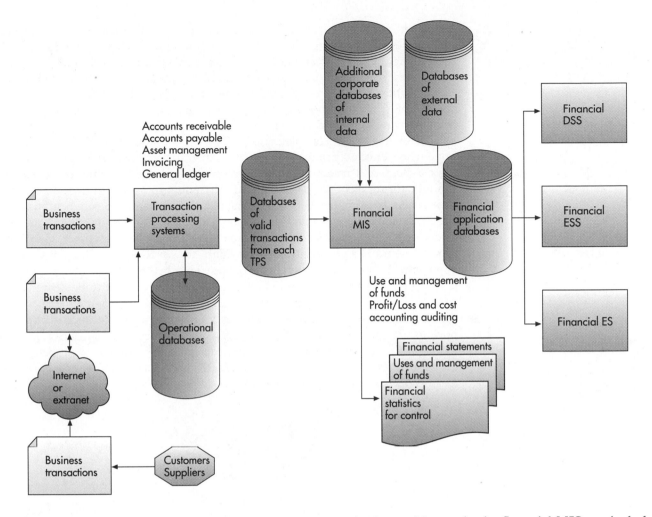

FIGURE 6.6

Overview of a Financial MIS

Depending on the organization and its needs, the financial MIS can include both internal and external systems that assist in acquiring, using, and controlling cash, funds, and other financial resources. These subsystems of the financial MIS have a unique role in adding value to a company's business processes. For example, a real estate development company might use a financial MIS subsystem to aid it in using and managing funds. Suppose the firm takes $10,000 deposits on condominiums in a new development. Until construction begins, the company will be able to invest these surplus funds. By using reports produced by the financial MIS, finance staff can analyze investment alternatives. The company might invest in new equipment or purchase global stocks and bonds. The profits generated from the investment can then be passed along to customers in different ways. The company can pay stockholders dividends, buy higher-quality materials, or sell the condominiums at a lower cost.

Important financial subsystems include the use and management of funds, profit/loss and cost accounting, and auditing. Each subsystem interacts with the TPS in a specialized way and has information reports that help financial managers make better decisions. These reports include financial statements, uses and management of funds, and financial statistics for control.

A Manufacturing Management Information System

More than any other functional area, manufacturing has been revolutionized by advances in technology. As a result, many manufacturing operations have been dramatically improved over the last decade. Also, with businesses' emphasis on greater quality and productivity, it is becoming even more critical for them to have an efficient and effective manufacturing process. All levels of manufacturing use computerized systems—from the shop floor to the executive suite. The use of the Internet has also streamlined all aspects of manufacturing.[4] Figure 6.7 gives an overview of some of the manufacturing MIS inputs, subsystems, and outputs.

The subsystems and outputs of the manufacturing MIS monitor and control the flow of materials, products, and services through the organization. The subsystems include design and engineering, production scheduling, inventory control, and material requirements planning (MRP). The basic goal of MRP is to determine when finished products are needed, then to work backward in determining deadlines and resources needed to complete the final product on schedule. Manufacturing resource planning (MRPII) refers to an integrated, company-wide system based on network scheduling that enables people to run their business with a high level of customer service and productivity, while lowering costs and inventories. Other subsystems include the just-in-time (JIT) inventory approach, process control, and quality control. With the just-in-time (JIT) inventory approach, inventory and materials are delivered just before they are used in a product. A JIT inventory system would arrange for a car windshield to be delivered to the assembly line only a few moments before it is secured to the automobile, rather than having it sitting around the manufacturing facility while the car's other components are being assembled. The outputs include reports to help control quality, monitor the manufacturing process, provide inventory when it is needed (JIT inventory), plan the use of all materials (MRP), schedule production, and design new products and parts (computer-assisted design, or CAD, output).

FIGURE 6.7

Overview of a
Manufacturing MIS

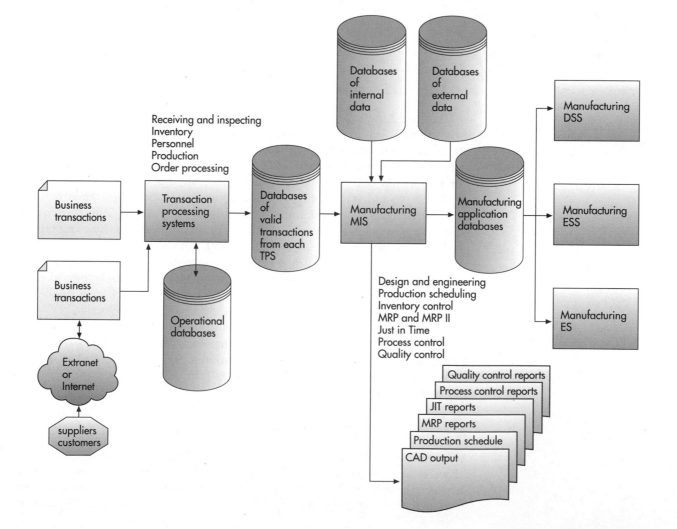

Management information and decision support systems can be used to help managers control all aspects of business. From order processing to product delivery, they can reveal new revenue sources and slash costs. Controlling inventory is perhaps one of the most complex and difficult aspects of managing any company. Some companies control inventory internally, but others look to outside companies and consultants for help.

Manugistics offers software products to help companies control inventory. Using the Internet, Manugistics has helped companies such as General Electric and Wal-Mart with all aspects of inventory control. Manugistics specializes in raw materials, production, and manufacturing. Manugistics also helps with shipping finished inventory to customers. In a joint effort with Burlington Northern Santa Fe (BNSF), Manugistics is giving customers the tools they need to help them decide between truck or rail in shipping products to their warehouses and retail stores. BNSF is also involved in FreightWise, an online e-commerce site to help companies ship products to market.

Manugistics is also developing a system for an e-commerce site that will allow companies to bid on transportation alternatives. The system operates like an auction site. Shipping companies will have the ability to post their excess shipping capacity on the Internet. A trucking company, for example, will be able to advertise truck availability on-line. Railroad companies can advertise available space on rail cars on the site. Companies that need to ship products can then bid for this excess capacity.

Discussion Questions

1. Why is inventory control so important to companies?
2. How can an MIS be used in inventory control?

Critical Thinking Questions

3. Assume that you are the manager of a medium-sized manufacturing company. What features would you like to have in an on-line MIS for inventory control?
4. If you were a manager for Manugistics, would you try to integrate other systems, such as finance or marketing, into your inventory control system?

Sources: Adapted from Daniel Machalaba, "Manugistics Set to Expand E-Commerce Role," *The Wall Street Journal*, January 25, 2000, p. B4; "Supply Chain Vendor Feels the Pressure," *Computing*, September 23, 1999, p. 14; and Clinton Wilder et al., "Companies Invest in Business-to-Business Marketplaces," *InformationWeek*, December 20, 1999.

The objective of the manufacturing MIS is to produce products that meet customer needs—from the raw materials provided by suppliers to finished goods and services delivered to customers—at the lowest possible cost. The activities of the manufacturing MIS subsystems support value-added business processes. As raw materials are converted to finished goods, the manufacturing MIS monitors the process at almost every stage. Take a car manufacturer that converts raw steel, plastic, and other materials into a finished automobile. In this case, the manufacturer has added thousands of dollars of value to the raw materials used in assembling the car. If the manufacturing MIS also lets the manufacturer provide customized paint colors on any of its models, it has further added value (although less tangible) by ensuring a direct customer fit. In doing so, the MIS helps provide the company the edge that can differentiate it from competitors. The success of an organization can depend on the manufacturing function. See the "E-Commerce" box above for examples of how companies are using inventory control systems.

A Marketing Management Information System

marketing MIS

an information system that supports managerial activities in product development, distribution, pricing decisions, promotional effectiveness, and sales forecasting

A **marketing MIS** supports managerial activities in product development, distribution, pricing decisions, promotional effectiveness, and sales forecasting. Marketing functions are increasingly being performed on the Internet.[5] Knight Ridder, a newspaper and publishing company, for example, places content and classified advertisements on the Internet. In addition, a number of companies are developing Internet marketplaces to advertise and sell products.[6] Customer relationship management (CRM) programs, available from some ERP vendors, help a company manage all aspects of customer encounters.[7] CRM can help a company collect customer data, contact customers, educate customers on new products, and sell products to customers through an Internet site. Yet not all marketing sites on the Internet are successful.[8] Rubbermaid, a company that makes plastic containers to store leftover food and other items, could not successfully sell products on its Internet site. Instead, the company stopped trying to sell products and developed a site that helps customers build a wish list of products that can be taken to a normal retail store. Figure 6.8 shows the inputs, subsystems, and outputs of a typical marketing MIS.

Subsystems for the marketing MIS include marketing research, product development, promotion and advertising, and product pricing. These subsystems and their outputs help marketing managers and executives increase sales, reduce marketing expenses, and develop plans for future products and services to meet the changing needs of customers. The specific reports include an analysis of sales (sales by customer, sales by salesperson, and sales by product), an analysis of possible prices for products and services (pricing reports), a summary of sales and service calls made to customers, and an analysis of customer satisfaction.

human resource MIS

an information system that is concerned with activities related to employees and potential employees of an organization

A Human Resource Management Information System

A **human resource MIS**, also called the personnel MIS, is concerned with activities related to employees and potential employees of the organization. Because the personnel function relates to all other functional areas in the business, the human resource MIS plays a valuable role in ensuring organizational success. The activities performed by this important MIS are workforce analysis

FIGURE 6.8

Overview of a Marketing MIS

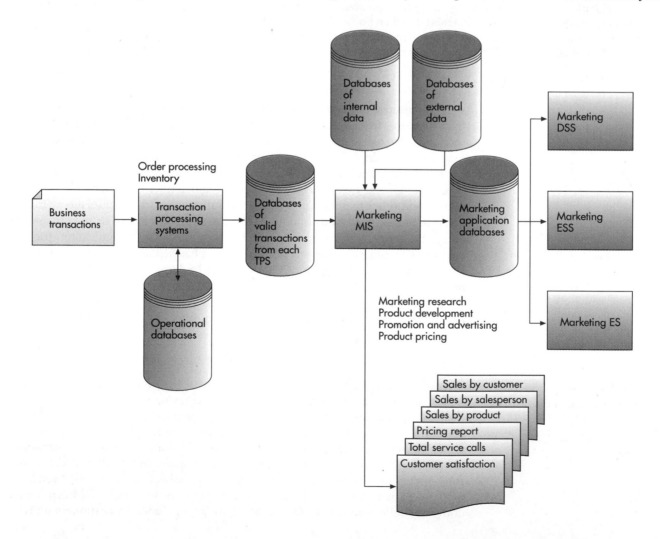

and planning, hiring, training, job and task assignment, and many other personnel-related issues. Personnel issues can include offering new hires attractive stock option and incentive programs.[9] Interwoven, for example, is offering new engineers a two-year lease on a sporty BMW Z3 roadster as a signing bonus. An effective human resource MIS will allow a company to keep personnel costs at a minimum while serving the required business processes needed to achieve corporate goals. Figure 6.9 shows some of the inputs, subsystems, and outputs of the human resource MIS.

Human resource subsystems and outputs range from the determination of human resource needs and hiring through retirement and outplacement. The subsystems include needs and planning assessments, recruiting, training and skills development, scheduling and assignment, and employee benefits. Most medium and large organizations have computer systems to assist with human resource subsystems. Outputs of the human resource MIS include benefit reports, salary surveys, scheduling reports, training test scores, job applicant profiles, and needs and planning reports.

Other Management Information Systems

FIGURE 6.9

Overview of a Human
Resource MIS

In addition to finance, manufacturing, marketing, and human resource MISs, some companies have other functional management information systems. For example, most successful companies have well-developed accounting functions and a supporting accounting MIS. Also, many companies use geographic information systems to present data in a useful form.

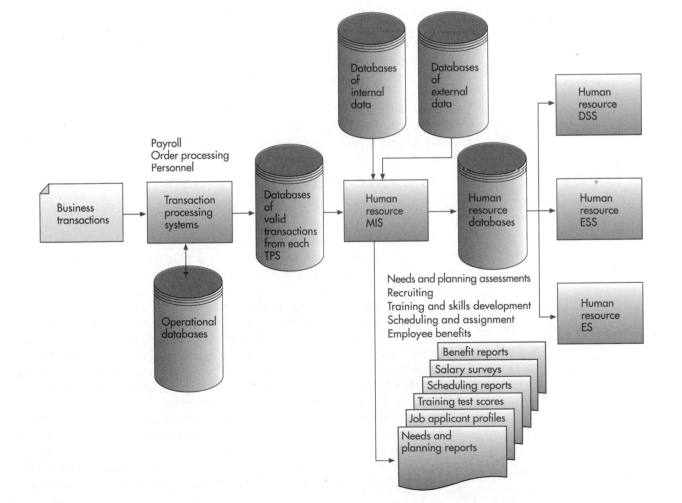

FIGURE 6.9

Overview of a Human Resource MIS

accounting MIS

an information system that provides aggregate information on accounts payable, accounts receivable, payroll, and many other applications

Accounting MISs

In some cases, accounting works closely with financial management. An **accounting MIS** performs a number of important activities, providing aggregate information on accounts payable, accounts receivable, payroll, and many other applications. The organization's TPS captures accounting data, which is also used by most other functional information systems.

Some smaller companies hire outside accounting firms to assist them with their accounting functions. These outside companies produce reports for the firm using raw accounting data. In addition, many excellent integrated accounting programs are available for personal computers in small companies. Depending on the needs of the small organization and its personnel's computer experience, using these computerized accounting systems can be a very cost-effective approach to managing information.

geographic information system (GIS)

a computer system capable of assembling, storing, manipulating, and displaying geographic information (i.e., data identified according to its location)

Geographic Information Systems

Increasingly, managers want to see data presented in graphical form. A **geographic information system (GIS)** is a computer system capable of assembling, storing, manipulating, and displaying geographically referenced information—that is, data identified according to its location. A GIS enables users to pair maps or map outlines with tabular data to describe aspects of a particular geographic region. For example, sales managers may want to plot total sales for each county in the states they serve. Using a GIS, they can shade each county to indicate the relative amount of sales—no shading or light shading represents no or little sales and deeper shading represents more sales. Because the GIS works with any data represented in tabular form, GIS capability is finding its way into spreadsheets. For example, Excel and Lotus include a mapping tool that lets you plot spreadsheet data as a demographic map. Such applications show up frequently in scientific investigations, resource management, and development planning. Retail, government, and utility organizations are also frequent users of GISs. Retail chains, for example, need to determine where potential customers are located and where their competition is.

We saw earlier in this chapter that MISs provide useful summary reports to help solve structured and semistructured business problems. Decision support systems (DSSs) offer the potential to also assist in solving these problems.

AN OVERVIEW OF DECISION SUPPORT SYSTEMS

A DSS is an organized collection of people, procedures, software, databases, and devices used to support problem-specific decision making and problem solving. The focus of a DSS is on decision-making effectiveness when faced with unstructured or semistructured business problems. Decision support systems offer the potential to generate higher profits, lower costs, and better products and services. For example, healthcare organizations use DSSs to track and reduce costs. Visteon, a transportation equipment company, uses a decision support system at its Sterling plant to help the self-directed teams that oversee the design and development of front axles for Ford Expeditions and Explorers.[10] As a result, productivity has increased by 30%. The U.S. Forest Service uses a DSS optimization model to help it determine land usage.[11] As with a TPS and an MIS, a DSS should be designed, developed, and used to help the organization achieve its goals and objectives.

Decision support systems, although skewed somewhat toward the top levels of management, are used at all levels. To some extent, managers at all levels are faced with less structured, nonroutine problems, but the quantity and magnitude of these decisions increase as a manager rises higher in an organization. Many organizations face a bureaucracy of complex rules, procedures, and decisions.

DSSs are used to bring more structure to these problems to aid in decision making. In addition, because of the inherent flexibility of decision support systems, managers at all levels can use DSSs to assist in some relatively routine, programmable decisions in lieu of more formalized management information systems.

Capabilities of a Decision Support System

Developers of decision support systems strive to make them more flexible than management information systems and to give them the potential to assist decision makers in a variety of situations. See Table 6.2 for a few DSS applications. DSSs can assist with all or most problem-solving phases, decision frequencies, and different degrees of problem structure. DSS approaches can also help at all levels of the decision-making process. In this section we investigate these DSS capabilities. Remember, though, that a particular DSS may provide only a few of these capabilities, depending on its uses and scope.

Support for Problem-Solving Phases

The objective of most decision support systems is to assist decision makers with the phases of the problem-solving process. As previously discussed, these phases include intelligence, design, choice, implementation, and monitoring. A specific DSS might support only one or a few problem-solving phases.

Support for Different Decision Frequencies

Decisions can range from one-of-a-kind to repetitive decisions. One-of-a-kind decisions are typically handled by an **ad hoc DSS**. An ad hoc DSS is concerned with situations or decisions that come up only a few times during the life of the organization; in small businesses, they may happen only once. For example, a company might be faced with a decision on whether to build a new manufacturing facility in another area of the country. Repetitive decisions are addressed by an institutional DSS. An **institutional DSS** handles situations or decisions that occur more than once, usually several times a year. An institutional DSS is used repeatedly and refined over the years. Examples of institutional DSSs include systems that support portfolio and investment decisions and production scheduling. These decisions may require decision support numerous times during the year. Between these two extremes are decisions managers make several times, but not regularly or routinely.

ad hoc DSS

a DSS concerned with situations or decisions that come up only a few times during the life of the organization

institutional DSS

a DSS that handles situations or decisions that occur more than once, usually several times a year

TABLE 6.2

Selected DSS Applications

Company or Application	Description
Cinergy Corporation	The electric utility developed a DSS to reduce lead time and effort required to make decisions in purchasing coal.
RCA	The company developed a DSS called Industrial Relation Information System (IRIS) to help solve personnel problems and issues.
U.S. Army	It developed a DSS to help recruit, train, and educate enlisted forces. The DSS uses simulation that incorporates what-if features.
National Audubon Society	It developed a DSS called Energy Plan (EPLAN) to analyze the impact of U.S. energy policy on the environment.
Hewlett-Packard	The computer company developed a DSS called Quality Decision Management to help improve the quality of its products and services.
Virginia	The state of Virginia developed the Transportation Evacuation Decision Support System (TEDSS) to determine the best way to evacuate people in case of a nuclear disaster at its nuclear power plants.

FIGURE 6.10

Decision-Making Level

Strategic-level managers are involved with long-term decisions, which are often made infrequently. Operational-level managers are involved with decisions that are made more frequently.

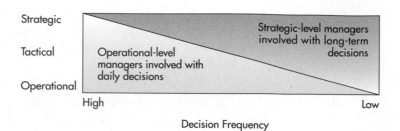

highly structured problems

problems that are straightforward and require known facts and relationships

semistructured or unstructured problems

more complex problems in which the relationships among the data are not always clear, the data may be in a variety of formats, and the data is often difficult to manipulate or obtain

Support for Different Problem Structures

As discussed previously, decisions can range from highly structured and programmed to unstructured and nonprogrammed. **Highly structured problems** are straightforward, requiring known facts and relationships. **Semistructured or unstructured problems**, on the other hand, are more complex. The relationships among the data are not always clear, the data may be in a variety of formats, and the data is often difficult to manipulate or obtain. In addition, the decision maker may not know the information requirements of the decision in advance.

Support for Various Decision-Making Levels

Decision support systems can offer help for managers at different levels within the organization. Operational-level managers can be assisted with daily and routine decision making. Tactical-level decision makers can be supported with analysis tools that assist in proper planning and control. At the strategic level, DSSs can help managers by providing analysis for long-term decisions requiring both internal and external information (Figure 6.10).

A Comparison of DSSs and MISs

A DSS differs from an MIS in numerous ways, including the type of problems solved; the support given to users; the decision emphasis and approach; and the type, speed, output, and development of the system used. Table 6.3 lists brief descriptions of these differences.

COMPONENTS OF A DECISION SUPPORT SYSTEM

dialogue manager

user interface that allows decision makers to easily access and manipulate the DSS and use common business terms and phrases

model base

part of a DSS that provides decision makers access to a variety of models and assists them in decision making

model management software

software that coordinates the use of models in a DSS

At the core of a DSS are a database and a model base. In addition, a typical DSS contains a **dialogue manager**, which allows decision makers to easily access and manipulate the DSS and use common business terms and phrases. External database access allows the DSS to tap into vast stores of information contained in the corporate database, letting the DSS retrieve information on inventory, sales, personnel, production, finance, accounting, and other areas. Finally, access to the Internet, networks, and other computer-based systems permits the DSS to tie into other powerful systems, including the TPS or function-specific subsystems. Internet software agents, for example, can be used in creating powerful decision support systems.[12] Figure 6.11 on page 22 shows a conceptual model of a DSS. Since database and database management system (DBMS) concepts were thoroughly covered in Chapter 3 and networks and the Internet were covered in Chapter 4, we begin with a discussion of the model base.

The Model Base

The purpose of the **model base** in a DSS is to give decision makers access to a variety of models and to assist them in the decision-making process. The model base can include **model management software (MMS)** that coordinates the

Factor	DSS	MIS
Problem Type	A DSS is good at handling unstructured problems that cannot be easily programmed.	An MIS is normally used only with more structured problems.
Users	A DSS supports individuals, small groups, and the entire organization. In the short run, users typically have more control over a DSS.	An MIS supports primarily the organization. In the short run, users have less control over an MIS.
Support	A DSS supports all aspects and phases of decision making; it does not replace the decision maker—people still make the decisions.	This is not true of all MIS systems—some make automatic decisions and replace the decision maker.
Emphasis	A DSS emphasizes actual decisions and decision-making styles.	An MIS usually emphasizes information only.
Approach	A DSS is a direct support system that provides interactive reports on computer screens.	An MIS is typically an indirect support system that uses regularly produced reports.
System	The computer equipment that provides decision support is usually on-line (directly connected to the computer system) and related to real time (providing immediate results). Computer terminals and display screens are examples— these devices can provide immediate information and answers to questions.	An MIS, using printed reports that may be delivered to managers once a week, may not provide immediate results.
Speed	Because a DSS is flexible and can be implemented by users, it usually takes less time to develop and is better able to respond to user requests.	An MIS's response time is usually longer.
Output	DSS reports are usually screen oriented, with the ability to generate reports on a printer.	An MIS, however, typically is oriented toward printed reports and documents.
Development	DSS users are usually more directly involved in its development. User involvement usually means better systems that provide superior support. For all systems, user involvement is the most important factor for the development of a successful system.	An MIS is frequently several years old and often was developed for people who are no longer performing the work supported by the MIS.

TABLE 6.3

Comparison of DSSs and MISs

financial models

models that provides cash flow, internal rate of return, and other investment analysis

use of models in a DSS, including financial, statistical analysis, graphical, and project management models. Depending on the needs of the decision maker, one or more of these models can be used.

Financial Models

Financial models provide cash flow, internal rate of return, and other investment analysis. Spreadsheet programs such as Excel are often used for this purpose. In addition, more sophisticated financial planning and modeling programs can be employed. Some organizations develop customized financial models to handle the unique situations and problems faced by the organization. However, as spreadsheet packages continue to increase in power, the need for sophisticated financial modeling packages may decrease. Individuals also develop financial models to help them plan for retirement or invest in the market. As seen in the "Ethical and Societal Issues" box (on page 231), however, using financial models for such activities as on-line trading can lead to problems.

statistical analysis models

models that can provide summary statistics, trend projections, hypothesis testing, and more

Statistical Analysis Models

Statistical analysis models can provide summary statistics, trend projections, hypothesis testing, and more. These programs are available on both personal and mainframe systems. Many software packages, including SPSS and

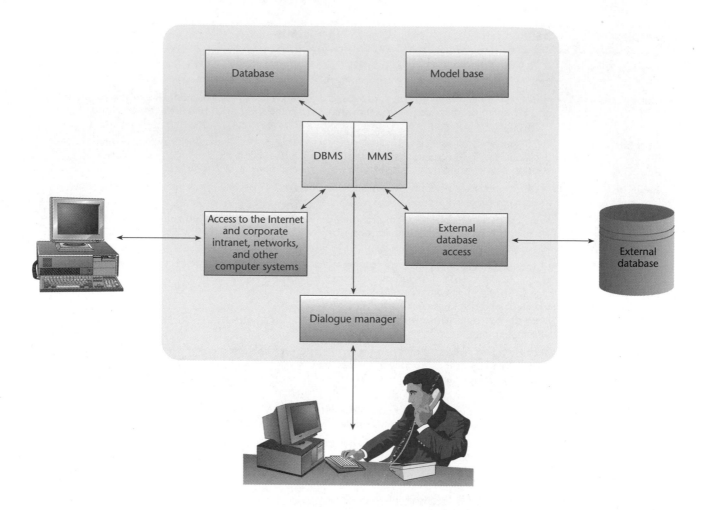

FIGURE 6.11

A Conceptual Model of a DSS

DSS components include a model base; database; external database access; access to the Internet and corporate intranet, networks, and other computer systems; and a dialogue manager.

graphical modeling program

software package that assists decision makers in designing, developing, and using graphic displays of data and information

project management model

model used to coordinate large projects and identify critical activities and tasks that could delay or jeopardize an entire project if they are not completed on time and cost-effectively

SAS, provide outstanding statistical analysis for organizations of all sizes. These statistical programs can compute averages, standard deviations, correlation coefficients, and regression analysis; do hypotheses testing; and use many more techniques. Some statistical analysis programs also have the ability to produce graphic displays that reveal the relationship between variables or quantities. Hyperion Solutions Corporation and Cognos help statistical analysts determine which reports and graphs will generate the most meaningful decision support.[13]

Graphical Models

Graphical modeling programs are software packages that assist decision makers in designing, developing, and using graphic displays of data and information. Personal computer programs that can perform this type of analysis, such as PowerPoint, are available. In addition, sophisticated graphical analysis, including computer-assisted design (CAD), is available.

Project Management Models

Project management models are used to coordinate large projects; they are also used to identify critical activities and tasks that could delay or jeopardize an entire project if they are not completed in a timely and cost-effective fashion. Some of these programs can also determine the best way to speed up a project by using additional resources, including cash, labor, and equipment. Project management models allow decision makers to keep tight control over projects of all sizes and types. These models can also help project managers prioritize user needs and requirements.[14]

ETHICAL AND SOCIETAL ISSUES
The Dark Side of E-Trading

For decades, financial institutions and brokerage firms have developed sophisticated decision support systems for trading stocks and bonds. Some DSSs use the futures market and options to help companies hedge their stock bets and avoid risk. Financial DSSs can provide a wealth of investment information and discounted stock trades for corporations, which individual investors did not have access to previously. But this situation is changing rapidly. Additional investment information and discounted trades are now affordable and available to investors with few assets or financial resources, and discount brokers are aggressively advertising their services to the general public. But does this additional investment data and inexpensive trades result in better investment decisions?

You turn on your TV and see a young man with his face pushed against the glass of a copy machine. He pushes the copy button and makes a copy of his face. The secretary looks on in disgust. The boss walks by and calls the young man in for a talk. The boss looks concerned. Will the young man lose his job? Inside the office, the mood changes. The boss wants the young man to help him open an Internet trading account. The young man shows his boss how it is done and invites the boss to a party. The boss buys some stock. He might come to the party. With this and similar ads for discounted trading companies, it can appear that making money using discounted trading firms and free investment information is easy. Yet there are many potential problems, including fraud and large losses.

In Los Angeles, an employee of a company posted false information on an investment Web site and drove up the company's stock price by more than 30 percent. Unsuspecting and inexperienced investors bought the stock on the way up and lost a substantial amount of money when the stock price plunged. The person posting the false information, like others involved with stock fraud, faces up to ten years in prison and up to $1 million in fines. Many stock frauds, however, go undetected, and investors go unprotected. In another case, an on-line trader lost about $750,000 through on-line trading. He was so upset that he took out a $500,000 life insurance policy on his wife and then pushed her off a balcony to collect the insurance money to help pay off his trading debts. When she didn't die, he climbed down and tried to suffocate her. The wife never knew of her husband's losses. After he landed in prison, he sued the on-line trading company.

Discussion Questions

1. How can a financial DSS be used to generate profits?
2. What are some of the potential problems of on-line trading for individual investors?

Critical Thinking Questions

3. What additional legislation, if any, would you suggest to help prevent stock fraud?
4. Is a trading company that aggressively advertises successful stock trading responsible to some extent for the trading activities of its clients?

Sources: Adapted from Rebbecca Buckman, "Heavy Losses," *The Wall Street Journal*, February 28, 2000, p. C1; David Rogers, "Online Stock Fraud," *The Wall Street Journal*, June 10, 1999, p. A6; and Michael Allen, "Bank Rule to Catch Money Launderers," *The Wall Street Journal*, March 23, 1999, p. A4.

The Dialogue Manager

The dialogue manager allows users to interact with the DSS to obtain information. It assists with all aspects of communications between the user and the hardware and software that constitute the DSS. In a practical sense, to most DSS users, the dialogue manager is the DSS. Upper-level decision makers are often less interested in where the information came from or how it was gathered than the fact that the information is both understandable and accessible.

With today's emphasis on teamwork for all types of business processes, decisions are often made by groups, rather than individual employees. Also, with global business trends, co-workers are often spread around the world and find it more convenient to communicate and make collaborative decisions with information systems. Decision support systems have evolved to assist such group decision making, and we now look at the major components of these systems.

THE GROUP DECISION SUPPORT SYSTEM

The DSS approach has resulted in better decision making for all levels of individual users. However, many DSS approaches and techniques are not suitable for a group decision-making environment. Although not all workers and managers

group decision support system (GDSS)

software application that consists of most elements in a DSS, plus software needed to provide effective support in group decision making

delphi approach

a decision-making approach in which group decision makers are geographically dispersed; this approach encourages diversity among group members and fosters creativity and original thinking in decision making

brainstorming

decision-making approach that often involves members offering ideas "off the top of their heads"

group consensus approach

decision-making approach that forces members in the group to reach a unanimous decision

nominal group technique

decision-making approach that encourages feedback from individual group members; the final decision is made by voting, similar to the way public officials are elected

are involved in committee meetings and group decision-making sessions, some tactical and strategic-level managers can spend more than half their decision-making time in a group setting. Such managers need effective approaches to assist with group decision making. A **group decision support system (GDSS)**, also called a *group support system* or a *computerized collaborative work system*, consists of most of the elements in a DSS, plus GDSS software needed to provide effective support in group decision-making settings (Figure 6.12).

Characteristics of a GDSS

A GDSS has a number of unique characteristics that go beyond the traditional DSS. Developers of these systems try to build on the advantages of individual support systems while realizing that additional approaches are needed for group decision making. For example, some GDSSs can allow groups to exchange information and expertise without meetings or direct face-to-face interaction.[15] The following are some characteristics of a typical GDSS.

Special design. The GDSS approach acknowledges that special procedures, devices, and approaches are needed in group settings. These procedures must foster creative thinking, effective communications, and good group decision-making techniques.

Ease of use. Like an individual DSS, a GDSS must be easy to learn and use. Systems that are complex and hard to operate will seldom be used. Many groups have less tolerance for poorly developed systems than do individual decision makers. In addition, the skill levels of individuals can vary greatly.

Flexibility. Two or more decision makers working on the same problem may have different decision-making styles and preferences. Each manager makes decisions in a unique way, in part because of different experiences and cognitive styles. An effective GDSS not only has to support the different approaches that managers use to make decisions but also must find a means to integrate their different perspectives into a common view of the task at hand.

Decision-making support. A GDSS can support different decision-making approaches, including the **delphi approach**, in which group decision makers are geographically dispersed throughout the country or the world. This approach encourages diversity among group members and fosters creativity and original thinking in decision making. Another approach, called **brainstorming**, which often involves members offering ideas "off the top of their heads," fosters creativity and free thinking. The **group consensus approach** forces members in the group to reach a unanimous decision. With the **nominal group technique**, each decision maker can participate; this technique encourages feedback from individual group members, and the final decision is made by voting, similar to a system for electing public officials.

Anonymous input. Many GDSSs allow anonymous input, where the person giving the input is not known to other group members. For example, some organizations use a GDSS to help rank the performance of managers. Anonymous input allows the group decision makers to concentrate on the merits of the input without considering who gave it. In other words, input given by a top-level manager is given the same consideration as input from lower-level employees or other members of the group. Some studies have shown that groups using anonymous input can make better decisions and have superior results compared with groups that do not use anonymous input. Anonymous input, however, can result in flaming, where an unknown team member posts insults or even obscenities on the GDSS system.[16]

Computer-assisted design (CAD) programs are powerful tools in developing products.
(Source: Bob Krist/Corbis)

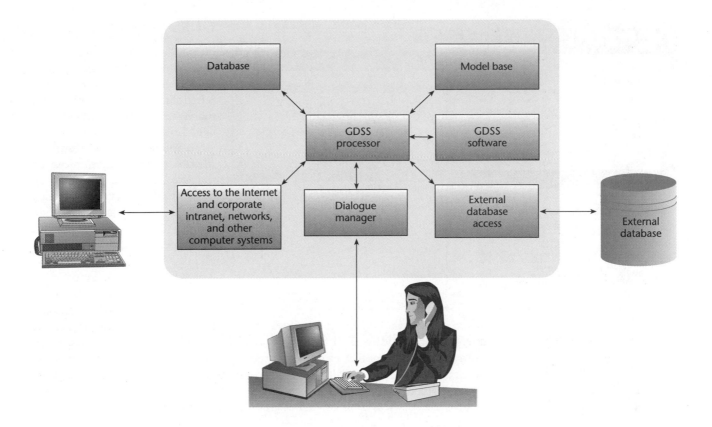

FIGURE 6.12

Configuration of a GDSS

A GDSS contains most of the elements found in a DSS, plus GDSS software to facilitate group member communications.

Reduction of negative group behavior. One key characteristic of any GDSS is the ability to suppress or eliminate group behavior that is counter-productive or harmful to effective decision making. In some group settings, dominant individuals can take over the discussion, which can prevent other members of the group from presenting creative alternatives. In other cases, one or two group members can sidetrack or subvert the group into areas that are nonproductive and do not help solve the problem at hand. Other times, members of a group may assume they have made the right decision without examining alternatives—a phenomenon called *groupthink*. If group sessions are poorly planned and executed, the result can be a tremendous waste of time. Today, many GDSS designers are developing software and hardware systems to reduce these types of problems. Procedures for effectively planning and managing group meetings can be incorporated into the GDSS approach. A trained meeting facilitator is often employed to help lead the group deci-sion-making process and to avoid groupthink.

Parallel communication. With traditional group meetings, people must take turns addressing various issues. One person normally talks at a time. With a GDSS, it is possible for every group member to address issues or make com-ments at the same time by entering them into a PC or workstation. These com-ments and issues are displayed on every group member's PC or workstation immediately. Parallel communication can speed meeting times and result in better decisions.

Automated record keeping. Most GDSSs have the ability to keep detailed records of a meeting automatically. Each comment that is entered into a group member's PC or workstation can be anonymously recorded. In some cases, lit-erally hundreds of comments can be stored for future review and analysis. In addition, most GDSS packages have automatic voting and ranking features. After group members vote, the GDSS records each vote and makes the appro-priate rankings.

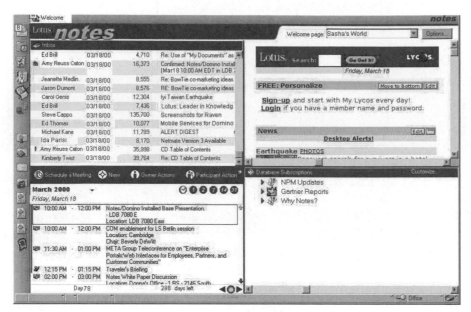

Workgroup software, such as Lotus Notes, allows people located around the world to work on the same project, documents, and files efficiently and at the same time.
(Source: Courtesy of Lotus Development Corporation.)

decision room

a room that supports decision making, with the decision makers in the same building, combining face-to-face verbal interaction with technology to make the meeting more effective and efficient

FIGURE 6.13

GDSS Alternatives

This figure demonstrates that the decision room may be the best alternative for group members who are located physically close together and who need to make infrequent decisions as a group. By the same token, group members who are situated at distant locations and who frequently make decisions together may require a wide area decision network to accomplish their goals.

GDSS Software

GDSS software, often called *groupware* or *workgroup software*, helps with joint work group scheduling, communication, and management. One popular package, Lotus Notes, can capture, store, manipulate, and distribute memos and communications that are developed during group projects. Microsoft's NetMeeting product supports application sharing in multiparty calls. It gains performance by sharing one task at a time. The user designated to be the "operator" must select one previously launched application, and each participant has to choose whether to "collaborate." Any collaborating participant can assume mouse control and work in the shared application while others watch. Exchange from Microsoft is another example of groupware. This software allows users to set up electronic bulletin boards, schedule group meetings, and use e-mail in a group setting. Other GDSS software packages include Collabra Share, OpenMind, and TeamWare. All these tools can aid in group decision making.[17]

GDSS Alternatives

Group decision support systems can take on a number of alternative network configurations, depending on the needs of the group, the decision to be supported, and the geographic location of group members. The frequency of GDSS use and the location of the decision makers are two important factors (Figure 6.13).

The Decision Room

The **decision room** is ideal for situations in which decision makers are located in the same building or geographic area and the decision makers are occasional users of the GDSS approach. In these cases, one or more decision rooms or facilities can be set up to accommodate the GDSS approach. Groups, such as marketing research teams, production management groups, financial control teams, and quality-control committees, can use the decision

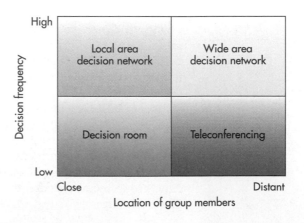

The GDSS Decision Room

For group members in the same physical location, the decision room is an optimal GDSS alternative. By use of networked computers and computer devices, such as project screens and printers, the meeting leader can pose questions to the group, instantly collect their feedback, and, with the help of the governing software loaded on the control station, process this feedback into meaningful information to aid in the decision-making process.

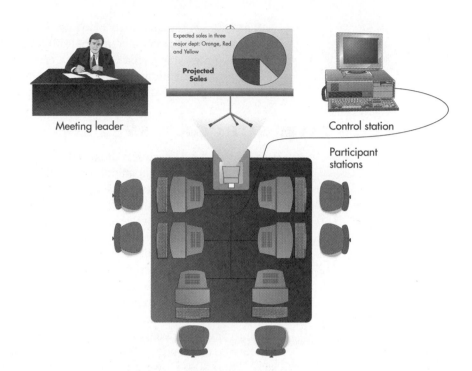

rooms when needed. The decision room alternative combines face-to-face verbal interaction with technology-aided formalization to make the meeting more effective and efficient. A typical decision room is shown in Figure 6.14.

The Local Area Decision Network

The local area decision network can be used when group members are located in the same building or geographic area and under conditions in which group decision making is frequent. In these cases, the technology and equipment of the GDSS approach is placed directly into the offices of the group members. Usually this is accomplished via a local area network (LAN).

The Teleconferencing Alternative

The teleconferencing alternative is used for situations in which the decision frequency is low and the location of group members is distant. These distant and occasional group meetings can tie together multiple GDSS decision-making rooms across the country or around the world. Using long-distance communications technology, these decision rooms are electronically connected in teleconferences and videoconferences. This alternative can offer a high degree of flexibility. The GDSS decision rooms can be used locally in a group setting or globally when decision makers are located throughout the world. GDSS decision rooms are often connected through the Internet.

The Wide Area Decision Network

The wide area decision network is used for situations in which the decision frequency is high and the location of group members is distant. In this case, the decision makers require frequent or constant use of the GDSS approach. This situation requires decision makers located throughout the country or the world to be linked electronically through a wide area network (WAN). The group facilitator and all group members are at geographically dispersed locations. In some cases, the model base and database are also geographically dispersed. This GDSS alternative allows people to work in **virtual workgroups**, where teams of people located around the world can work on common problems.

virtual workgroups

teams of people located around the world working on common problems

The Internet is increasingly being used to support wide area decision networks. As discussed in Chapter 4, a number of technologies—including videoconferencing, instant messaging, chat rooms, and telecommuting—can be used to assist the GDSS process. In addition, many specialized wide area decision networks use the Internet for group decision making and problem solving.

THE EXECUTIVE SUPPORT SYSTEM

executive support system (ESS), or executive information system (EIS)

specialized DSS that includes all hardware, software, data, procedures, and people used to assist senior-level executives within the organization

Because top-level executives often require specialized support when making strategic decisions, many companies have developed systems to assist executive decision making. This type of system, called an **executive support system (ESS)**, is a specialized DSS that includes all hardware, software, data, procedures, and people used to assist senior-level executives within the organization. In some cases, an ESS, also called an **executive information system (EIS)**, supports the actions of members of the board of directors, who are responsible to stockholders. These top-level decision-making strata are shown in Figure 6.15.

An ESS can also be used by individuals further down in the organizational structure. Once targeted at the top-level executive decision makers, ESSs are now marketed to—and used by—employees at other levels in the organization. In the traditional view, ESSs give top executives a means of tracking critical success factors. Today, all levels of the organization share information from the same databases. However, for our discussion, we will assume that ESSs remain in the upper management levels, where they present important corporate issues, indicate new directions the company may take, and help executives monitor the company's progress.

Executive Support Systems in Perspective

An ESS is a special type of DSS, and like a DSS, it is designed to support higher-level decision making. The two systems are, however, different in important ways. DSSs provide a variety of modeling and analysis tools to enable users to thoroughly analyze problems—that is, they allow users to *answer* questions. ESSs present structured information about aspects of the organization that executives consider important—in other words, they allow executives to *ask* the right questions.[18]

The following are general characteristics of ESSs.

* *Tailored to individual executives.* ESSs are typically tailored to individual executives; DSSs are not tailored to particular users. Because of this customization, ESSs truly represent the overall objective of information systems: to deliver the right information to the right person at the right time in the right format. An ESS is an interactive, hands-on tool that allows an executive to focus, filter, and organize data and information.
* *Easy to use.* A top-level executive's most critical resource can be his or her time. Thus, an ESS must be easy to learn and use and not overly complex. Early ESSs were noted for their extreme ease of use, since they were targeted for decision makers who were often not technically oriented. The extensive use of color and graphics is common with an ESS.
* *Have drill down abilities.* An ESS allows executives to "drill down" into the company to determine how certain data was produced. Drill down allows an executive to get more detailed information if needed. For example, after seeing a report summarizing the progress and costs of a large project, an executive might want to drill down to see the status of a particular activity within the project.

FIGURE 6.15

The Layers of Executive Decision Making

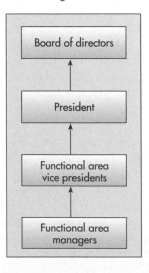

• *Support the need for external data.* The data needed to make effective top-level decisions is often external—information from competitors, the federal government, trade associations and journals, consultants, and so on. This data is often "soft" as well. An effective ESS can extract data useful to the decision maker from a wide variety of sources, including the Internet and other electronic publishing sources such as Lexis/Nexis. Often they employ intelligent bots to search the Web for new information about competitors, suppliers, customers, and key business trends. Typically, decisions made at top levels are also complex and unstructured. Traditional computer programs are normally ineffective for these types of decisions. As a result, more advanced ESSs are required to help support all types of strategic decisions.

• *Can help with situations that have a high degree of uncertainty.* There is a high degree of uncertainty with most executive decisions. What will happen if a new plant is started? What will be the consequences of a potential merger? How will union negotiators take a new proposal? The answers to these questions are not known with certainty. Handling these unknown situations using modeling and other ESS procedures helps top-level managers measure the amount of risk in a decision.

• *Have a future orientation.* Executive decisions are future oriented, meaning that decisions will have a broad impact for years or decades. The information sources to support future-oriented decision making are usually informal—from golf partners to members of social clubs or civic organizations.

• *Are linked with value-added business processes.* Like other information systems, executive support systems are linked with executive decision making about value-added business processes. For instance, executive support systems can be used by car rental companies to analyze trends. By detecting which firms generate enough business to be worth a certain discount, car-rental executives can ask questions to determine why these firms create high revenue and how this business can be continued or bettered. They might also use these factors to identify similar firms with business potential. Strategic planning of this type is just one of the functions that can be aided by an ESS.

Capabilities of an Executive Support System

The responsibility given to top-level executives and decision makers brings unique problems and pressures to their jobs. Following is a discussion of some of the characteristics of executive decision making that are supported through the ESS approach. As you will note, most of these characteristics are related to an organization's overall profitability and direction. An effective ESS should be able to support executive decisions with many of these capabilities, such as strategic planning and organizing, crisis management, and more.

Support for defining an overall vision. One of the key roles of senior executives is to provide a broad vision for the entire organization. This vision includes the organization's major product lines and services, the types of businesses it supports today and in the future, and its overriding goals.

Support for strategic planning. ESSs also support strategic planning. **Strategic planning** involves determining long-term objectives by analyzing the strengths and weaknesses of the organization, predicting future trends, and projecting the development of new product lines. It also involves planning the acquisition of new equipment, analyzing merger possibilities, and making difficult decisions concerning downsizing and the sale of assets if required by unfavorable economic conditions.

strategic planning

determining long-term objectives by analyzing the strengths and weaknesses of the organization, predicting future trends, and projecting the development of new product lines

Support for strategic organizing and staffing. Top-level executives are concerned with organization structure. For example, decisions concerning the creation of new departments or downsizing the labor force are made by top-level managers. Should the information systems department be placed under new leadership? Should the accounting and finance departments be joined under a new vice president of financial services? Should the marketing department be divided along the company's two major product lines? These and similar questions can profoundly affect the overall effectiveness of the organization and should be supported by an ESS.

Overall direction for staffing decisions and effective communication with labor unions are major decision areas for top-level executives. Middle- and lower-level managers make staffing decisions, but general decisions about the types and number of employees for various departments in the organization are determined at the top level of the organization. In addition, top-level managers are responsible for labor negotiation. ESSs can be employed to help analyze the impact of staffing decisions, potential pay raises, changes in employee benefits, and new work rules.

Support for strategic control. Another type of executive decision relates to strategic control, which involves monitoring and managing the overall operation of the organization. Effective ESS approaches can help top-level managers make the most of their existing resources and control all aspects of the organization.

Support for crisis management. Even with careful strategic planning, a crisis can occur. Major disasters, including hurricanes, tornadoes, floods, earthquakes, fires, and sabotage, can totally shut down major parts of the organization. Handling these emergencies is another responsibility for top-level executives. In many cases, strategic emergency plans can be put into place with the help of an ESS. These contingency plans help organizations recover quickly if an emergency or crisis occurs.

Decision making is a vital part of managing businesses strategically. IS systems such as decision support, group decision support, and executive support systems help employees by tapping existing databases and providing them with current, accurate information. The increasing integration of all business information systems—from TPSs to MISs to DSSs—can help organizations monitor their competitive environment and make better-informed decisions.

INFORMATION SYSTEMS IN ACTION

Instant Messaging for the Network-Supported Course

Gerald M. Santoro, *School of Information Sciences and Technology, Pennsylvania State University*

As we explore models for distance education and network-supported residential instruction, communication tools prove to be crucial components. These tools provide the connectivity that allow instructors to guide students through predeveloped learning environments. They also allow students to work together in problem-solving groups.

The trend has been toward asynchronous modes, such as electronic mail and bulletin board systems, or toward scheduled real-time meetings in chat rooms. Opportunities are disappearing for a quick "drop-in" meeting, where the student happens to be near the instructor's office and drops in for a quick question.

A nice solution to this problem has emerged in the form of instant messaging (IM). With this service, users register themselves with a central network server and are then available for on-demand chat. For the past two years, I have been giving students my instant messaging address to see (1) if they would use IM to contact me with questions and (2) if the use of IM would prove beneficial to the learning process. The results have been very positive on both counts.

I have found that students will contact me via IM an average of 2 to 3 times each day (total), with more contacts immediately preceding important due dates for assignments and exams. As an example, following is a portion of an IM transcript:

rzr99 (10:50:24 AM): Hey doc
gmsantoro (10:54:54 AM): hello
rzr99 (10:57:55 AM): Should we make the database to contain the entire text in the database or should we use links in the database?
gmsantoro (10:58:25 AM): you mean for pbl4?
rzr99 (10:59:04 AM): yes
rzr99 (10:59:13 AM): to complete the assignment website.
gmsantoro (11:03:07 AM): have you looked at the rubric?
rzr99 (11:03:31 AM): I haven't to be honest going to now. Been working on the pages with the database.
gmsantoro (11:04:49 AM): ok - i think it answers all questions

rzr99 (11:04:56 AM): ok thanks
gmsantoro (11:04:59 AM): you should only produce stuff relevant to the rubric
gmsantoro (11:05:06 AM): i didn't see a database listed there at all
gmsantoro (11:05:20 AM): pbl4 is all about multimedia
rzr99 (11:05:29 AM): Oh ok
gmsantoro (11:05:42 AM): you produce a document that describes your MM plans
gmsantoro (11:05:54 AM): to a detail such that you could hire an artist to do it
rzr99 (11:06:03 AM): Oh ok
gmsantoro (11:07:34 AM): good thing you asked :-)
rzr99 (11:09:33 AM): Thanks doc. Too many things on my mind this weekend and today.
gmsantoro (11:09:41 AM): no prob - see you tonight

This transcript demonstrates a few of the advantages of IM over competing technologies, such as the telephone. Because IM is Internet-based, students do not have to know where I am physically located to reach me. I can easily move between my home and campus offices and still be available.

Since most IM systems are text-based (input is typed and displayed in a window for reading), transcripts are easy to make. I typically use the cut-and-paste method to copy text from an IM window and paste it into an e-mail note to myself. These transcripts can serve as "minutes" of the virtual meeting.

Many IM systems also allow for group meetings. This way, I can meet with a group of students who could themselves be meeting either physically or virtually. I can provide them with answers to their questions, and possibly observe their discussions, without otherwise interfering.

I have found that providing students with a way to reach me easily for a quick Q-and-A session is also a great student motivator. Students are impressed by my willingness to be contacted in real time when they feel a need to contact me. They understand that I will usually not be able to socialize on-line because I am working—and as a result there is a genuine effort to make the sessions efficient.

IM is a great tool. It is effective for learning support and students like it. I encourage all instructors to give it a try.

● SUMMARY

PRINCIPLE: • **The management information system (MIS) must provide the right information to the right person in the right fashion at the right time.**

Problem solving begins with decision making. A well-known model developed by Herbert Simon divides the decision-making phase of the problem-solving process into three stages: intelligence, design, and choice. The three phases of decision making are augmented by implementation and monitoring to result in problem solving.

The first stage in the decision-making phase of the problem-solving process is the intelligence stage. During this stage, potential problems and opportunities are identified and defined. Information is gathered that relates to the cause and scope of the problem. Constraints on the possible solution and the problem environment are investigated. In the design stage, alternative solutions to the problem are developed. In addition, the feasibility and implications of these alternatives are evaluated. The last stage of the decision-making phase, the choice stage, requires selecting a course of action.

Problem solving includes and goes beyond decision making. It also includes the implementation stage, when the solution is put into effect. The final stage of the problem-solving process is the monitoring stage. In this stage, the decision makers evaluate the implementation of the solution to determine whether the anticipated results were achieved and to modify the process in light of new information learned during the implementation stage.

• • •

A management information system is an integrated collection of people, procedures, databases, and devices that provide managers and decision makers with information to help achieve organizational goals. An MIS can help an organization achieve its goals by providing managers with insight into the regular operations of the organization so that they can control, organize, and plan more effectively and efficiently. The primary difference between the reports generated by the TPS and those generated by the MIS is that MIS reports support managerial decision making at higher levels of management.

Data that enters the MIS originates from both internal and external sources. The output of most management information systems is a collection of reports that are distributed to managers. These reports include scheduled reports, key-indicator reports, demand reports, exception reports, and drill down reports. Scheduled reports are produced periodically, or on a schedule, such as daily, weekly, or monthly. A key-indicator report is a special type of scheduled report. Demand reports are developed to give certain information at a manager's request. Exception reports are automatically produced when a situation is unusual or requires management action. Drill down reports provide increasingly detailed data about situations.

• • •

Most MISs are organized along the functional lines of an organization. Typical functional management information systems include accounting, manufacturing, marketing, and human resources. Each system is composed of inputs, processing subsystems, and outputs. The primary sources of input to functional MISs include the corporate strategic plan, data from the TPS, information from other functional areas, and external sources, including the Internet. The primary output of these functional MISs are summary reports that assist in managerial decision making.

A financial management information system provides financial information to all financial managers within an organization, including the chief financial officer (CFO). Important financial subsystems include the use and management of funds, profit/loss and cost accounting, and auditing. Reports from the financial MIS include financial statements, use and management of funds, and financial statistics for control.

A manufacturing MIS accepts inputs from the strategic plan, the TPS, and external sources. The manufacturing MIS subsystems include design and engineering, production scheduling, inventory control, material requirements planning (MRP), manufacturing resource planning II (MRPII), just-in-time (JIT) inventory, process control, and quality control. The outputs include reports to help control quality, monitor the manufacturing process, provide inventory when it is needed (JIT inventory), plan the use of all materials (MRP), schedule production, and design new products and parts (computer-assisted design, or CAD, output).

A marketing MIS supports managerial activities in product development, distribution, pricing decisions, promotional effectiveness, and sales forecasting. Subsystems for the marketing MIS include marketing research, product development, promotion and advertising, and product pricing. The specific reports include an analysis of sales (sales by customer, sales by salesperson, and sales by product), an analysis of possible prices for products and services (pricing reports), a summary of sales and service calls made to customers, and an analysis of customer satisfaction.

A human resource MIS is concerned with activities related to employees of the organization. The subsystems include needs and planning assessments, recruiting, training and skills development, scheduling and assignment, and employee benefits. Outputs of the human resource MIS include benefit reports, salary surveys, scheduling reports, training test scores, job applicant profiles, and needs and planning reports.

In addition, there are other management information systems. An accounting MIS performs a number of important activities, providing aggregate information on accounts payable, accounts receivable, payroll, and many other applications. The organization's TPS captures accounting data, which is also used by most other functional information systems. Geographic information systems provide regional data in graphical form.

PRINCIPLE: ● **Decision support systems (DSSs) are used when the problems are more unstructured.**

A decision support system (DSS) is an organized collection of people, procedures, software, databases, and devices working to support managerial decision making. DSSs provide assistance through all phases of the decision-making process. The degree of problem structure and scope contributes to the complexity of the decision support system. Problems may be highly structured or unstructured, infrequent or routine. An ad hoc DSS addresses unique, infrequent decision situations; an institutional DSS handles routine decisions. A common database is often the link that ties together a company's TPS, MIS, and DSS.

● ● ●

The components of a DSS are the database, model base, dialogue manager, and a link to external databases, the Internet, the corporate intranet, extranets, networks, and other systems. The model base contains the models used by the decision maker, such as financial, statistical, graphical, and project management models. The dialogue manager provides a dialogue management facility to assist in communications between the system and the user. Access to other computer-based systems permits the DSS to tie into other powerful systems, including the TPS or function-specific subsystems.

PRINCIPLE: ● **Specialized support systems, such as group decision support systems (GDSSs) and executive support systems (ESSs), use the overall approach of a DSS in situations such as group and executive decision making.**

A group decision support system (GDSS), also called a group support system or a computerized collaborative work system, consists of most of the elements in a DSS, plus GDSS software needed to provide effective support for group decision making. GDSSs are typically easy to learn and use and can offer specific or general decision-making support. GDSS software, also called groupware, is specially designed to help generate lists of decision alternatives and perform data analysis. These packages let people work on joint documents and files over a network.

The frequency of GDSS use and the location of the decision makers will influence the GDSS alternative chosen. The decision room alternative supports users in a single location that meet infrequently. Local area networks can be used when group members are located in the same geographic area and users meet regularly. Teleconferencing is used when decision frequency is low and the location of group members is distant. A wide area network is used for situations where the decision frequency is high and the location of group members is distant.

● ● ●

Executive support systems (ESSs) are specialized decision support systems designed to meet the needs of senior management. They present issues of importance to the organization, indicate new directions the company may take, and help executives monitor the company's progress. ESSs are typically easy to use, offer a wide range of computer resources, and handle a variety of internal and external data. In addition, the ESS performs sophisticated data analysis, offers a high degree of specialization, and provides flexibility and comprehensive communications abilities. An ESS also supports individual decision-making styles. Some of the major decision-making areas that can be supported through an ESS are providing an overall vision, strategic planning and organizing, staffing and labor relations, crisis management, and strategic control.

● REVIEW QUESTIONS

1. What are the five stages of problem solving?
2. Define *management information system (MIS)*.
3. What are the four basic kinds of reports produced by an MIS?
4. What guidelines should be followed in developing reports for management information systems?
5. Identify the functions performed by all MIS systems.
6. What are the functions performed by a financial MIS?
7. Describe the functions of a manufacturing MIS.
8. What is a human resource MIS? What are its outputs?
9. List and describe some other types of MISs.
10. What is a geographic information system?
11. Describe the difference between a structured and an unstructured problem and give an example of each.

12. Define *decision support system*. What are its characteristics?
13. What are the components of a decision support system?
14. Describe four models used in decision support systems.
15. What is meant by *groupthink?*
16. State the objective of a group decision support system (GDSS) and identify three characteristics that distinguish it from a DSS.
17. Identify three group decision-making approaches often supported by a GDSS.
18. What is an executive support system? Identify three fundamental uses for such a system.

● DISCUSSION QUESTIONS

1. What is the relationship between an organization's transaction processing systems and its management information systems? What is the primary role of management information systems?
2. How can management information systems be used to support the objectives of the business organization?
3. Describe a financial MIS for a Fortune 1000 manufacturer of food products. What are the primary inputs and outputs? What are the subsystems?
4. How can a strong financial MIS provide strategic benefits to a firm?
5. Imagine that you are the CFO for a services organization. You are concerned with the integrity of the firm's financial data. What steps might you take to ascertain the extent of problems?

6. How does a DSS differ from an MIS?
7. How is decision making in a group different from individual decision making, and why are information systems that assist groups different? What are the advantages and disadvantages of making decisions as a group?
8. The use of ESSs should not be limited to the executives of the company. Do you agree or disagree? Why?
9. Imagine that you are the vice president of manufacturing for a Fortune 1000 manufacturing company. Describe the features and capabilities of your ideal ESS.

● PROBLEM-SOLVING EXERCISES

 1. You have been asked to develop your company's employee training program. Your company sells personal computers and software to university students. Make a list of the kind of information that must be available. Make a second list of the services that should be provided to the employee. Make a third list of the information you would provide managers, summarizing test results and performance information of employees taking the training course. Use your word processor or other software to describe the training program and all lists and reports.

⭐ 2. Review the summarized consolidated statement of income for the manufacturing company described next. Use graphics software to prepare a set of bar charts that shows data for this year compared with data for last year.
a. Operating revenues increase by 3.5% while operating expenses increase by 2.5%.
b. Other income and expenses decrease to $13,000.
c. Interest and other charges increase to $265,000.

Operating Results (in millions)

Operating Revenues	$2,924,177
Operating Expenses (including taxes)	2,483,687
Operating Income	440,490
Other Income and Expenses	13,497
Income before Interest and Other Charges	453,987
Interest and Other Charges	262,845
Net Income	191,142
Average Common Shares Outstanding	147,426
Earnings per Share	$1.30

If you were a financial analyst tracking this company, what detailed data might you need to perform a more complete analysis? Write a brief memo summarizing your data needs.

 3. As the head buyer for a major supermarket chain, John is constantly being asked by manufacturers and distributors to stock their new products. More than ten new items are introduced each week. Develop a spreadsheet that lists the costs, sales price, and expected sales in units each week for ten new products. Use your spreadsheet to compute net profit for each unit and expected total net profits for the week for the ten new products.

● TEAM ACTIVITIES

1. Divide the class into teams of three or four classmates and use the following role-playing scenario: Your consulting team is designing a management information system for a small manufacturing organization that produces ten different models of high-performance bicycles. These cycles are sold to distributors and bicycle shops throughout North America. One of the major problems this company faces is poor inventory control. It always seems that there are too many bikes in stock and yet the "hottest selling" model is out of stock. The owner has suggested that your team look at the feasibility of implementing a manufacturing MIS to help deal with this problem. Prepare a brief memo that describes the subsystems and outputs associated with the typical manufacturing MIS. Outline how these subsystems need to be integrated. What are some of the issues that will need to be addressed to develop a single integrated MIS to meet the needs of this organization? What additional benefits, besides better inventory control, can be expected?

2. Have your team work together in making a group decision, such as where to eat or the best classes to take next semester. After the meeting, have your team describe the most important features you would include in a software package to support the group's decision. What features described in this chapter would be the least important to your team in making its group decision?

3. Imagine that you and your team have decided to develop an ESS software product to support senior executives in the music recording industry. What are some of the key decisions these executives must make? Make a list of the capabilities that such a system must provide to be useful. Identify at least six sources of external information that will be useful to its users.

● WEB EXERCISES

1. Most companies typically have a number of functional MISs, such as finance. Find the sites of two finance companies, such as a bank or a brokerage company. Compare these sites. Which one do you prefer? How could these sites be improved? (*Hint:* If you are having trouble, try Yahoo. It should have a listing for "Business and Economy" on its home page. From there you can go to "Companies" and then "Finance." There will be several menu choices from there.) You may be asked to develop a report or send an e-mail message to your instructor about what you found.

2. FedEx and UPS both offer DSSs to help customers track their shipments. Research one of these companies using the Web and describe the Web pages and their features that can help customers track packages as they are shipped from one location to another. You may be asked to develop a report or send an e-mail message to your instructor about what you found.

3. Group decision support systems (GDSS) permit collaboration and joint work. Groupware, an important software component, allows people to communicate and work on joint projects. Using an Internet search engine, find one or more companies that make groupware. Describe the features of these groupware products. You may be asked to develop a report or send an e-mail message to your instructor about what you found.

● CASES

 Collaborative Work Gives a Competitive Edge

In addition to superior people skills, good project management skills are the hallmark of a good manager. Controlling costs and delivering a project on time is no simple task. But some managers have to juggle hundreds of projects at the same time. Tom Liddell, vice president of operations at Medical Manager Midwest, for example, has about 400 projects involving more than 150 people in over ten states. Other managers face similar project management workloads. How do managers control all these projects?

As discussed in this chapter, project management models can help a manager monitor and control projects. Software can help a manager determine the best and least expensive way to reduce the time it takes to complete a project. Often called *project crashing*, effectively reducing project completion time can involve optimization and decision support capabilities. Yet good project management systems are typically not enough for many project managers. Good group support systems are also needed in many cases.

Because most projects involve meetings, collaboration is essential. As with Medical Manager Midwest, projects can involve more than a hundred people who are dispersed over wide areas, requiring a wide area decision network or a teleconferencing approach to group support. Fortunately, software products combine some of the features of project management and group support packages. Medical Manager Midwest, for example, uses Project Home Page and ActionPlan products to help manage projects using group support capabilities. Both products are available through Netmosphere. These programs link project teams, give them access to centralized information, and use the Internet to allow project members to work in different geographic locations, permitting real-time communication and discussion. According to Netmosphere CEO Kevin Nickles, "Most of the traditional views of project management are not collaborative." Project Home Page and ActionPlan attempt to combine the best of project management and group decision support systems.

Discussion Questions

1. Why is project management so important?
2. What features of a GDSS would be important to help manage projects?

Critical Thinking Questions

3. If you were managing a large project with people in the United States and France, what capabilities would you like to see in software to help you manage the project?
4. Search the Internet for project management software. Describe the features you found and the features that are lacking.

Sources: Adapted from Helen Johnson, "A Competitive Edge in Collaborative Work," *Computerworld*, January 31, 2000, p. 68; Andy Donoghue, "Lutus Is Banking on Raven," *Network News*, February 2, 2000, p. 12; and "Project Tracker," *InformationWeek*, February 7, 2000, p. 137.

2 Marketing Research on the Internet

A marketing MIS is a key competitive tool for most companies. In addition to developing products, advertising products, and determining product prices, an effective marketing MIS requires good marketing research. Traditionally, companies have used surveys, questionnaires, pilot studies, and interviews to determine problems with existing products and services and to explore the potential of proposed products and services. Data collected from these marketing research approaches are typically entered into sophisticated statistical analysis programs. In the past, marketing research was done manually. For example, surveys and questionnaires were mailed to customers or potential customers. After a few weeks, the returned surveys and questionnaires were analyzed. Interviews were also conducted by companies that sent staff to shopping malls or other locations where they were likely to find people. Interviewing a few hundred people could take weeks and thousands of dollars. As with surveys and questionnaires, the data would then be entered into statistical analysis programs. Today, some companies are performing marketing research on the Internet.

Internet marketing research seems to be a winning strategy. It can be fast and inexpensive. Responses can take only a few days, and the raw data from the Internet site can be directly entered into statistical programs for analysis. With traditional surveys and questionnaires, the data often has to be manually entered into a computer. In addition to being fast, Internet marketing research can be inexpensive or even free. Questionnaires and surveys can be put onto Web pages. Mailing costs are eliminated for surveys and questionnaires, and a team of interviewers is not needed. With all these advantages, why would any company use more traditional ways to perform marketing research?

Polaroid Corporation decided to use the Internet to perform marketing research. The company asked more than 1,500 people about their use of Polaroid cameras and image scanners. The results were amaz-ing. About one fourth of the people responding to the survey said they had scanned a Polaroid picture. That seemed to be wonderful news until the company uncovered a very disturbing fact when the results were forecast to the entire population. The director of marketing research for Polaroid, Bruce Godfrey, soon realized that "that is more than the number of people who have even taken a Polaroid picture last year!" Something was terribly wrong. People responding to the Internet survey either didn't understand the question or lied. In addition, for any survey to be accurate, it must be representative of the general population. Most people agree that Web surfers are not typical consumers. Most Internet users tend to be richer and more technically skilled than the population in general. Marketing research performed on the Internet could then be seriously biased. To overcome these problems, some companies try to weight Internet results to make them more accurate. To others, however, weighting bad data only results in bad analysis.

Discussion Questions

1. What are the advantages of using the Internet to perform marketing research?
2. What are some of the disadvantages?

Critical Thinking Questions

3. If you were a marketing manager, under what circumstances would you use the Internet to perform marketing research?
4. Comment on the following: "The problems of performing marketing research on the Internet can be reduced by combining the results from the Internet with results from traditional marketing research methods."

Sources: Adapted from Erin White, "Market Research on the Internet Has Its Drawbacks," *The Wall Street Journal,* March 2, 2000, p. B4; "Outsourcing E-Commerce Website for Polaroid," *Washington Technology,* January 24, 2000, p. S4; and "Polaroid, Avery Form Marketing Alliance," *Photo Marketing Newsline,* January 19, 2000, p. 1.

3 Investment Information from Financial Web Sites

The Internet has penetrated most aspects of business and marketing. As discussed in the opening vignette, the "Ethical and Societal Issues" box, and throughout the chapter, financial information is now available over the Internet. The Internet excels at providing and analyzing data. Although general news, sports, and entertainment normally contain pictures and depend on sometimes quirky, subjective personalities, financial information is based on large amounts of more objective raw data and the power to analyze it. These requirements for financial information are ideal for delivery on the Internet, and a number of companies specialize in providing this type of information.

Larry Kramer was tired of the newspaper business and believed the entire industry was in decline. In 1994, he decided to launch an Internet site specializing in providing financial information. The original Internet site was a part of Data Broadcasting, which provided real-time stock quotes. Data Broadcasting gave a 38 percent stake of the company to CBS in return for the CBS brand name and $30 million in promotional advertising on CBS's TV, radio, and billboard operations. Kramer's upstart Internet site became CBS MarketWatch. CBS MarketWatch then made a $21 million deal with America Online to get an even bigger audience. The strategy appears to have worked. Today, CBS MarketWatch offers hard news on stocks, bonds, and the market in general. CBS MarketWatch is the most popular investment information site on the Internet, with more visitors per month than other financial information Internet sites. In December 1999, for example, about 3.5 million people visited the site at least once. With more than 70 reporters and editors, CBS MarketWatch attempts to provide fast, reliable information. Although the market share for CBS MarketWatch has soared, its stock price has had a few bumps along the way. The initial public offering in 1999 started at $17 per share. The share price then climbed to $120 per share before declining to about $40 per share.

In addition to CBS MarketWatch, a number of other Internet sites offer financial information. The second most visited site is The Motley Fool, which is known for its commentary and investment chat rooms. The Motley Fool had about two million visitors in December 1999. CNN FN, with about 1.3 million visitors, and TheStreet.com, with about 900,000 visitors, are next in terms of visitors. SmartMoney, CNBC, and The Wall Street Journal Online also have a significant number of visitors each month. Although some sites are free—notably,

CBS MarketWatch and The Motley Fool—others charge for their news and financial information (e.g., The Wall Street Journal Online). Like many Internet companies, it is not unusual for financial information companies to lose money each year. The hope is to gain market share that one day will turn into profits.

The competition among these firms is fierce. One company charged that one of its competitors tried to drive its stock price down to buy it at a lower price. This charge, however, has not been substantiated.

Discussion Questions

1. Why is financial information ideally placed on the Internet compared with other information and news?
2. What information would you like to see on a financial information Internet site?

Critical Thinking Questions

3. Do you think some financial information Internet companies are trying to drive the price of their competitors' stocks down by reporting misleading or false information? How would you determine whether this was happening?
4. Use the Internet to research several of the companies described in this case. Which one would be best for your financial needs? Describe the features you like about this site and which features you would like to see in the future.

Sources: Adapted from Marc Gunther, "The Top Financial Website Is Nobody's Fool," *Fortune*, February 21, 2000, p. 54; Marcia Vickers, "Call Him The Streetfighter.com," *Business Week*, March 6, 2000, p. 8; and "Motley Falls at Starting Line on Survey," *Money Marketing*, November 11, 1999, p. 1.

● NOTES

Sources for the opening vignette on p. 000: Adapted from Richard Morais, "Instihot," *Forbes,* March 6, 2000, p. 68; "Setting Up Shop," *The Economist,* February 12, 2000; and Stanley Reed, "Reuters Jumps into the Net," *Business Week,* February 21, 2000, p. 55.

1. Maria Trombly, "New Bank's Net-Only Vision," *Computerworld,* April 3, 2000, p. 12.

2. Chip Cummins, "Deutsche Bank, Chase Manhattan Join Move toward Online Currency Dealing," *The Wall Street Journal,* April 3, 2000, p. A43C.

3. Marcia Stepanek, "Webling," *Business Week,* April 3, 2000, p. EB26.

4. Janet Ginsburg, "Selling Sofas Online Is No Snap," *Business Week,* April 3, 2000, p. 96.

5. Craig Stedman, "Online Exchange to Sell Oracle Software," *Computerworld,* March 13, 2000, p. 4.

6. Julia King, "Businesses Weigh Pros and Cons of Web Marketplaces," *Computerworld,* March 13, 2000, p. 28.

7. Robin Robinson, "Customer Relationship Management," *Computerworld,* February 28, 2000, p. 67.

8. Neil Weinberg, "Not.Coms," *Forbes,* April 17, 2000, p. 424.

9. Joan Hamilton, "The Panic over Hiring," *Business Week,* April 3, 2000, p. EB130.

10. Pfeil et al., "Visteon's Sterling Plant Uses Simulation-Based Decision Support," *Interfaces,* January 2000, p. 115.

11. Richard Curch et al., "Support System Development for Forest Ecosystem Management," *European Journal of Operations Research,* March 1, 2000, p. 247.

12. Traci Hess et al., "Using Autonomous Software Agents to Create the Next Generation of Decision Support Systems," *Decision Sciences,* Winter 2000, p. 1.

13. Peggy King, "Decision Support Grows Up and Out," *CIO,* November 15, 1999, p. 88.

14. Ed Yourdan, "The Value of Triage," *Computerworld,* March 20, 2000. p. 40.

15. Milam Aiken et al., "An Abductive Model of Group Support Systems," *Information & Management,* March 1, 2000, p. 87.

16. Milam Aiken et al., "Flaming among First-Time Group Support System Users," *Information Management,* March 1, 2000, p. 95.

17. David Essex, "Teamware Offering That's Tailor-Made," *Computerworld,* April 10, 2000, p. 90.

18. Betty Vandenbosch, "An Empirical Analysis of the Association between the Use of Executive Support Systems and Perceived Organizational Competitiveness," *Accounting, Organizations and Society,* January 1999, p. 7.

Specialized Business Information Systems: Artificial Intelligence, Expert Systems, and Virtual Reality

> *A*ccepting what a computer tells you, instead of thinking, is a big mistake.
>
> — Herbert W. Lovelace, "But the Computer Said," *Information Week*, November 8, 1999

Principles	Learning Objectives
Artificial intelligence systems form a broad and diverse set of systems that can replicate human decision making for certain types of well-defined problems.	• *Define the term* artificial intelligence *and state the objective of developing artificial intelligence systems.* • *List the characteristics of intelligent behavior and compare the performance of natural and artificial intelligence systems for each of these characteristics.* • *Identify the major components of the artificial intelligence field and provide one example of each type of system.*
Expert systems can enable a novice to perform at the level of an expert but must be developed and maintained very carefully.	• *List the characteristics and basic components of expert systems.* • *Identify the individuals involved in the development of an expert system.* • *Identify the benefits associated with the use of expert systems.*
Virtual reality systems have the potential to reshape the interface between people and information technology by offering new ways to communicate information, visualize processes, and express ideas creatively.	• *Define the term* virtual reality *and provide three examples of virtual reality applications.*

Ask Jeeves

Search Engine Helps "Humanize" On-Line Experience

Ask Jeeves operates a consumer and corporate Web search service that allows users to seek information on the Internet using standard English questions. In addition, Microsoft, Dell, and other corporate customers employ Ask Jeeves to provide customer support and answer specific questions about their products and services. Using the Web site provides a customized, outsourced service for corporations that want an easier, more intuitive, and more intelligent way of interacting with customers. Ask Jeeves at Ask.com is the thirteenth most visited Web property, according to Media Metrix, a Web industry watcher. Ask.com answers three million questions each day and receives more than 12 million unique visitors per month, presenting a tremendous opportunity for advertisers to tap into Ask Jeeves's broad and diverse user base.

Once the user asks a question of Jeeves, the process goes like this:

1. Ask Jeeves attempts to understand the precise nature of the question by using a question-processing engine. Ask Jeeves has been programmed to understand and respond to statements made in English, so it determines both the meaning of the words in the question (semantic processing) and the meaning in the grammar of the question (syntactic processing).

2. Ask Jeeves's answer-processing engine provides a question template response—a list of questions that users see after they ask Jeeves a question. When users click on a question that closely matches theirs, the answer-processing engine retrieves an answer template that contains links to locations that can answer the question.

The Ask Jeeves knowledge base contains links to more than seven million answers, which contain information about the most frequently asked questions on the Internet. Smart lists allow one question template to point to many answers (e.g., "What is the population of <city name>?").

To assure users that they'll get the best answers to their questions, Ask Jeeves also has partners who provide answers to supplement Jeeves's own answers. This additional resource guarantees that the user can find the best answer; it includes the top ten answers from other leading search engines.

As it does its searches, Ask Jeeves captures customer intelligence from the questions they ask, the language they use to ask those questions, and the items they select in the Ask Jeeves system. This information is used to better understand what customers need and want, what is successful on its system, and what Ask Jeeves can't provide. So Ask Jeeves is constantly improving its responses to people's questions.

A new extension of the technology is a decision support engine that enables users to solve complex problems such as buying a digital camera, selecting annuities, and choosing among healthcare programs. The decision support engine engages in a dialogue concerning preferences leading beyond simple questions and answers and document location.

As you read this chapter, consider the following:

- What does it take to design and implement an effective artificial intelligence system?

- What business needs can be addressed by an expert system?

- What are some practical applications of virtual reality?

At a Dartmouth College conference in 1956, John McCarthy proposed the use of the term *artificial intelligence (AI)* to describe computers with the ability to mimic or duplicate the functions of the human brain. Many AI pioneers attended this first conference; a few predicted that computers would be as "smart" as people by the 1960s. The prediction has not yet been realized, but the benefits of artificial intelligence in business and research can be seen today. Advances in AI have led to many practical applications of systems (like the one used at Ask Jeeves) that are capable of making complex decisions.

AN OVERVIEW OF ARTIFICIAL INTELLIGENCE

Science fiction novels and popular movies have featured scenarios of computer systems and intelligent machines taking over the world. Computer systems such as Hal in the classic movie *2001: A Space Odyssey* are futuristic glimpses of what might be. These accounts are fictional, but we see the real application of many computer systems that use the notion of AI. These systems help to make medical diagnoses, explore for natural resources, determine what is wrong with mechanical devices, and assist in designing and developing other computer systems. In this chapter we explore the exciting applications of artificial intelligence and virtual reality and look at what the future really might hold.

Artificial Intelligence in Perspective

artificial intelligence systems

people, procedures, hardware, software, data, and knowledge needed to develop computer systems and machines that demonstrate characteristics of intelligence

Artificial intelligence systems include the people, procedures, hardware, software, data, and knowledge needed to develop computer systems and machines that demonstrate characteristics of intelligence. Researchers, scientists, and experts on how humans think are often involved in developing these systems. The objective in developing contemporary AI systems is not to replace human decision making completely but to replicate it for certain types of well-defined problems. As with other information systems, the overall purpose of artificial intelligence applications in business is to help an organization achieve its goals.

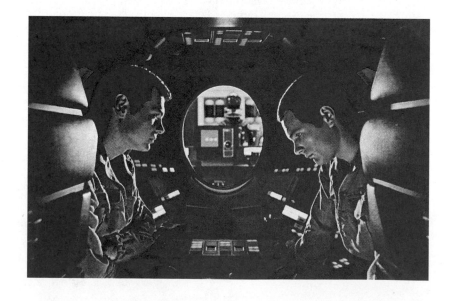

Science fiction movies give us a glimpse of the future, but many practical applications of artificial intelligence exist today, among them medical diagnostics, mechanical diagnostics, and development of computer systems.
(Source: Photofest)

The Nature of Intelligence

intelligent behavior

the ability to learn from experience and apply knowledge acquired from experience, handle complex situations, solve problems when important information is missing, determine what is important, and react quickly and correctly to a new situation

From the early AI pioneering stage, the research emphasis has been on developing machines with **intelligent behavior**. Some of the specific characteristics of intelligent behavior include the ability to do the following:

Learn from experience and apply the knowledge acquired from experience. Being able to learn from past situations and events is a key component of intelligent behavior and is a natural ability of humans, who learn by trial and error. However, learning from experience is not natural for computer systems. This ability must be carefully programmed into the system. Today, researchers are developing systems that have this ability. For instance, computerized AI chess games can learn to improve their game while they play human competitors (Figure 7.1).

In addition to learning from experience, people apply what they have learned to new settings and circumstances. Often, individuals take what they learn and succeed with in one endeavor and apply it to another. For example, a company that developed a dish-washing product effective for cleaning greasy dishes developed a variation of the product for use in cleaning up messy highway spills. Although humans have the ability to apply what they have learned to new settings, this characteristic is not automatic with computer systems. Developing computer programs to allow computers to apply what they have learned can be difficult.

Handle complex situations. Humans are involved in complex situations. World leaders face difficult political decisions regarding conflict, global economic conditions, hunger, and poverty. In a business setting, top-level managers and executives must handle a complex market, difficult and challenging competitors, intricate government regulations, and a demanding workforce. Even human experts make mistakes in dealing with these situations. Developing computer

FIGURE 7.1

Computers like Deep Blue attempt to learn from past chess moves. The powerful supercomputer's logic system was able to calculate the ramifications of up to 100 billion chess maneuvers within the allotted time for each move.
(Source: Photo courtesy of the Association for Computing Machinery.)

systems that can handle perplexing situations requires careful planning and elaborate computer programming.

Solve problems when important information is missing. The essence of decision making is dealing with uncertainty. Quite often, decisions must be made even when we lack information or have inaccurate information, because obtaining complete information is too costly or impossible. You have probably seen movies in which computers have responded to human commands with statements like "Does not compute" and "Insufficient information." Today, AI systems can make important calculations, comparisons, and decisions even when missing information.

Determine what is important. Knowing what is truly important is the mark of a good decision maker. Every day we are bombarded with facts and must process large amounts of data, filtering out what is unnecessary. Determining which items are crucial can make the difference between good decisions and those that ultimately lead to problems or failures. Computers, on the other hand, do not have this natural ability. Developing programs and approaches to allow computer systems and machines to identify important information is not a simple task.

React quickly and correctly to a new situation. A small child can look over a ledge or a drop-off and know not to venture too close. The child reacts quickly and correctly to a new situation. Computers, on the other hand, do not have this ability without complex programming.

Understand visual images. Interpreting visual images can be extremely difficult, even for sophisticated computers. People and animals can look at objects interacting in our environment and understand exactly what is going on. For instance, we can see a man sitting at a table and know that he has legs and feet that we cannot see. Being able to understand and correctly interpret visual images is an extremely complex process for computer systems. Moving through a room of chairs, tables, and other objects can be trivial for people but extremely complex for machines, robots, and computers. Such machines require an extension of understanding visual images, called a **perceptive system**. Having a perceptive system allows a machine to approximate the way a human sees, hears, and feels objects.

perceptive system

a system that approximates the way a human sees, hears, and feels objects

Process and manipulate symbols. People see, manipulate, and process symbols every day. Visual images provide a constant stream of information to our brains. By contrast, computers have difficulty handling symbolic processing and reasoning. Although computers excel at numerical calculations, they aren't as good at dealing with symbols and three-dimensional objects. Recent developments in machine-vision hardware and software, however, allow some computers to process and manipulate symbols on a limited basis.

Be creative and imaginative. Throughout history, some people have turned difficult situations into advantages by being creative and imaginative. For instance, when shipped a lot of defective mints with holes in the middle, an enterprising entrepreneur decided to market these new mints as Lifesavers instead of returning them to the manufacturer. Ice cream cones were invented at the St. Louis World's Fair when an imaginative store owner decided to wrap ice cream with a waffle from his grill for portability. Developing new and exciting products and services from an existing (perhaps negative) situation is a human characteristic. Few computers have the ability to be truly imaginative or creative in this way, although software has been developed to enable a computer to write short stories.

Use heuristics. With some decisions, people use heuristics (rules of thumb arising from experience) or even guesses. In searching for a job, we may decide to rank companies we are considering according to profits per employee. Companies making more profits might pay their employees more. In a manufacturing setting, a corporate president may decide to look at only certain locations for a new plant. We make these types of decisions using general rules of thumb, without completely searching all alternatives and possibilities. Today, some computer systems also have this ability. They can, given the right programs, obtain good solutions that use approximations instead of trying to search for an optimal solution, which would be technically difficult or too time-consuming.

This list of traits only partially defines intelligence. Unlike virtually every other field of information systems research in which the objectives can be clearly defined, the term *intelligence* is a formidable stumbling block. One of the problems in artificial intelligence is arriving at a working definition of real intelligence against which to compare the performance of an artificial intelligence system.

The Difference between Natural and Artificial Intelligence

Since the term *artificial intelligence* was defined in the 1950s, experts have disagreed about the difference between natural and artificial intelligence. For instance, is there a difference between carbon life (human or animal life) and silicon life (a computer chip) in terms of behavior? Profound differences exist, but they are declining in number (Table 7.1). One of the driving forces behind AI research is an attempt to understand how humans actually reason and think. Many experts believe that the ability to create machines that can reason will be possible only once we truly understand our own mental processes.

The Major Branches of Artificial Intelligence

AI is a broad field that includes several specialty areas, such as expert systems, robotics, vision systems, natural language processing, learning systems, and neural networks (Figure 7.2). Many of these areas are related; advances in one can occur simultaneously with or result in advances in others.

Expert Systems

An *expert system* consists of hardware and software that stores knowledge and makes inferences, similar to a human expert. Because of their many business applications, expert systems are discussed in more detail in the next several sections of the chapter.

TABLE 7.1

A Comparison of Natural and Artificial Intelligence

Attributes	Natural Intelligence (Human)	Artificial Intelligence (Machine)
The ability to use sensors (eyes, ears, touch, smell)	High	Low
The ability to be creative and imaginative	High	Low
The ability to learn from experience	High	Low
The ability to be adaptive	High	Low
The ability to afford the cost of acquiring intelligence	High	Low
The ability to use a variety of information sources	High	High
The ability to acquire a large amount of external information	High	High
The ability to make complex calculations	Low	High
The ability to transfer information	Low	High
The ability to make a series of calculations rapidly and accurately	Low	High

FIGURE 7.2

A Conceptual Model of
Artificial Intelligence

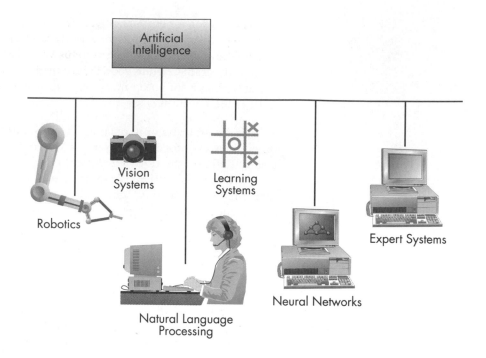

robotics

mechanical or computer devices
that perform tasks requiring a high
degree of precision or that are
tedious or hazardous for humans

FIGURE 7.3

Robots can be used in situations
that are hazardous, repetitive,
or difficult to do in other ways.
(Source: Courtesy of ABB Flexible
Automation.)

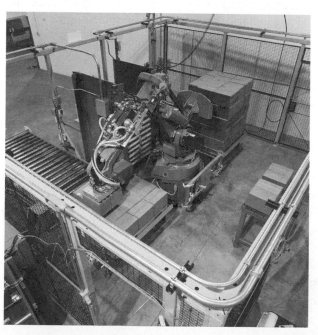

Robotics

Robotics involves developing mechanical or computer devices that can paint cars, make precision welds, and perform other tasks that require a high degree of precision or are tedious or hazardous for humans. Contemporary robotics combines both high-precision machine capabilities with sophisticated controlling software. The controlling software in robots is what is most important in terms of AI. The brain in an advanced industrial robot today works at about ten million instructions per second (MIPS)—no smarter than an insect. To achieve anything even approaching human intelligence, the robot brain must achieve 100 trillion operations per second.[1]

Many applications of robotics exist, and research into these unique devices continues (Figure 7.3). Manufacturers use robots to assemble and paint products. Welding robots have enabled firms to manufacture top-quality products and reduce labor costs while shortening delivery time to their customers.

In addition to their use in supporting manufacturing operations, robots have been applied in more unusual situations. Standardized Teleoperation System (STS) is a modular kit of robotic controls for converting a vehicle to unmanned operation. STS controls permit normal (or manual) vehicle operation and unmanned operation with the flick of a switch. The strap-on kit feature converts existing vehicles; vehicles can also be delivered with the STS installed. STS controls are used in many types and classes of vehicles, including bulldozers, tanks, and trucks used for bomb and landmine detection, neutralization, detonation, clearing, and route proofing. Some of the more unusual applications of robotics include a Ping-Pong ball server for the U.S. Olympic team, a toy soldier marching machine, and a bug sorter for museum display.

Although robots are essential components of today's automated manufacturing systems, future robots will find wider applications outside the factory in banks, restaurants, homes, and hazardous working environments such as nuclear stations. A robot must not only execute tasks programmed by the user but also be able to interact with its environment through its sensors and actuators, sense and avoid unforeseen obstacles, and perform its duties much the same way as humans.

Vision Systems

vision system

the hardware and software that permit computers to capture, store, and manipulate visual images and pictures

Another area of AI involves vision systems. **Vision systems** include hardware and software that permit computers to capture, store, and manipulate visual images and pictures. The U.S. Justice Department uses vision systems to perform fingerprint analysis, with almost the same level of precision as human experts. The speed with which the system can search through the huge database of fingerprints has brought a quick resolution to many long-standing mysteries. Vision systems are also effective at identifying people based on facial features.

Vision systems can be used in conjunction with robots to give these machines "sight." Robots such as those used in factory automation typically perform mechanical tasks with little or no visual stimuli. Robotic vision extends the capability of these systems, allowing the robot to make decisions based on visual input. Generally, robots with vision systems can recognize black and white and some gray-level shades but do not have good color or three-dimensional vision. Other systems concentrate on only a few key features in an image, ignoring the rest. It may take years before a robot or other computer system can "see" in full color and draw conclusions from what it sees, the way humans do.

Natural Language Processing

natural language processing

processing that allows the computer to understand and react to statements and commands made in a "natural" language, such as English

Natural language processing allows a computer to understand and react to statements and commands made in a "natural" language, such as English. There are three levels of voice recognition: command (recognizes dozens to hundreds of words), discrete (recognizes dictated speech with pauses between words), and continuous (recognizes natural speech). For example, a natural language processing system can be used to retrieve important information without typing in commands or searching for key words. With natural language processing, it is possible to speak into a computer's microphone and have the computer convert the electrical impulses generated from the voice into text files or program commands. With some simple natural language processors, you say a word into a microphone and type the same word on the keyboard. The computer then matches the sound with the typed word. With more advanced natural language processors, recording and typing words is not necessary.

Dragon Systems' Naturally Speaking Preferred Edition 5 uses continuous voice recognition or natural speech, allowing the user to speak to the computer at a normal pace without pausing between words. The spoken words are transcribed immediately onto the computer screen.
(Source: Courtesy of Dragon Systems, Inc., a Lemout & Hauspie Company)

Brokerage services are a perfect fit for voice-recognition technology to replace the existing "press 1 to buy or sell a stock" touch-pad telephone menu system. People buying and selling stock use a vocabulary too varied for easy access through menus and touch pads but small enough for software to process in real time. Several brokerages—including Charles Schwab, Fidelity Investments, DLJdirect, and TD Waterhouse Group—offer voice-recognition services. One of the big advantages is that the number of calls routed to the customer service department drops considerably once the new voice features are added. That is desirable to brokerages because market volatility means that the call centers are often either overstaffed or understaffed.

Within the next few years, voice recognition is expected to spread to all automated phone-answering systems. The main holdup is price and ease of installation—and the major vendors are working on both of these issues.[2]

Learning Systems

Another part of AI deals with **learning systems**, a combination of software and hardware that allows the computer to change how it functions or reacts to situations based on feedback it receives. For example, some computerized games have learning abilities. If the computer does not win a particular game, it remembers not to make the same moves under the same conditions. Learning systems software requires feedback on the results of its actions or decisions. At a minimum, the feedback needs to indicate whether the results are desirable (winning a game) or undesirable (losing a game). The feedback is then used to alter what the system will do in the future.

Neural Networks

An increasingly important aspect of AI involves neural networks. A **neural network** is a computer system that can simulate the functioning of a human brain. The systems use massively parallel processors in an architecture that is based on the human brain's own meshlike structure. In addition, neural network software can be used to simulate a neural network using standard computers. Neural networks can process many pieces of data at once and learn to recognize patterns. The systems then program themselves to solve related problems on their own. Some of the specific features of neural networks include the following:

- The ability to retrieve information even if some of the neural nodes fail
- Fast modification of stored data as a result of new information
- The ability to discover relationships and trends in large databases
- The ability to solve complex problems for which all the information is not present

Neural networks excel at pattern recognition. For example, neural network computers can be used to read bank check bar codes despite smears or poor-quality printing. Some hospitals use neural networks to determine a patient's likelihood of contracting cancer or other diseases. Neural nets work particularly well when analyzing detailed trends. Large amusement parks and banks use neural networks to figure out staffing needs based on customer traffic—a task that requires precise analysis, down to the half-hour. Increasingly, businesses are firing up neural nets to help them navigate ever-thicker forests of data and make sense of myriad customer traits and buying habits. Computer Associates has developed Neugents, neural intelligence agents, which "learn" patterns and behaviors and predict what will happen next. For example, Neugents can be used to track the habits of insurance customers and predict which ones will not renew, say, an automobile policy. They can then suggest to an insurance agent what changes might be made in the policy to get the consumer to renew it. The technology also can be employed to track individual users and their on-line preferences so that users at e-commerce sites don't have to input the same information each time they log on—their purchasing history and other data will be factored in each time they access a Web site.[3] Standard & Poor's (S&P) began using a neural network system called Decider from United Kingdom–based Neural Technologies to launch a credit-rating system on the Internet. The system, called CreditModel, allows S&P customers (banks and asset managers) to identify the firm for which they want a rating and then to launch the Decider engine, which develops a credit "score" on that company. It analyzes a list of variables, such as the company's debt load, revenue, and credit ratings that S&P collected in its files over the years— all at a fraction of the cost of a full rating.[4]

Read the "Ethical and Societal Issues" box to learn how neural networks might help disabled people.

ETHICAL AND SOCIETAL ISSUES
Neural Networks Provide Hope

It has been estimated that computers that can exhibit humanlike intelligence—including musical and artistic aptitude, creativity, the ability to move physically through the world, and emotional responsiveness—require processing power of 20 million billion calculations per second. Even more challenging than developing computer hardware with such capacity is developing software to mimic a human. Maybe it will be accomplished through essentially reverse-engineering the human brain, but we won't be able to crack the human brain's complexity anytime soon. It will be at least 2030 before we can design such computers. Much more promising is the use of neural network technology, which is being applied to assist disabled people.

The disabling effects of spinal injury or degenerative disease on voluntary movement can be permanent because damaged nerve cells and their "wiring" cannot regenerate. However, the motor areas of the brain that control body movements are usually left intact. Research scientists are exploring whether the activity of these motor areas can be used to operate robotic limbs.

John Chapin and his colleagues at MCP Hahnemann Medical College in Philadelphia trained a rat to press a lever for a food reward that was delivered by a robotic device. A 16-electrode array implanted in the rat's brain recorded the activity of 30 neurons that controlled the muscles involved in the lever press. The team then used these recordings to "train" a neural network to recognize the brain activity patterns that occur during a lever press. Control of the robotic device was switched from the lever to the neural network driven by the rat's thought patterns. Eventually the rat learned that pressing the lever with its paw was unnecessary and stopped its paw movements. Its brain activity continued to drive the robotic arm, bypassing nerves and muscles altogether.

The next stage of research involves the use of monkeys rather than rats and larger electrode arrays to record up to 130 movement-related neurons simultaneously. Such simultaneous recording is critical, because a neuron's activity is not specific to a particular muscle contraction, so it cannot give complete directions for movements by itself. If that experiment can succeed, the team could encode directions for a more complex robotic device.

Although it now seems possible that direct brain control of robotic actions could assist those disabled by spinal damage, transferring the technology from animal experiments to human use will be extremely difficult. Use of a neural network to control a robotic limb would require paralyzed patients to learn, through trial and error, how to shape brain activity appropriate for driving it. Another complication is that although electrodes can be anchored to the skull, they cannot be "hardwired" to the neurons themselves. Without permanent connections, the electrode tips and neurons could move slightly relative to one another and short-circuit the whole system. Perhaps the biggest obstacle to overcome in creating neural signal-based actuators is the design of multielectrode arrays that are both stable and safe for humans over the long term.

Discussion Questions

1. How probable do you think it is that we will soon see a humanlike robot such as those depicted in the *Terminator* movies?
2. What are the biggest barriers to development of successful direct brain control of robotic actions?

Critical Thinking Questions

3. Would mastery of direct brain control of robots lead to artificial limbs stronger and faster than natural limbs? Why or why not?
4. Would there be a need for an "emergency short-circuit" of direct brain control robots in the event that their owners lost their temper and had brief flashes of terrible thoughts?

Sources: Adapted from Ray Kurzweil, "Will My PC Be Smarter Than I Am?" *Time*, June 19, 2000, pp. 82–84; and Mimi Zucker, "Mind over Matter," *Scientific American*, November 1999, accessed at http://www.sciam.com/1999/1199issue/1199techbus2.html.

AN OVERVIEW OF EXPERT SYSTEMS

As we mentioned earlier, an expert system behaves similarly to a human expert in a particular field. Computerized expert systems have been developed to diagnose problems, predict future events, and solve energy problems. They have also been used to design new products and systems, determine the best use of lumber, and increase the quality of healthcare. Like human experts, computerized expert systems use heuristics, or rules of thumb, to arrive at conclusions or make suggestions. Expert systems have also been used to determine credit limits for credit cards (Figure 7.4). The research conducted in AI during the past two decades is resulting in expert systems that explore new business possibilities, increase overall profitability, reduce costs, and provide superior service to customers and clients.

Characteristics of an Expert System

Expert systems have a number of characteristics and capabilities, including the following:

Can explain their reasoning or suggested decisions. A valuable characteristic of an expert system is the capability to explain how and why a decision or solution was reached. For example, the expert system can explain the reasoning behind the conclusion to approve a particular loan application. The ability to explain its reasoning processes can be the most valuable feature of a computerized expert system. The user of the expert system can then understand the reasoning behind the conclusion.

Can display "intelligent" behavior. Considering a collection of data, an expert system can propose new ideas or approaches to problem solving. A few of the applications of expert systems are an imaginative medical diagnosis based on a patient's condition, a suggestion to explore for natural gas at a particular location, and providing job counseling for workers.

Can draw conclusions from complex relationships. Expert systems can evaluate complex relationships to reach conclusions and solve problems. For example, one proposed expert system will work with a flexible manufacturing system to determine the best use of tools. Another expert system can suggest ways to improve quality control procedures.

Can provide portable knowledge. One unique capability of expert systems is that they can be used to capture human expertise that might otherwise be lost. A classic example is the expert system called DELTA (Diesel Electric Locomotive Troubleshooting Aid), which was developed to preserve the expertise of the retiring David Smith, the only engineer competent to handle many highly technical repairs of such machines.

Can deal with uncertainty. One of an expert system's most important features is its ability to deal with knowledge that is incomplete or not completely accurate. The system deals with this problem through the use of probability, statistics, and heuristics.

Even though these characteristics of expert systems are impressive, other characteristics limit their current usefulness. Many of these limiting characteristics are related to concerns of cost, control, and complexity. Some of these characteristics are as follows:

Not widely used or tested. Even though successes occur, expert systems are not used in a large number of organizations. In other words, they have not been widely tested in corporate settings.

Difficult to use. Some expert systems are difficult to control and use. In some cases, the assistance of computer personnel or individuals trained in the use of expert systems is required to help the user get the most from these systems. Today's challenge is to make expert systems easier to use by decision makers who have limited computer programming experience.

Limited to relatively narrow problems. Although some expert systems can perform complex data analysis, others are limited to simple problems. Also, many problems solved by expert systems are not that beneficial in business settings. An expert system designed to provide advice on how to repair a machine, for example, cannot assist in decisions about when or whether to repair it. In general, the narrower the scope of the problem, the easier it is for an expert system to solve it.

Cannot readily deal with "mixed" knowledge. Expert systems cannot easily handle a knowledge base that has a mixed representation. Knowledge can be represented through defined rules, through comparison to similar cases, and in various other ways. An expert system in one application might not be able to deal with knowledge that combined both rules and cases.

Possibility of error. Although some expert systems have limited abilities to learn from experience, the primary source of knowledge is a human expert. If this knowledge is incorrect or incomplete, it will affect the system negatively. Other development errors involve programming. Because expert systems are more complex than other information systems, the potential for such errors is greater.

Cannot refine own knowledge base. Expert systems are not capable of acquiring knowledge directly. A programmer must provide instructions to the system that determine how the system is to learn from experience. Also, some expert systems cannot refine their own knowledge bases—such as eliminating redundant or contradictory rules.

Difficult to maintain. Related to the preceding point is that expert systems can be difficult to update. Some are not responsive or adaptive to changing conditions. Adding new knowledge and changing complex relationships may require sophisticated programming skills. In some cases, a spreadsheet can be used in conjunction with other expert system tools to modify the system. In others, upgrading an expert system can be too difficult for the typical manager or executive. Future expert systems are likely to be easier to maintain and update.

expert system shell

a collection of software packages and tools used to develop expert systems

May have high development costs. Expert systems can be expensive to develop with traditional programming languages and approaches. Development costs can be greatly reduced through the use of software for expert system development. **Expert system shells**, a collection of software packages and tools used to develop expert systems, can be implemented on most popular PC platforms to reduce development time and costs.

Raise legal and ethical concerns. People who make decisions and take action are legally and ethically responsible for their behavior. For example, a person can be taken to court and punished for a crime. But when expert systems are used to make decisions or help in the decision-making process, who is legally and ethically responsible? The human experts used to develop the knowledge base, the expert system developer, the user, or someone else? For example, if a doctor uses an expert system to make a diagnosis and the diagnosis is wrong, who is responsible? These legal and ethical issues have not been completely resolved.

When to Use Expert Systems

Sophisticated expert systems can be difficult, expensive, and time-consuming to develop, especially for large expert systems implemented on mainframes. Thus, it is important for an organization to make sure that the potential benefits are worth the effort and that various expert system characteristics are balanced in terms of cost, control, and complexity.

Following is a list of factors that normally make expert systems worth the expenditure of time and money:

- Provide a high potential payoff or significantly reduced downside risk
- Can capture and preserve irreplaceable human expertise
- Can develop a system more consistent than human experts
- Can provide expertise needed at a number of locations at the same time or in a hostile environment that is dangerous to human health
- Can provide expertise that is expensive or rare
- Can develop a solution faster than human experts can
- Can provide expertise needed for training and development to share the wisdom and experience of human experts with a large number of people

Components of Expert Systems

An expert system consists of a collection of integrated and related components, including a knowledge base, an inference engine, an explanation facility, a knowledge base acquisition facility, and a user interface. A diagram of a typical expert system is shown in Figure 7.5. In this figure, the user interacts with the user interface, which interacts with the inference engine. The inference engine interacts with the other expert system components. These components work together in providing expertise.

The Knowledge Base

knowledge base

a component of an expert system that stores all relevant information, data, rules, cases, and relationships used by the expert system

The **knowledge base** stores all relevant information, data, rules, cases, and relationships used by the expert system. A knowledge base must be developed for each unique application. For example, a medical expert system contains facts about diseases and symptoms. The knowledge base can include generic knowledge from general theories that have been established over time and specific knowledge that comes from more recent experiences and rules of thumb. Knowledge bases, however, go far beyond simple facts, also storing relationships, rules or frames, and cases. For example, certain telecommunications network problems may be related or linked; one problem may cause another. In other cases, rules suggest certain conclusions, based on a set of given facts. In many instances, these rules are stored as **if-then statements**, such as "If a certain set of network conditions exists, then a certain network problem diagnosis is appropriate." Cases can also be used. This technique involves finding instances, or cases, that are similar to the current problem and modifying the solutions to these

if-then statements

rules that suggest certain conclusions

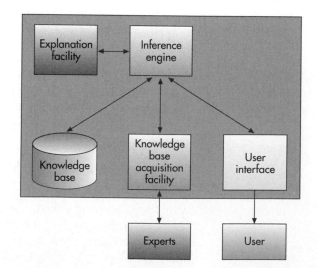

FIGURE 7.5

Components of an Expert System

cases to account for any differences between the previously solved cases stored in the computer and the current situation or problem.

Purpose of a knowledge base. The overall purpose of the knowledge base is to hold the relevant facts and information for the specific expert system. A knowledge base is similar to the sum of a human expert's knowledge and experience gained through years of work in a specific area or discipline. The goal of the system is to capture as much experience and knowledge as possible.

Consider an expert system that locates hardware problems for a large mainframe computer system. A human expert knows a large number of facts about the system and will also look for specific problems that have occurred frequently in the past. Human experts use a number of rules to help locate problems. If the malfunctioning computer displays certain types of behaviors, then the human inspects certain parts of the hardware for potential problems. A knowledge base developed to identify hardware problems also contains important facts on the hardware, information on frequent problems, and relationships between computer performance and what may be wrong.

Assembling human experts. One challenge in developing a knowledge base is to assemble the knowledge of multiple human experts. Typically, the objective in building a knowledge base is to integrate the knowledge of individuals with similar expertise (e.g., many doctors may contribute to a medical diagnosis knowledge base). A knowledge base that contains information from numerous experts can be extremely powerful and accurate. Unfortunately, human experts can disagree on important relationships and interpretations of data, presenting a dilemma for designers and developers of knowledge bases and expert systems in general. Some human experts are more expert than others; their knowledge, experience, and information are better developed and more accurately represent reality. When human experts disagree on important points, it can be difficult for expert systems developers to determine which rules and relationships to place in the knowledge base.

The use of fuzzy logic. Another challenge for expert system designers and developers is capturing knowledge and relationships that are not precise or exact. Computers typically work with numerical certainty; certain input values will always result in the same output. In the real world, as you know from experience, this is not always the case. To handle this dilemma, a specialty research area in computer science, called **fuzzy logic**, has been developed. Research into fuzzy logic has been going on for decades, but application to expert systems is just beginning to show results in a variety of areas.

fuzzy logic
a special research area in computer science that allows shades of gray and does not require conditions to be black/white, yes/no, or true/false

Instead of the usual black/white, yes/no, or true/false conditions of typical computer decisions, fuzzy logic allows shades of gray, or what are known as "fuzzy sets." The criteria on whether a subject or situation fits into a set are given in percentages or probabilities. For example, a weather forecaster might state that "if it is very hot with high humidity, the likelihood of rain is 75 percent." The imprecise terms of "very hot" and "high humidity" are what fuzzy logic must determine to formulate the chance of rain. Fuzzy logic rules help computers evaluate the imperfect or imprecise conditions they encounter and make "educated guesses" based on the likelihood or probability of correctness of the decision. This ability to estimate whether a condition fits a situation more closely resembles the judgment a person makes when evaluating situations.

Fuzzy logic is used in embedded computer technology—for example, in autofocus cameras, medical equipment that monitors patients' vital signs and makes automatic corrections, and temperature sensors attached to furnace controls.

Fuzzy logic was first applied in Japan. Seiji Yasunobu used it in an automatic control system for the city of Sendai's subway system. Even in a country famed for the precision of its underground railways, Sendai's is impressive. Each train stops to within 7 cm (3 in.) of the right spot on the platform. In addition, the trains travel more smoothly and use about 10 percent less energy than their human-controlled equivalents. The person in the driver's compartment is there for little more than reassurance. DaimlerChrysler used fuzzy logic to optimize the design process of Mercedes Benz truck components, such as gear boxes, axles, and steering. And in a pilot study, a U.S. hospital used fuzzy logic to estimate the length of patients' hospital stays. The system uses the information that is provided by the doctor who admits the patient to the hospital, considering the diagnosis, the patient's general condition, the likelihood of complications, the patient's previous medical history (if available), and other information. In yet another application of fuzzy logic, data analysis, the system must decide whether two similar entries in a database really represent the same person. This decision is harder than it appears at first glance because entries can vary in many ways—minor differences in spelling, order and grouping of words, and typos. Most humans, based on their experience comparing addresses, can come up with a pretty good guess whether two addresses in a database belong to the same person, but a mathematical model for the similarity of two addresses is hard to define.

rule

a conditional statement that links given conditions to actions or outcomes

The use of rules. A **rule** is a conditional statement that links given conditions to actions or outcomes. As we saw earlier, a rule is constructed using if-then statements. If certain conditions exist, then specific actions are taken or certain conclusions are reached. In an expert system for a weather forecasting operation, for example, the rules could state that if certain temperature patterns exist with a given barometric pressure and certain previous weather patterns over the last 24 hours, then a specific forecast will be made, including temperatures, cloud coverage, and the wind-chill factor. Rules are often combined with probabilities, such as if the weather has a particular pattern of trends, then there is a 65 percent probability that it will rain tomorrow. Likewise, rules relating data and conclusions can be developed for any knowledge base. Most expert systems prevent users from entering contradictory rules. Figure 7.6 shows the use of expert system rules in helping to determine whether a person should receive a mortgage

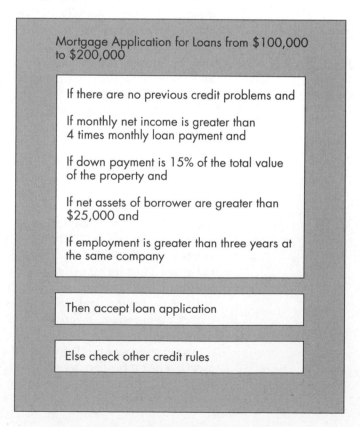

Mortgage Application for Loans from $100,000 to $200,000

If there are no previous credit problems and

If monthly net income is greater than 4 times monthly loan payment and

If down payment is 15% of the total value of the property and

If net assets of borrower are greater than $25,000 and

If employment is greater than three years at the same company

Then accept loan application

Else check other credit rules

FIGURE 7.6

Rules for a Credit Application

loan from a bank. In general, as the number of rules an expert system knows increases, the precision of the expert system increases.

The use of cases. As mentioned previously, an expert system can use cases in developing a solution to a current problem or situation. This process involves (1) finding cases stored in the knowledge base that are similar to the problem or situation at hand and (2) modifying the solutions to the cases to fit the current problem or situation. Cases stored in the knowledge base can be identified and selected by comparing the parameters of the new problem with the cases stored in the computer system. For example, a company may be using an expert system to determine the best location of a new service facility in the state of New Mexico. Labor and transportation costs may be the most important factors. The expert system may identify two cases involving the location of a service facility where labor and transportation costs were also important—one in the state of Colorado and the other in the state of Nevada. The expert system will modify the solution to these two cases to determine the best location for a new facility in New Mexico. The result might be to locate the new service facility in the city of Santa Fe.

The Inference Engine

inference engine

part of the expert system that seeks information and relationships from the knowledge base and provides answers, predictions, and suggestions the way a human expert would

The overall purpose of an **inference engine** is to seek information and relationships from the knowledge base and to provide answers, predictions, and suggestions the way a human expert would. In other words, the inference engine is the component that delivers the expert advice.

The process of retrieving relevant information and relationships from the knowledge base is not simple. As you have seen, the knowledge base is a collection of facts, interpretations, and rules. The inference engine must find the right facts, interpretations, and rules and assemble them correctly. In other words, the inference engine must make logical sense out of the information contained in the knowledge base, the way the human mind does when sorting out a complex situation. The inference engine has a number of ways of accomplishing its tasks, including backward and forward chaining.

backward chaining

the process of starting with conclusions and working backward to the supporting facts

Backward chaining. **Backward chaining** is the process of starting with conclusions and working backward to the supporting facts. If the facts do not support the conclusion, another conclusion is selected and tested. This process is continued until the correct conclusion is identified.

Consider an expert system that forecasts product sales for next month. With backward chaining, we start with a conclusion, such as "Sales next month will be 25,000 units." Given this conclusion, the expert system searches for rules in the knowledge base that support the conclusion, such as "IF sales last month were 21,000 units and sales for competing products were 12,000 units, THEN sales next month should be 25,000 units or greater." The expert system verifies the rule by checking sales last month for the company and its competitors. If the facts are not true—in this case, if last month's sales were not 21,000 units or 12,000 units for competitors—the expert system would start with another conclusion and proceed until rules, facts, and conclusions matched.

forward chaining

the process of starting with the facts and working forward to the conclusions

Forward chaining. **Forward chaining** starts with the facts and works forward to the conclusions. Consider the expert system that forecasts future sales for a product. With forward chaining, we start with a fact, such as "The demand for the product last month was 20,000 units." With the forward-chaining approach, the expert system searches for rules that contain a reference to product demand. For example, "IF product demand is over 15,000 units, THEN check the demand for competing products." As a result of this process, the expert system might use information on the demand for competitive products. Next, after searching additional rules, the expert system might use information on personal income or inflation on a national basis. This process continues until the expert system can reach a conclusion using the data supplied by the user and the rules that apply in the knowledge base.

   A comparison of backward and forward chaining. Forward chaining can reach conclusions and yield more information with fewer queries to the user than backward chaining, but this approach requires more processing and a greater degree of sophistication. Forward chaining is often used by more expensive expert systems. It is also possible to use mixed chaining, which is a combination of backward and forward chaining.

The Explanation Facility

explanation facility

component of an expert system that allows a user or decision maker to understand how the expert system arrived at certain conclusions or results

An important part of an expert system is the **explanation facility**, which allows a user or decision maker to understand how the expert system arrived at certain conclusions or results. A medical expert system, for example, may have reached the conclusion that a patient has a defective heart valve given certain symptoms and the results of tests on the patient. The explanation facility allows a doctor to find out the logic or rationale of the diagnosis made by the expert system. The expert system, using the explanation facility, can indicate all the facts and rules that were used in reaching the conclusion. This facility allows doctors to determine whether the expert system is processing the data and information in a correct and logical fashion.

The Knowledge Acquisition Facility

A difficult task in developing an expert system is the process of creating and updating the knowledge base. In the past, when more traditional programming languages were used, developing a knowledge base was tedious and time-consuming. Each fact, relationship, and rule had to be programmed into the knowledge base. In most cases, an experienced programmer was required to create and update the knowledge base.

knowledge acquisition facility

part of the expert system that provides convenient and efficient means of capturing and storing all the components of the knowledge base

   Today, specialized software allows users and decision makers to create and modify their own knowledge bases through the knowledge acquisition facility (Figure 7.7). The overall purpose of the **knowledge acquisition facility** is to provide a convenient and efficient means for capturing and storing all components of the knowledge base. Knowledge acquisition software can present users and decision makers with easy-to-use menus. After filling in the appropriate attributes, the knowledge acquisition facility correctly stores information and relationships in the knowledge base, making the knowledge base easier and less expensive to set up and maintain. Knowledge acquisition can be a manual process or a mixture of manual and automated procedures. Regardless of how the knowledge is acquired, it is important to validate and update the knowledge base frequently to make sure it is still accurate.

The User Interface

Specialized user interface software is employed for designing, creating, updating, and using expert systems. The main purpose of the user interface is to make the development and use of an expert system easier for users and decision makers. At

FIGURE 7.7

The knowledge acquisition facility acts as an interface between experts and the knowledge base.

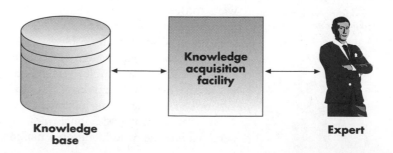

Knowledge base   **Knowledge acquisition facility**   **Expert**

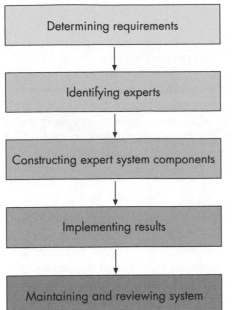

FIGURE 7.8

Steps in the Expert System Development Process

domain expert

the individual or group whose expertise or knowledge is captured for use in the expert system

one time, skilled computer personnel created and operated most expert systems; today, the user interface permits decision makers to develop and use their own expert systems. Because expert systems place more emphasis on directing user activities than do other types of systems, text-oriented user interfaces (using menus, forms, and scripts) may be more common in expert systems than the graphical interfaces often used with DSSs.

Expert Systems Development

Like other computer systems, expert systems require a systematic development approach for best results. This approach includes determining the requirements for the expert system, identifying one or more experts in the area or discipline under investigation, constructing the components of the expert system, implementing the results, and maintaining and reviewing the complete system (Figure 7.8). First we describe the participants who are involved in expert system development for an organization.

Participants in Developing and Using Expert Systems

Typically, several people are involved in developing and using an expert system, chiefly domain experts, knowledge engineers, and knowledge users (Figure 7.9).

The domain expert. Because of the time and effort involved in the task, an expert system is developed to address only a specific area of knowledge. This area of knowledge is called the domain. An organization must carefully evaluate the domain of the expert system to determine its stability and longevity and weigh that against its cost to implement. Many domains, such as the design of microcomputer chips, change quickly in their content and structure. Rapid changes in the knowledge or rules used to make decisions will quickly invalidate the system. On the other hand, an expert system should be built to be flexible so that new rules and knowledge can be added to the system—in effect, permitting the system to learn. The **domain expert** is the individual or group who has the expertise or knowledge an organization is trying to capture in the expert system. In most cases, the domain expert is a group of human experts. The domain expert (individual or group) usually has the ability to do the following:

- Recognize the real problem
- Develop a general framework for problem solving
- Formulate theories about the situation
- Develop and use general rules to solve a problem
- Know when to break the rules or general principles

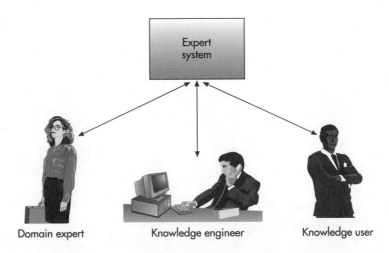

FIGURE 7.9

Participants in Expert Systems Development and Use

- Solve problems quickly and efficiently
- Learn from past experience
- Know what is and is not important in solving a problem
- Explain the situation and solutions of problems to others

knowledge engineer

an individual who has training or experience in the design, development, implementation, and maintenance of an expert system

knowledge user

the individual or group who uses and benefits from the expert system

The knowledge engineer and knowledge users. A **knowledge engineer** is an individual who has training or experience in the design, development, implementation, and maintenance of an expert system, including training or experience with expert system shells. The **knowledge user** is the individual or group who uses and benefits from the expert system. Knowledge users do not need any previous training in computers or expert systems.

Expert Systems Development Alternatives

Expert systems can be developed from scratch by using an expert system shell or by purchasing an existing expert system package. The approach selected depends on the system benefits compared with the cost, control, and complexity of each alternative. A graph of the general cost and time of development is shown in Figure 7.10. It is usually faster and less expensive to develop an expert system using an existing package or an expert system shell. If the organization does not already have this type of software, then there will be an additional cost to develop or acquire it.

In-house development: develop from scratch. Developing an expert system from scratch is usually more costly than the other alternatives, but an organization has more control over the features and components of the system. Such customization also has a downside; it can result in a more complex system, with higher maintenance and updating costs.

In-house development: develop from a shell. As discussed earlier, an expert system shell consists of one or more software products that assist in the development of an expert system. In some instances, the same shell can be used to develop many expert systems. An expert system developed from a shell

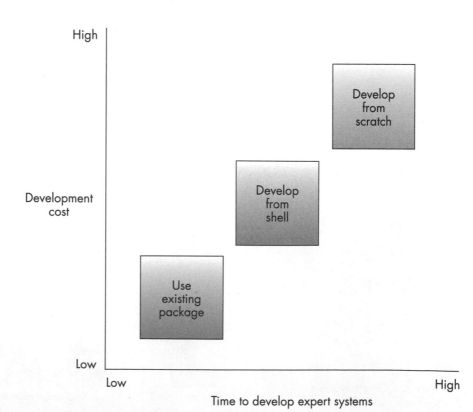

FIGURE 7.10

Some Expert Systems Development Alternatives and Their Relative Cost and Time Values can be less complex and easier to maintain than one developed from scratch. However, the expert system may need to be modified to tailor it to specific applications. In addition, the capabilities and features of the expert system can be more difficult to control.

Off-the-shelf purchase: use existing packages. Using an existing expert system package is the least expensive and fastest approach, in most cases. An existing expert system package is one that has been developed by a software or consulting company for a specific field or area, such as the design of a new computer chip or a weather forecasting and prediction system. The advantages of using an existing package can go beyond development time and cost. These systems can also be easy to maintain and update over time. A disadvantage of using an off-the-shelf package is that it may not be able to satisfy the unique needs of an organization.

Applications of Expert Systems and Artificial Intelligence

Expert systems and artificial intelligence are being used in a variety of ways. Read the "E-Commerce" box to on page 269 learn how bots are being used to support corporate buyers. Some of the other applications of these systems are summarized next.

Games. Some expert systems are used for entertainment. For example, Proverb is an expert system designed to solve standard American crosswords given the grid and clues.

Legal profession. SHYSTER is a case-based legal expert system, developed by James Popple. It provides advice in areas of case law that have been defined by a legal expert using a unique specification language.

AI and expert systems embedded in products. The antilock braking system on modern automobiles is an example of a rudimentary expert system. A processor senses when the tires are beginning to skid and releases the brakes for a fraction of a second to prevent the skid. AI researchers are also finding ways to use neural networks and robotics in everyday devices, such as toasters, alarm clocks, and televisions.

Plant layout. FLEXPERT is an expert system that uses fuzzy logic to perform plant layout. The software helps companies determine the best placement for equipment and manufacturing facilities.

Hospitals and medical facilities. Some hospitals use expert systems to determine a patient's likelihood of contracting cancer or other diseases. MYCIN is an expert system started at Stanford University to analyze blood infections. A medical expert system used by the Harvard Community Health Plan allows members of the HMO to get medical diagnoses via home personal computers. For minor problems, the system gives uncomplicated treatments; for more serious conditions, the system schedules appointments. The system is highly accurate, diagnosing 97 percent of the patients correctly (compared with the doctors' 78 percent accuracy rating). To help doctors in the diagnosis of thoracic pain, MatheMEDics has developed THORASK, a straightforward, easy-to-use program, requiring only the input of carefully obtained clinical information.

MatheMEDics' Easy Diagnosis is an on-line medical expert system. Patients answer a series of questions about their symptoms or signs, and the program presents the most likely conditions or diagnoses, in descending order of their probabilities.
(Source: Courtesy of MatheMEDics, Inc.)

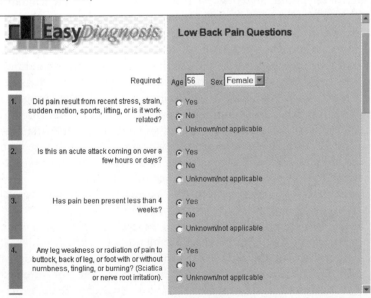

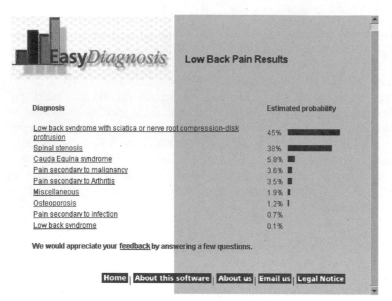

The program helps the less experienced to distinguish the three principal categories of chest pain from each other. It does what a true medical expert system should do without the need for complicated user input. You answer basic questions about the patient's history and directed physical findings, and the program immediately displays a list of diagnoses. The diagnoses are presented in decreasing order of likelihood, together with their estimated probabilities. The program also provides concise descriptions of relevant clinical conditions and their presentations as well as brief suggestions for diagnostic approaches. For record keeping, documentation, and data analysis, there are options for saving and printing cases.

Help desks and assistance. Expert systems are used by customer service help desks to provide timely and accurate assistance. Kaiser Permanente, a large HMO, uses an expert system and voice response to automate its help desk function. The automated help desk frees up staff to handle more complex needs, while still providing more timely assistance for routine calls.

Employee performance evaluation. An expert system by Austin-Hayne, called Employee Appraiser, provides managers with expert advice for use in employee performance reviews and career development.

Loan analysis. KPMG Peat Marwick uses an expert system called Loan Probe to review loan loss reserves to determine whether sufficient funds have been set aside to cover the risk that some loans will be uncollectible.

Virus detection. IBM is using neural network technology to help create more advanced software for eradicating computer viruses, a major problem in American businesses. IBM's neural network software deals with "boot sector" viruses, the most prevalent type, using a form of artificial intelligence that mimics the human brain and generalizes by looking at examples. It requires a vast number of training samples, which in the case of antivirus software are three-byte virus fragments.

Repair and maintenance. ACE is an expert system used by AT&T to analyze the maintenance of telephone networks. Nynex (New York and New England Telephone Exchange) has expert systems to help its workers locate and solve customer-related phone problems. IET-Intelligent Electronics uses an expert system to diagnose maintenance problems related to aerospace equipment. In the airline industry, prognosis can help reduce the high costs of unscheduled major component removals (such as engines) or failures in flight (such as in-flight engine shut-downs). Assessment of equipment health also supports early and better identification and planning for optimal maintenance. General Electric Aircraft Engine Group uses an expert system to enhance maintenance performance levels at all sites, improve diagnostic accuracy and reduce ambiguity of the fault, advise on real-time repair action, provide clues to failures, and access and display relevant maintenance information.

Marketing. CoverStory is an expert system that extracts marketing information from a database and automatically writes marketing reports.

Warehouse optimization. United Distillers uses an expert system to determine the best combinations of stocks to produce its blends of Scottish whiskey. This information is then supplemented with information about location of the casks for each blend. The system then optimizes the selection of required casks, keeping to a minimum the number of "doors" (warehouse sections)

E-COMMERCE

Bots Automate Corporate Purchasing

Bots (short for "knowledge robots") are programs that help users accomplish certain tasks. Bots can be written in a variety of programming languages and feature triggers that allow them to execute without human intervention. Most bots also can adapt to learn users' tendencies and preferences and so become personalized based on what they learn about users.

There are plenty of bots in use today. The beady-eyed paper clip seen in Windows 98 applications is an example of a bot. The most basic type of bot simply does the same routine task every day, such as backing up a hard drive for a server every night. More powerful bots can roam the Internet on your behalf and automatically perform intelligent searches, answer questions, tell you when an event occurs, provide individualized news delivery, and comparison shop. In fact, experts see procurement as the killer application of business-to-business bots.

One comparison shopping robot, MySimon (http://www.mysimon.com/index.anml), doesn't sell anything; instead, it compares millions of products and prices at thousands of on-line stores that offer a wide range of products—appliances, credit cards, chocolates, electronics, and toys. Type in "Paul Garmirian cigars", and it lists 25 merchants. The results can be sorted by merchant, manufacturer, model, or price. You can also find out a product's price history. There are buyer guides for in-depth information and product suggestions based on your preferences. Now extend the MySimon model to corporate purchasing and procurement, and you can see why bots are a hot issue for businesses. Such databots use neural networks and exhibit characteristics usually associated with intelligent life forms, such as learning and creativity.

Much work remains before bots realize their potential. If bots are to collaborate and form a vast network of super corporate buyer agents, standards must be developed and adopted to address communication and security issues. The Defense Advanced Research Projects Agency is working on developing the Knowledge Query and Manipulation Language (KQML), which is both a message format and a message-handling protocol to support knowledge sharing among agents. In other words, KQML can be used as a language for an application program to interact with a bot or for two or more bots to share knowledge for cooperative problem solving. KQML provides a basic architecture for knowledge sharing through a special class of agent called *communication facilitators*, which coordinate the interactions of other bots.

In addition to complicated technical problems, there are legal and ethical issues as well. For example, in May 2000, U.S. District Court Judge Ronald Whyte barred Burlington, Massachusetts–based Bidder's Edge from using an automated system to search eBay on the grounds that it could slow the auction giant's site. The decision could also have broader implications for the openness of the Internet, because it relies on laws against trespass, not copyright infringement.

Discussion Questions

1. How might a communication facilitator bot assist a procurement bot in carrying out its role?
2. Why is it important for a procurement bot to exhibit the ability to learn?

Critical Thinking Questions

3. Why is it important for bots to be able to communicate with one another? What would be the impact of a "misinformation" robot that, when unleashed, provides incorrect information to other bots?
4. If the use of procurement bots proves highly successful, how might the role of the human purchasing agent change? Can you identify a set of products or services for which bots would not be effective?

Sources: Adapted from Steve Ulfelder, "Undercover Agents," *Computerworld*, June 5, 2000, p. 85; George Lawton, "Putting Agents to Work," *Knowledge Management*, November 1999, pp. 68–73; "What Is KQML," KQML Lab Web site, http://www.csee. umbc.edu/kqml/whats-kqml.html, accessed June 16, 2000; and Linda Rosencrance and Melissa Solomon, "U.S. Judge Blocks Web Bot from eBay Site," *Computerworld*, May 26, 2000, http://www.computerworld.com/.

from which the casks must be taken and the number of casks that need to be moved to clear the way. Other constraints must be satisfied, such as the current working capacity of each warehouse and the maintenance and restocking work that may be in progress.

VIRTUAL REALITY

virtual reality system

a system that enables one or more users to move and react in a computer-simulated environment

The term *virtual reality* was initially coined by Jaron Lanier, founder of VPL Research, in 1989. Originally, the term referred to *immersive virtual reality,* in which the user becomes fully immersed in an artificial, three-dimensional world that is completely generated by a computer. A **virtual reality system** enables one or more users to move and react in a computer-simulated environment.

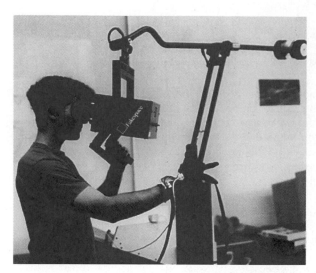

FIGURE 7.11

The BOOM, a Head-Coupled Display Device
(Source: Courtesy of University of Michigan Virtual Reality Laboratory)

Virtual reality simulations require special devices that transmit the sights, sounds, and sensations of the simulated world to the user. These devices can also record and send the speech and movements of the participants to the simulation program. Thus, users can sense and manipulate virtual objects much as they would real objects. This natural style of interaction gives participants the feeling that they are immersed in the simulated world.

Interface Devices

To see in a virtual world, often the user wears a head-mounted display (HMD) with screens directed at each eye. The HMD also contains a position tracker to monitor the location of the user's head and the direction in which the user is looking. Using this information, a computer generates images of the virtual world—a slightly different view for each eye—to match the direction in which the user is looking and displays these images on the HMD. With current technology, virtual-world scenes must be kept relatively simple so that the computer can update the visual imagery quickly enough (at least ten times a second) to prevent the user's view from appearing jerky and from lagging behind the user's movements.

Alternative concepts—BOOM and CAVE—were developed for immersive viewing of virtual environments to overcome the often uncomfortable intrusiveness of a head-mounted display.

The BOOM (Binocular Omni-Orientation Monitor) from Fakespace Labs is a head-coupled stereoscopic display device (Figure 7.11). Screens and optical systems are housed in a box that is attached to a multilink arm. The user looks into the box through two holes, sees the virtual world, and can guide the box to any position within the virtual environment. Head tracking is accomplished via sensors in the links of the arm that holds the box.

The Electronic Visualization Laboratory at the University of Illinois at Chicago introduced a room constructed of large screens on which the graphics are projected onto the three walls and the floor (Figure 7.12). The CAVE, as this room is called, provides the illusion of immersion by projecting stereo images on the walls and floor of a room-sized cube. Several persons wearing lightweight stereo glasses can enter and walk freely inside the CAVE. A head-tracking system continuously adjusts the stereo projection to the current position of the leading viewer.

FIGURE 7.12

Viewing the Detroit Midfield Terminal in an Immersive CAVE System
(Source: Courtesy of University of Michigan Virtual Reality Laboratory.)

Users hear sounds in the virtual world through earphones. The information reported by the position tracker is also used to update audio signals. When a sound source in virtual space is not directly in front of or behind the user, the computer transmits sounds to arrive at one ear a little earlier or later than at the other and to be a little louder or softer and slightly different in pitch.

The *haptic* interface, which relays the sense of touch and other physical sensations in the virtual world, is the least developed and perhaps the most challenging to create. Currently, with the use of a glove and position tracker, the computer locates the user's hand and measures finger movements. The user can reach into the virtual world and handle objects; however, it is difficult to generate the sensations that are felt when a person taps a hard surface, picks up an object, or runs a finger across a textured surface. Touch sensations also have to be synchronized with the sights and sounds users experience.

Immersive Virtual Reality

In immersive virtual reality, the virtual world is presented in full scale and relates properly to humans' size. It may represent any three-dimensional setting, real or abstract, such as a building, an archaeological excavation site, the human anatomy, a sculpture, or a crime scene reconstruction. Virtual worlds can be animated, interactive, and shared. Through immersion, the user can gain a deeper understanding of the virtual world's behavior and functionality.

Other Forms of Virtual Reality

Virtual reality can also refer to applications that are not fully immersive, such as mouse-controlled navigation through a three-dimensional environment on a graphics monitor, stereo viewing from the monitor via stereo glasses, stereo projection systems, and others. Some virtual reality applications allow views of real environments with superimposed virtual objects. Motion trackers monitor the movements of dancers or athletes for subsequent studies in immersive virtual reality. Telepresence systems (e.g., telemedicine, telerobotics) immerse a viewer in a real world that is captured by video cameras at a distant location and allow for the remote manipulation of real objects via robot arms and manipulators. Many believe that virtual reality will reshape the interface between people and information technology by offering new ways to communicate information, visualize processes, and express ideas creatively.

Useful Applications

There are literally hundreds of applications of virtual reality, with more and more being developed as the cost of hardware and software declines and people's imaginations are opened to its potential. Here is a summary of some of the more interesting applications.

Medicine

Surgeons in France performed the first successful closed-chest coronary bypass operation in 1999. Instead of cutting open the patient's chest and breaking his breastbone, as is usually done, surgeons used a virtual reality system that enabled them to operate through three tiny half-inch holes between the patient's ribs. They inserted thin tubes to tunnel to the operating area and protect the other body tissue. Then three arms were inserted into the tubes. One was for a 3-D camera; the other two held tiny artificial wrists to which a variety of tools—scalpels, scissors, needle—were attached. The virtual reality system mimicked the movements of the surgeon's shoulders, elbows, and wrists. The surgeon sits at a computer workstation several feet from the operating table and moves instruments that control the ones inside the patient. The instruments inside the patient mimic the motion of the surgeon's hands so accurately that it is possible to sew up a coronary artery as thin as a thread. The surgeon watches his progress on a screen that enlarges the artery in 3-D to the size of a garden hose.[5]

Doctors use a technique called *palpation* to feel with their fingertips for telltale signs of abdominal injury or disease. By feeling the shape of the organs under the skin and noting how the skin springs back from the fingers' touch, skilled physicians can gain information they can't get from other instruments. In yet another medical application of virtual reality, researchers are trying to translate that expertise into biomechanical measurements that can be stored in a computer database. Data from actual patients will be gathered from sensors mounted on the fingertips of virtual reality gloves to create a database that will be made available to examining physicians. Upon examination of a new patient, a feedback system hooked up to the virtual reality gloves would capture sensations felt during the examination and compare them with data in the database. This approach could allow surgeons to go right into the operating room without having to obtain a CAT scan, thus saving time—and with many injuries, that's absolutely critical.

Education

Virtual environments are used in education to bring exciting new resources into the classroom. Students can stroll among digital bookshelves, learn anatomy on a simulated cadaver, or participate in historical events, all in virtual reality.

Third-grade students at John Cotton Tayloe School in Washington, North Carolina, can take an exciting virtual trip down the Nile for an integrated-curriculum lesson on ancient Egypt. This interactive virtual reality computer lesson integrates social studies, geography, music, art, science, math, and language arts. The software used to design the lesson was 3-D Website Builder (created by the Virtus Corporation), which allowed the designers to create a desert landscape complete with an oasis, camels, and a pyramid. Students could view the scenes from all angles, including front and top views, and could even enter a pyramid and view the sarcophagus holding the mummy in the middle of a room containing different Egyptian items. On one wall was a hieroglyphic message that was part of the lesson. On another wall was artwork depicting life in ancient Egypt, and against another were pieces of Egyptian furniture and a harp.[6]

Virtual technology has also been applied to aircraft maintenance training. The trainer uses virtual reality to simulate the aircraft. Feedback gives the user a sense of touch, while computer graphics give the senses of sight and sound. The user sees, touches, and manipulates the various parts of the virtual aircraft during training. The Virtual Aircraft Maintenance System simulates real-world maintenance tasks that are routinely performed on the AV8B vertical takeoff and landing aircraft used by U.S. Marines.

Real Estate Marketing

Virtual reality has been used to increase real estate sales in several powerful ways. From Web publishing to laptop display to a potential buyer, virtual reality provides excellent exposure for properties and attracts potential clients. Clients can take a virtual walk through properties and eliminate wrong choices without wasting valuable time. Virtual walk-throughs can be mailed on diskettes or posted on the Web as a convenience for nonlocal clients. A CD-ROM containing all virtual reality homes can also be sent to clients and other agents. Realatrends Real Estate Service, offering homes for sale in Orange County, California (http://www.realatrends.com/virtual_tours.htm), is just one of many real estate firms offering this service.

FIGURE 7.13

Computer-generated image technology is used widely in sports simulations. Shown here is NFL GameDay from Sony.
(Source: Courtesy of Sony Computer Entertainment.)

Computer-Generated Images

Computer-generated image technology, or CGI, has been around since the 1970s. A number of movies released in 2000 used this technology to bring realism to the silver screen. A team of artists rendered the roiling seas and crashing waves of *The Perfect Storm* almost entirely on computers using weather reports, scientific formulas, and their imagination. There was also *Dinosaur* with its realistic talking reptiles, *Titan A.E.'s* beautiful 3-D space-scapes, and the casts of computer-generated crowds and battles in *Gladiator* and *The Patriot*. CGI can also be used for sports simulation, as Figure 7.13 shows, to enhance the viewers' knowledge and enjoyment of a game.

INFORMATION SYSTEMS IN ACTION
Artificial Neural Network Satisfies Customers

Kenneth Baldauf,
Florida State University

Think back to the last time that you needed to communicate with a manufacturer or service provider. Perhaps you had problems with your new computer or a question about your credit card balance. Chances are you picked up a phone, dialed an 800 number, and spoke with a customer service representative. For many corporations, the corporate call center is the primary point of communication with their customers. A well-managed and efficient call center will impress customers, increase a company's customer base with return business and customer referrals, and give the company a competitive advantage. A poorly run call center that keeps customers waiting on hold or employs unhelpful agents will frustrate customers and build a bad reputation.

NeuralAct Inc. (http://www.neuralact.com) offers a software solution that allows call center managers to organize their workforce and human resource management systems more efficiently with the assistance of an artificial neural network. The key to agent scheduling is not to staff so sparely that customers run out of patience on hold and not to overstaff so that call center dollars are wasted. Neural Act's Workforce Management software uses an artificial neural network (ANN), a form of artificial intelligence that can "think" to identify trends in call center traffic and predict them into the future. The software takes into account the history of a call center and current and future external influences (holidays,

time of day, time of year, day of the week, for example); then it predicts call volume, the number of agents needed to efficiently handle the calls, and the skills those agents will need.

NeuralAct offers a second suite of applications, Call Center HR Management (CHRM), that complements the Workforce Management suite by assisting call centers in the recruitment, training, performance assessment, and retention of top agents. The software tracks a call center agent from application to exit interview. It provides tools for evaluating an agent's effectiveness and recommends additional training when needed.

The founders of NeuralAct recognized that the high turnover rate of call center agents was an issue that needed to be addressed. Turnover is costly in terms of time, money, and efficiency. NeuralAct focuses on creating happy and satisfied *super agents* who are well equipped to respond to any customer request. Through AI, they enable managers to get the most out of their agents without undue stress on the agent or customer.

Sources: Adapted from Ellen Muraskin, "Optimizing Agent Resources," *ComputerTelephony.com*, September 5, 2000, accessed at http://www.computertelephony.com/article/CTM20000829S0013; and "NeuralAct Uses Artificial Intelligence (AI) to Enhance Customer Service; Predict Call Center Volume Intelligently with AgentCARE™," *PR Newswire*, November 9, 2000.

● SUMMARY

PRINCIPLE • Artificial intelligence systems form a broad and diverse set of systems that can replicate human decision making for certain types of well-defined problems.

The term *artificial intelligence* is used to describe computers with the ability to mimic or duplicate the functions of the human brain. The objective of building AI systems is not to replace human decision making completely but to replicate it for certain types of well-defined problems.

* * *

Intelligent behavior encompasses several characteristics including the abilities to learn from experience and apply this knowledge to new experiences; handle complex situations and solve problems for which pieces of information may be missing; determine relevant information in a given situation, think in a logical and rational manner and give a quick and correct response; and understand visual images and processing symbols. The computer is better than humans at transferring information, making a series of calculations rapidly and accurately, and making complex calculations, but a human is better than a computer at all other attributes of intelligence.

* * *

Artificial intelligence is a broad field that includes several key components, such as expert systems, robotics, vision systems, natural language processing, learning systems, and neural networks. An expert system consists of the hardware and software to produce systems that behave like a human expert in a field or area (e.g., credit analysis). Robotics involves developing mechanical or computer devices to perform tasks that require a high degree of precision or are tedious or hazardous for humans (e.g., stacking cartons on a pallet). Vision systems include hardware and software that permit computers to capture, store, and manipulate images and pictures (face-recognition software). Natural language processing allows the computer to understand and react to statements and commands made in a "natural" language, such as English. Learning systems use a combination of software and hardware that allows the computer to change how it functions or reacts to situations based on feedback it receives (e.g., a computerized chess game). A neural network is a computer system that can simulate the functioning of a human brain (e.g., disease diagnostics system).

PRINCIPLE • Expert systems can enable a novice to perform at the level of an expert but must be developed and maintained very carefully.

Expert systems can explain their reasoning or suggested decisions, display intelligent behavior, manipulate symbolic information and draw conclusions from complex relationships, provide portable knowledge, and deal with uncertainty. They are not yet widely used; some are difficult to use, are limited to relatively narrow problems, cannot readily deal with mixed knowledge, present the possibility for error, cannot refine their own knowledge base, are difficult to maintain, and may have high development costs. Their use also raises legal and ethical concerns.

An expert system consists of a collection of integrated and related components, including a knowledge base, an inference engine, an explanation facility, a knowledge acquisition facility, and a user interface. The knowledge base contains all the relevant data, rules, and relationships used in the expert system. The rules are often composed of if-then statements, which are used for drawing conclusions. Fuzzy logic allows expert systems to incorporate facts and relationships into expert system knowledge bases that may be imprecise or unknown.

The inference engine performs the processing of the rules, data, and relationships stored in the knowledge base to provide answers, predictions, and suggestions the way a human expert would. Two common methods for processing include forward and backward chaining. Backward chaining starts with a conclusion, then searches for facts to support it; forward chaining starts with a fact, then searches for a conclusion to support it. Mixed chaining is a combination of backward and forward chaining.

The explanation facility of an expert system allows the user to understand what rules were used in arriving at a decision. The knowledge acquisition facility helps the user add or update knowledge in the knowledge base. The user interface makes it easier to develop and use the expert system.

* * *

The individuals involved in the development of an expert system include the domain expert, the knowledge engineer, and the knowledge users. The domain expert is the individual or group who has the expertise or knowledge being captured for the system. The knowledge engineer is the developer whose job is the extraction of the expertise from the domain expert. The knowledge user is the individual who benefits from the use of the developed system.

* * *

Following is a set of benefits that normally make expert systems worth the expenditure of time and money: a high potential payoff or significantly reduced downside risk, the ability to capture and preserve irreplaceable human expertise, the ability to develop a system more consistent than human experts, expertise needed at a number of locations at the same time, and expertise needed in a hostile environment that is dangerous to human health. The expert system solution can be developed faster than the solution from human experts. An ES also provides expertise needed for training and development to share the wisdom and experience of human experts with a large number of people.

PRINCIPLE • Virtual reality systems have the potential to reshape the interface between people and information technology by offering new ways to communicate information, visualize processes, and express ideas creatively.

A virtual reality system enables one or more users to move and react in a computer-simulated environment. Virtual reality simulations require special interface devices that transmit the sights, sounds, and sensations of the simulated world to the user. These devices can also record and send the speech and movements of the participants to the simulation program. Thus, users are able to sense and manipulate virtual objects much as they would real objects. This natural style of interaction gives the participants the feeling that they are immersed in the simulated world.

Virtual reality can also refer to applications that are not fully immersive, such as mouse-controlled navigation through a three-dimensional environment on a graphics monitor, stereo viewing from the monitor via stereo glasses, stereo projection systems, and others. Some virtual reality applications allow views of real environments with superimposed virtual objects.

● REVIEW QUESTIONS

1. Define the term *artificial intelligence*.
2. What is a vision system? Discuss two applications of such a system.
3. What is a perceptive system?
4. What is natural language processing? What are the three levels of voice recognition?
5. Describe three examples of the use of robotics.
6. What is a learning system? Give a practical example of such a system.
7. What is a neural network? Describe two applications of neural networks.
8. What is meant when it is said that neural networks learn to program themselves?
9. What are the capabilities of an expert system?
10. Under what conditions is the development of an expert system likely to be worth the effort?
11. Identify the basic components of an expert system and describe the role of each.

12. What are fuzzy sets and fuzzy logic?
13. How are rules used in expert systems?
14. Expert systems can be built based on rules or cases. What is the difference between the two?
15. Describe the roles of the domain expert, the knowledge engineer, and the knowledge user in expert systems.
16. What are the primary benefits derived from the use of expert systems?
17. Identify three approaches for developing an expert system.
18. Identify three special interface devices developed for use with virtual reality systems.
19. Identify and briefly describe three specific virtual reality applications.

● DISCUSSION QUESTIONS

1. What are the requirements for a computer to exhibit human-level intelligence? How long will it be before we have the technology to design such computers? Do you think we should push to try to accelerate such a development? Why or why not?

2. What are some of the tasks at which robots excel? Which human tasks are difficult for them to master? What fields of AI are required to develop a truly perceptive robot?

3. Accuracy slip occurs when there are significant changes in the real world, making an expert system less accurate. Identify at least three expert system applications where accuracy slip could lead to the loss of human life. What process can be put in place to safeguard against accuracy slip for these critical expert system applications?

4. You have been hired to capture the knowledge of a brilliant attorney who has an outstanding track record for selecting jury members favorable to her clients during the pretrial jury selection process. This knowledge will be used as the basis for an expert system to enable other attorneys to have similar success. Is this system a good candidate for an expert system? Why or why not?

5. Briefly explain why human decision making often does not lead to optimal solutions to problems.

6. What is the purpose of a knowledge base? How is one developed?

7. Describe an application that requires the concurrent use of more than one of the subfields of artificial intelligence.

8. Imagine that you are developing the rules for an expert system to select the strongest candidates for a medical school. What rules or heuristics would you include?

9. What skills does it take to be a good knowledge engineer? Would knowledge of the domain help or hinder the knowledge engineer in capturing knowledge from the domain expert?

10. Which interface is the least developed and most challenging to create in a virtual reality system? Why do you think this is so?

● PROBLEM-SOLVING EXERCISES

1. Imagine that you are a knowledge engineer and are developing an expert system to help consumers choose a used car that best meets their needs and their budget. You are going to your first interview with the owner of a used car lot who is the designated expert for this system. Use your word processing software and develop a list of questions that you would ask to begin to capture this individual's knowledge.

2. Assume you live in an area where there is a wide variation in weather from day to day. Develop a simple expert system to provide advice on the type of clothes to wear based on the weather. The system needs to help you decide which clothes and accessories (umbrella, boots, etc.) to wear for

sunny, snowy, rainy, hot, mild, or cold days. Key inputs to the system include last night's weather forecast, your observation of the morning temperature and clouds, yesterday's weather, and the activities you have planned for the day. Using your word processing program, create seven or more rules that could be used in such an expert system. Create five cases and use the rules you developed to determine the best course of action.

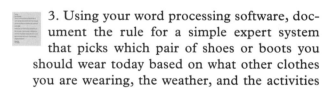

3. Using your word processing software, document the rule for a simple expert system that picks which pair of shoes or boots you should wear today based on what other clothes you are wearing, the weather, and the activities you have planned.

● TEAM ACTIVITIES

1. With two or three of your classmates, do research to identify three real examples of expert systems in use. Discuss the problems solved by each of these systems. Regardless of how the knowledge is acquired, it is important to validate and update the knowledge base frequently to make sure it is still accurate. Failure to do so will result in "knowledge creep," in which the knowledge base no longer matches reality. Identify any issues that may arise because of "knowledge creep." Choose which of the three systems provides the most benefit and state why you selected that system.

2. Form a team to debate other teams from your class on the following topic: "Are expert systems superior to humans when it comes to making objective decisions?" Develop several points supporting either side of the debate.

3. With members of your team, develop an expert system that makes suggestions about what to do if your car does not start. Develop a simple "dialogue" between the expert system and you, the end user. The expert system should suggest different actions to take in an attempt to diagnose and correct whatever may be wrong.

● WEB EXERCISES

1. Jess is an expert system shell written entirely in Java by Ernest Friedman-Hill. It enables you to build Java applets that have the capacity to "reason" based on knowledge you supply in the form of declarative rules. Visit the Jess Web site at http://herzberg.ca.sandia.gov/jess and read the FAQ page. Run the applet demo and see whether you can beat the expert system. Write a brief paper to your instructor summarizing your experience.

2. Visit the Web site describing artificial intelligence at http://ai.about.com/compute/ai/gi/dynamic/offsite.htm? and see how the business logic of lending companies can be expressed in plain-English sentences that are immediately translated into action in an expert system. Write a brief paper summarizing the advantages and disadvantages of using an expert system to make loan decisions.

3. Take the Fuzzy Shower Challenge; a challenge faced each morning by everyone—trying to survive controlling the shower temperature. Visit the Web site at http://ai.iit.nrc.ca/fuzzy/shower/ title.html and experience the use of a fuzzy logic system. Write a brief paper to your instructor summarizing your experience.

● CASES

Fuzzy Logic System Designed to Help Travelers

Yatra.net, a Web travel site launched in July 2000, claims it can provide better service to businesses than travel agents by using a proprietary system that employs fuzzy logic. Its target customers are the 35,000 small and medium-sized U.S. companies that spend $0.5 million to $7 million annually on air travel. Yatra's goal is to deliver significant savings through large buying volume and strategic sourcing capabilities.

At the heart of Yatra.net is its patent-pending Cognizer technology that enables travelers and travel suppliers to make complex purchasing decisions like never before. Cognizer was developed by a five-member team over a two-year period and uses fuzzy logic that takes into account corporate, traveler, and supplier preferences to create simple business rules and a dynamic travel policy. It uses information from multiple sources to analyze each purchasing event independently by weighing more than 50 corporate policies and traveler preferences across factors of convenience, service, and price. For instance, why should a salesperson's trip for an important sales call be handled in the same way as a staff accountant's travel to an internal meeting? Yatra believes it shouldn't. In one case, convenience and service should weigh heavily to ensure on-time arrival, while in the other, price should be the primary consideration. Cognizer ensures that every travel decision draws from all available knowledge to identify the best travel options.

If travelers fail to choose from the recommended itineraries, Cognizer instantly tracks the purchase and notifies the company's travel services manager of the decision.

Suppliers of travel services will also be able to create their own rules to target potential customers. For example, an airline can offer specials such as first-class upgrades to travelers who fly the San Diego to Hong Kong route a certain number of times a year and aren't part of its frequent-flier program.

Most industry analysts agree that the on-line travel market is already crowded. Yatra faces a major challenge—it has to convince customers that their current relationships with travel agencies aren't working out, and that may be hard to do. The biggest opportunity is in lower fees, but if agencies begin reducing fees for bookings made on-line, then Yatra loses its primary advantage.

Discussion Questions

1. Identify some potential business rules and traveler preferences that might be implemented to enable Cognizer to analyze each purchasing event independently.

2. What are some of the advantages and disadvantages of this system from the traveler's perspective?

Critical Thinking Questions

3. Access Yatra's Web site and do other research to learn how successful this new venture has been. Write a brief report summarizing your findings.

4. In many small companies, employees are used to booking their own travel arrangements. What sort of issues might be raised if such a company converts to Yatra?

Sources: Adapted from Cheryl Rosen, "Fuzzy Logic Provides Clear Solutions," *Information Week*, May 15, 2000, pg. 77; and Yatra home page, http://www.yatra.net/Default.htm_, accessed July 5, 2000.

2 Expert Systems Help Develop Business Plans

Unless you have easy access to lots of money, you usually must create a business plan to sell an internal project or launch a new business. Emerging enterprises, such as Internet start-ups, are doubly challenged—to generate financial backing, they must move quicker than established businesses, and they often do not have people experienced in developing such plans. Even managers in established companies must develop a business plan to gain senior management approval to launch a new product or service. Software has come to the rescue.

A number of software applications are designed to provide a standard template for data entry and guide you step by step in creating a business plan. As you proceed, the software offers additional assistance by providing descriptions of what is expected in each section and identifying issues to consider before writing a section. Use of such a product guarantees that you will produce a business plan with the key elements that most potential investors and senior managers expect to see; however, it provides no real feedback on the viability of the plan or the reasonableness of the business assumptions. Recognizing these shortcomings, a few software companies have developed an expert system that not only offers professional advice on how to create a viable business plan but also assesses the plan's probability for success.

Developing a plan with this software starts by answering a long series of questions asked by the inference engine. Some answers require a simple yes or no response; many of the questions relate to competitors and require a response on a scale of 1 to 10 to rate your relative strengths and weaknesses and other factors involved in the situation. When you are done answering the questions, the explanation facility provides a thumbnail analysis of your situation and some basic conclusions. It also provides a score of your chance of success and, most important, the option to trace its conclusions back to the answers you gave. The results are almost never based on a single answer

but on the way your answers interrelate. That is the power of expert system technology—it recognizes patterns in combinations of factors. You may next wish to review and revise some of your original assumptions and answers to key questions to evaluate their impact on the probability of success of the project.

A good business planning tool will not generate eccentric advice but will look at business problems in standard, proven ways to provide objective conclusions. Most users of such software think that answering the questions and examining the connections between answers and conclusions provides a deeper insight into the key factors and assumptions of their plan. Many times they realize that they simply do not have enough information about the situation and need to get more data before presenting a recommendation to management. The analysis delivered is often provocative and adds considerable value to initial planning.

Discussion Questions

1. What are the key advantages and disadvantages of using software such as that described in this case to evaluate business plans?
2. How might you choose between competing software packages to perform business planning?

Critical Thinking Questions

3. Imagine that you are the chief financial officer of a major corporation, and top executives have recommended use of business-planning software based on expert system technology to prepare all business proposals for more than $1 million. Would you support this recommendation? Why or why not?
4. How might the use of expert system business planning software speed up the approval of a business recommendation? Is it possible that it could slow the process down? Why or why not?

Sources: Adapted from Jeff Angus, "Expert System Acts as Outside Business Consultant," *Information Week,* November 8, 1999, accessed at http://www.informationweek.com/760/expert.htm; and Kathy Yakal, "Business Resources Software Plan Write for Business 5.0," *PC Magazine,* April 24, 2000, accessed at http://www.brs-inc.com/reviews.asp.

3 AI Software Used to Monitor and Control Networks

Companies have been quick to adopt distributed processing, as discussed in Chapter 4. So, many IT organizations manage and control a variety of applications, hardware, software, and data communications protocols for other companies. Computer Associates developed enterprise management software called Unicenter TNG to meet this need. Through a variety of integrated core and optional applications, Unicenter TNG can manage a technology infrastructure including networks, systems, desktops, applications, databases, point-of-sale devices, automated teller machines, manufacturing devices, environmental controls, hospital equipment, and power lines.

Neugents is Computer Associates' software that uses neural network technology to predict failure in the network or any of the devices connected to it. The software is offered as an option to the Unicenter TNG enterprise management software. Neugents analyzes system data collected by Unicenter and uses that data to learn what system behavior is acceptable and when a deviation in activity indicates a problem. For example, Neugents looks at 1,200 elements on a network server to determine what conditions foreshadow a failure. When similar conditions appear, Neugents predicts the likelihood of a slowdown or outage and sends Unicenter an alert when preselected thresholds are reached. For example, if Neugents' analysis indicates a 75 percent chance of a crash within an hour and 75 percent is the alert threshold, Neugents signal the Unicenter console. IT administrators can set Unicenter to respond automatically to Neugents' alerts.

Computer Associates promotes its Neugents technology to differentiate its support for e-business from its competitors. Computer Associates believes that Neugents will become even more critical as businesses become increasingly inundated with data and as back offices try to integrate seamlessly with Web storefronts.

NCI Information Systems typically required two engineers to support a 2,000-node network. After implementing Unicenter TNG and the Neugents application, the same two engineers can now support a 10,000-node network. At Allstate Insurance, Neugents traced a slowdown in a server's packet data transmissions to a bad network interface card that was then replaced before it could cause a system failure. Myers Industries, a plastic and rubber manufacturer, uses Neugents to review real-time information from its process controllers and production equipment on its manufacturing systems, giving the company early warnings of potential problems before faults begin corrupting processes.

To date, few Computer Associates clients have begun using Neugents to predict when various components of their network might fail because it is complicated to master.

Discussion Questions

1. What are some of the advantages of using neural network technology to monitor networks?
2. Why are effective and efficient network management and control key to the success of a company's e-commerce operation?

Critical Thinking Questions

3. Why do you think companies are finding it difficult to use software to predict when network components will fail?
4. Do a Web search to find at least two other companies that offer similar capabilities to manage and control networks. How are the products of these companies similar? How are they different?

Sources: Adapted from George V. Hulme, "CA Heads into the Next Dimension," *Information Week,* May 8, 2000, accessed at http://www.informationweek.com/785/vaca.htm; Catherine Shull, "Computer Associates Teaches Brigham Young University How to Enhance E-Business with the Power of Neugents," Computer Associates press release, March 13, 2000, accessed at http://www.ca.com/press/2000/03/byu_neugents.htm; and Amy K. Larsen, "CA Offers Free Neugent Trial," *Information Week,* March 15, 1999, accessed at http://www.informationweek.com/story/IWK19990315S0007.

● NOTES

Sources for the opening vignette on p. 249: Adapted from Dave Marino-Nachison, "The Motley Fool Interview with Ask Jeeves President and CEO Rob Wrubel," December 22, 1999, at http://www.fool.com/foolaudio/transcripts/1999/stocktalk991230_askj.htm; and "What Is Ask Jeeves," Ask Jeeves Web site at http://www.ask.com/docs/about/whatIsAskJeeves.html, accessed June 14, 2000.

1. Gray H. Anthes, "The Robots Are Coming," *Computerworld,* May 22, 2000, p. 82.
2. Maria Trombly, "Voice Recognition Eases Call-In Trading," *Computerworld,* May 22, 2000, accessed at http://www.computerworld.com.
3. Nancy Weil, "CA World to Focus on E-Commerce," *Computerworld,* April 10, 2000, accessed at http://www.computerworld.com/cwi/.
4. Thomas Hoffman, "Neural Nets Spot Credit Risks," *Computerworld,* July 26, 1999, accessed at http://www.computerworld.com/cwi/story/0,1199, NAV47_STO36440,00.html.
5. Jane Ellen Stevens, "Virtual Reality Hearts and Minds," accessed at http://www.msnbc.com/news, July 9, 2000.
6. Lynne Cox, "Cruising Down the Nile," "VR in the Schools," Vol. 4, no. 3, accessed at http://www.ecu.edu/vr/vrits/4-3Cox.htm.

PART IV

Systems Development and Social Issues

CHAPTER 8

Systems Development

*L*arge-scale software is being delivered on time and under budget. More important, a development project now sets 90 days for the key first deliverable [part of a new system] and no more than six months for full implementation.

— Peter Keen, consultant

Principles

Learning Objectives

Principles	Learning Objectives
Effective systems development requires a team effort, and it starts with careful planning.	• *Identify the key participants in the systems development process and discuss their roles.* • *Define the term* information systems planning *and list several reasons for initiating a systems project.*
Systems development often uses a predetermined methodology to select, implement, and monitor projects.	• *Discuss the key features of the traditional, prototyping, rapid application development, and end-user systems development life cycles.*
Systems development starts with investigation and analysis of existing systems.	• *State the purpose of systems investigation.* • *State the purpose of systems analysis and discuss some of the tools and techniques used in this phase.*
Designing new systems or modifying existing ones should always be aimed at helping an organization achieve its goals.	• *State the purpose of systems design and discuss the differences between logical and physical systems design.* • *Outline key steps taken during the design phase.* • *Define the term* RFP *and discuss the importance of this document.*
The primary emphasis of systems implementation is to make sure that the right information is delivered to the right person in the right format at the right time.	• *State the purpose of systems implementation and discuss the various activities associated with this phase.* • *Describe options for acquiring software.*
Maintenance and review add to the useful life of a system but can consume large amounts of resources.	• *State the importance of systems and software maintenance and discuss the activities involved.* • *Describe the systems review process.*

AT&T

Renting Software to Business Clients

AT&T is primarily known for its long-distance telephone services. But in the year 2000, the company decided to get into the business of developing and renting software to others. The company plans to invest about $250 million in the new business. AT&T's strategy is to rent software over the Internet. The new system, called Ecosystem for ASPs, will use AT&T's long-distance network, the largest in the world. AT&T is hoping to provide fast Internet access with off-site data storage capabilities for companies that rent software on-line through AT&T. The storage capabilities being offered by AT&T will allow companies to save by not having to invest in additional hard disks for new software. Referring to the new system, Kathleen Earley, president of AT&T's Data and Internet Services, stated, "This announcement is profound. This is the first time you will see all of the players at the table." AT&T plans to work with IBM, EMC Corporation, and Cisco Systems in developing the software rental business.

AT&T will have plenty of competition, though. How big is the rental market? A spokesman for Sun Microsystems predicts that it could be as high as $23 billion by the year 2003. That is one reason so many companies are now getting into the ASP business, including SAP, PeopleSoft, Hewlett-Packard, and Microsoft. For software companies, the ASP approach can mean shorter software release cycles, easier upgrades, and larger potential markets. According to Sanjav Sharma of IBM, "Intense competition in the software application market is driving vendors toward the ASP model to bring down the cost to customers."

As you read this chapter, consider the following:

- How can a company acquire software?

- Why is renting software becoming more attractive?

Sometime during your career, you will probably be directly or indirectly involved in systems development. As an individual, you will frequently use word processing programs and other software to write reports, make financial calculations, and store and retrieve information. As a member of a group, you will likely use e-mail programs and group decision support capabilities. As an employee of an organization, you will use corporate programs to help you serve customers, order raw materials from suppliers, and make important decisions that could affect the entire organization. In some cases, you will work directly with IS personnel in developing programs that are specifically designed to help you with some aspect of your job. In other cases, you may be indirectly involved in the development of a computer system. All of these programs and systems are created using systems development.

Systems development is the activity of creating or modifying existing business systems. It refers to all aspects of the process—from identifying problems to be solved or opportunities to be exploited to implementing and refining the chosen solution. Because these business systems resulting from systems development are critical to your career and the success of the organization, it is important that you understand your involvement and ways you can get the most from systems development efforts.

As we saw in the opening vignette, many companies are launching systems development efforts. When it comes to developing Internet applications, multiple companies are often involved. For example, Equiva Trading Company, a joint venture of Shell Oil Company, Texaco, and Saudi Arabian Oil Company, has invested more than $6 million into a systems development effort to build an oil-trading site to reduce oil costs.[1] Developing new business applications for the Internet is only one part of systems development, as we'll see.

AN OVERVIEW OF SYSTEMS DEVELOPMENT

Understanding systems development is important to all professionals, not just those in the field of information systems. In today's businesses, managers and employees in all functional areas work together and use business information systems. As a result, users of all types are helping with development and, in many cases, leading the way. This chapter will provide you with a deeper appreciation of the systems development process and help you avoid costly failures.

Participants in Systems Development

Effective systems development requires a team effort. The team usually consists of stakeholders, users, managers, systems development specialists, and various support personnel.[2] This team, called the development team, is responsible for determining the objectives of the information system and delivering a system that meets these objectives to the organization. In the context of systems development, **stakeholders** are individuals who, either themselves or through the area of the organization they represent, ultimately benefit from the systems development project. **Users** are individuals who will interact with the system regularly. They can be employees, managers, customers, or suppliers. For large-scale systems development projects, where the investment in and value of a system can be quite high, it is common to have senior-level managers, including the company president and functional vice presidents (of finance, marketing, and so on), be part of the development team.

Depending on the nature of the systems project, the development team might include systems analysts and programmers, among others. A **systems analyst** is a professional who specializes in analyzing and designing business systems. Systems analysts play various roles while interacting with the stakeholders and users, management, vendors and suppliers, external companies, software programmers, and other IS support personnel (Figure 8.1). Like an architect developing blueprints for a new building, a systems analyst develops detailed plans for the new or modified

stakeholders

individuals who, either themselves or through the area of organization they represent, ultimately benefit from the systems development project

users

individuals who will interact with the system regularly

systems analyst

a professional who specializes in analyzing and designing business systems

FIGURE 8.1

The systems analyst plays an important role in the development team and is often the only person who sees the system in its totality. The one-way arrows in this figure do not mean that there is no direct communication between other team members. Instead, they indicate the pivotal role of the systems analyst—an individual who is often called upon to be a facilitator, moderator, negotiator, and interpreter for development activities.

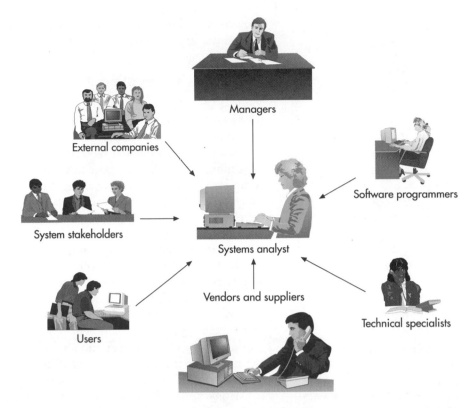

Managers

External companies

System stakeholders

Systems analyst

Software programmers

Users

Vendors and suppliers

Technical specialists

programmer

a specialist responsible for modifying or developing programs to satisfy user requirements

system. The **programmer** is responsible for modifying or developing programs to satisfy user requirements. Like a contractor constructing a new building or renovating an existing one, the programmer takes the plans from the systems analyst and builds or modifies the necessary software.

The other support personnel on the development team are usually technical specialists, including database and telecommunications experts, hardware engineers, and supplier representatives. One or more of these roles may be outsourced to nonemployees or consultants. Depending on the magnitude of the systems development project and the number of IS systems development specialists on the team, the team may also include one or more IS managers. The composition of a development team may vary over time and from project to project. For small businesses, the development team may consist of a systems analyst and the business owner as the primary stakeholder. For larger organizations, formal IS staff can include hundreds of people involved in a variety of IS activities, including systems development. Every development team should have a team leader. As shown in Table 8.1, different types of team leaders should also be considered for different

TABLE 8.1

Team Leaders for Different Systems Development Projects

Characteristics of Systems Development Project	Examples of Appropriate Team Leaders
Project involves new and advanced technology.	Individual from IS department
Impact will force critical changes in a functional area of the business.	Manager from that functional area
Project is extremely large and complex.	Specialist in project management
Project will have dramatic impact on personnel.	Individual from human resource department
Project will have a broad and important impact on the entire organization.	Senior management; leader should build a development team that includes personnel skilled in all affected areas
Project will involve two or more companies, such as an Internet exchange.	Senior executive or project manager with experience in starting and managing a new company; this team leader is often hired from outside

FIGURE 8.2

Information systems planning transforms organizational goals outlined in the strategic plan into specific system development activities.

information systems planning

the translation of strategic and organizational goals into systems development initiatives

projects and development teams. Senior-level team leaders can help translate a company's strategic plan into an information systems plan.

Information Systems Planning

Because an organization's strategic plan contains both organizational goals and a broad outline of steps required to reach them, the strategic plan affects the type of system an organization needs. For example, a strategic plan may identify as organizational goals a doubling of sales revenue within five years, a 20 percent reduction of administrative expenses over three years, acquisition of at least two competing companies within a year, or the capture of market leadership in a particular product category. Organizational commitments to policies such as continuous improvement are also reflected in the strategic plan. Such goals and commitments set broad outlines of system performance. As seen in the "E-Commerce" box (on page 288), information systems planning can often involve two or more companies.

The term **information systems planning** refers to the translation of strategic and organizational goals into systems development initiatives (Figure 8.2). The Marriott hotel chain, for example, invites its chief information officer to board meetings and other top-level management meetings. Proper IS planning ensures that specific systems development objectives support organizational goals.[3] Not translating the strategic plan into a specific IS plan can be disastrous.[4]

One of the primary benefits of IS planning is a long-range view of information technology use in the organization.[5] Specific systems development initiatives may spring from the IS plan, but the plan must also provide a broad framework for future success. The IS plan should provide guidance on how the information systems infrastructure of the organization should be developed over time. Another benefit of IS planning is that it ensures better use of information systems resources, including funds, IS personnel, and time for scheduling specific projects. The steps of IS planning are shown in Figure 8.3. After the IS planning steps are complete, a systems development method can be selected.

SYSTEMS DEVELOPMENT LIFE CYCLES

Organizations can select among four basic systems development life cycles, or methods, to develop new systems: traditional, prototyping, rapid application development (RAD), and end-user development. In some companies, these approaches are formalized and documented so that system developers have a well-defined process to follow; in others, less formalized approaches are used.

The Traditional Systems Development Life Cycle

Traditional systems development efforts can range from a small project, such as purchasing an inexpensive computer program, to a major undertaking. The steps of traditional systems development may vary from one company to the next, but most approaches have five common phases: investigation, analysis, design, implementation, and maintenance and review (Figure 8.4).

In the **systems investigation** phase, potential problems and opportunities are identified and considered in light of the goals of the business. Systems investigation attempts to answer the question "What is the problem, and is it worth solving?" The primary result of this phase is a defined information system project for which business problems or opportunity statements have been created, to which some organizational resources have been committed and for which systems analysis is recommended. **Systems analysis** attempts to answer the question "What must the information system do to solve the problem?" This phase involves the study of existing

systems investigation

the systems development phase during which problems and opportunities are identified and considered in light of the goals of the business

systems analysis

the systems development phase involving the study of existing systems and work processes to identify strengths, weaknesses, and opportunities for improvement

FIGURE 8.3

The Steps of IS Planning

Some projects are identified through overall IS objectives, whereas additional projects, called unplanned projects, are identified from other sources. All identified projects are then evaluated in terms of their organizational priority.

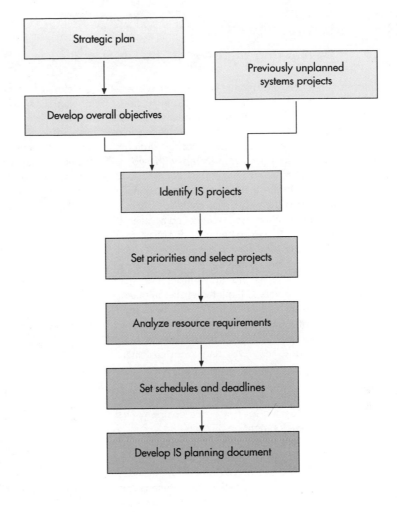

FIGURE 8.4

The Traditional Systems Development Life Cycle

Sometimes, information learned in a particular phase requires cycling back to a previous phase.

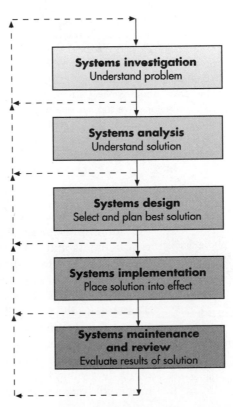

E-COMMERCE
Developing an Internet Site for Suppliers and Dealers

Good information systems planning is a key for successful systems development. In most cases, IS planning starts with a strategic plan. The strategic perspective guides general IS planning, which in turn can be converted into specific systems development initiatives. Today, strategic planning is often started by a partnership with two or more companies.

To stay competitive in the Internet age, Ford originally decided to form an alliance with Cisco Systems. Cisco is one of the largest and most successful Internet companies; it provides the infrastructure for a large number of e-commerce operations. The strategic plan for the Ford-Cisco alliance was to build a comprehensive Internet site for suppliers. The plan called for Cisco to aid in wiring more than 40,000 suppliers and dealers to the Internet site. In return, Cisco would obtain an equity stake in the new auto exchange, along with Oracle, a database company. According to Sue Bostrom, vice president of Internet business solutions group at Cisco, "We hear from customers across all industries. They want to move as quickly as possible to get connected." Because of the rush to e-commerce, the number of systems development initiatives will be huge. According to a spokesperson for Cisco, the company has never done systems development on this scale.

More recently, the Ford-Cisco alliance has been joined by other companies and given an official name, Covisint. In addition to Ford, the Covisint exchange now includes General Motors, DaimlerChrysler, Nissan, and Renault. It is hoped that Covisint will save the auto industry billions of dollars and allow designers to save months in preproduction planning. The production cost savings per car could be more than $2,000 on average if Covisint achieves its goals. Eventually, Covisint could allow a customer to custom order a car and receive it within weeks after ordering. "GM says that when somebody in a dealership orders a car with leather seats, the cow should wince," says Doug Grimm, a representative of an automotive supply company.

The Covisint systems development effort reveals a problem often faced by systems developers. Covisint can develop a fast, efficient Internet site, but will the thousands of suppliers also modify their systems? Although slow modems will be able to access the Covisint site, the real benefits to the auto companies and their suppliers will be achieved only if suppliers also upgrade their systems, including their connection to the Internet. If the suppliers fail to upgrade, the new system will be underutilized. According to Mark Duhaime, director of program release for the auto exchange, "There are a lot of suppliers that are using the site in a 'kick the tires' model. This announcement is an evolution from fingers and a browser to having system-to-system connectivity." The question remains, however, whether suppliers will get on the Internet bandwagon and modify their systems to take full advantage of the new and impressive systems development effort by Covisint.

Discussion Questions

1. What is the relationship between strategic planning and systems development initiatives?
2. What are some of the challenges facing systems developers when several companies form a partnership trying to connect thousands of suppliers to a comprehensive Internet site?

Critical Thinking Questions

3. Assume that you are in charge of this massive systems development effort. How would you ensure that suppliers upgraded their systems to take full advantage of the new auto exchange?
4. What types of company-supplier issues would have to be resolved in the IS planning stage to ensure a successful systems development effort?

Sources: Adapted from Fara Warner, "Ford, Cisco Team Up," *The Wall Street Journal*, February 10, 2000, p. A5; "Cisco Joins Oracle in Ford Exchange," *Computing*, February 17, 2000, p. 2; "Where E-Commerce Wins Hands Down," *The Economist*, February 26, 2000; and Steve Konicke, "Covisint's Rough Road," *Information Week*, August 2, 2000.

systems design

the systems development phase that defines how the information system will do what it must do to obtain the problem solution

systems implementation

the systems development phase that involves creating or acquiring various system components detailed in the systems design, assembling them, and placing the new or modified system into operation

systems and work processes to identify strengths, weaknesses, and opportunities for improvement. The major outcome of systems analysis is a list of requirements and priorities. **Systems design** seeks to answer the question "How will the information system do what it must do to obtain the problem solution?" The primary result of this phase is a technical design that either describes the new system or describes how existing systems will be modified. The system design details system outputs, inputs, and user interfaces; specifies hardware, software, database, telecommunications, personnel, and procedure components; and shows how these components are related. **Systems implementation** involves creating or acquiring the various system components detailed in the systems design, assembling them, and placing the new or modified system into operation. An important task during this phase is to train the users. Systems implementation results in an installed, operational information system that meets the business needs for which it was developed. The purpose of **systems maintenance and review** is to ensure that the system operates and to modify the system so that it continues to meet changing business needs. As shown in Figure 8.4, a system under development moves from one phase of the traditional SDLC to the next.

Prototyping

Prototyping takes an iterative approach to the systems development process. During each iteration, requirements and alternative solutions to the problem are identified and analyzed, new solutions are designed, and a portion of the system is implemented. Users are then encouraged to try the prototype and provide feedback (Figure 8.5). Prototyping begins with the creation of a preliminary model of a major subsystem or a scaled-down version of the entire system. For example, a prototype might be developed to show sample report formats and input screens.[6] Once developed and refined, the prototypical reports and input screens are used as models for the actual system, which may be developed using an end-user programming language such as SAS, Focus, or Visual Basic. The first preliminary model is refined to form the second- and third-generation models, and so on until the complete system is developed (Figure 8.6).

systems maintenance and review

the systems development phase ensures that the system operates and modifies the system so that it continues to meet changing business needs

prototyping

an iterative approach to the systems development process

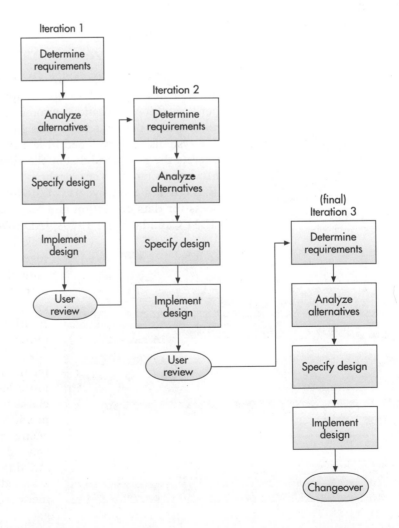

FIGURE 8.5

Prototyping Is an Iterative Approach to Systems Development

FIGURE 8.6

Prototyping is a popular technique in systems development. Each generation of prototype is a refinement of the previous generation based on user feedback.

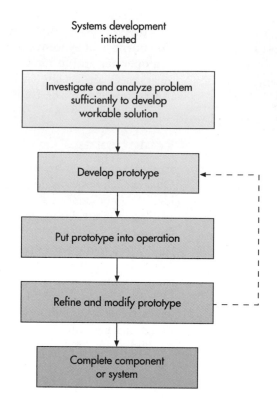

rapid application development (RAD)

a systems development approach that employs tools, techniques, and methodologies designed to speed application development

joint application development (JAD)

a process for data collection and requirements analysis

PowerBuilder, a RAD tool from Sybase, is used in both the public and private sectors. (Source: Courtesy of Sybase, Inc.)

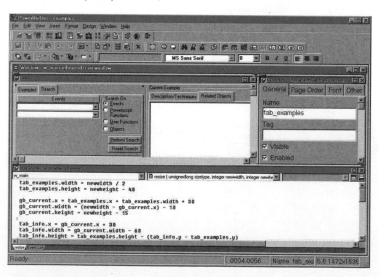

Rapid Application Development and Joint Application Development

Rapid application development (RAD) employs tools, techniques, and methodologies designed to speed application development.[7] RAD reduces paper-based documentation, automates program source code generation, and facilitates user participation in design and development activities. With RAD, entire systems are developed in less than six months. The ultimate goal is to accelerate the process so that applications can go into production much sooner than when using other approaches.

RAD makes extensive use of the **joint application development (JAD)** process for data collection and requirements analysis. Originally developed by IBM Canada in the 1970s, JAD involves group meetings in which users, stakeholders, and IS professionals work together to analyze existing systems, propose possible solutions, and define the requirements of a new or modified system. JAD groups consist of both problem holders and solution providers. A group normally requires one or more top-level executives who initiate the JAD process, a group leader for the meetings, potential users, and one or more individuals who act as secretaries and clerks to record what is accomplished and to provide general support for the sessions. Many companies have found that groups can develop better requirements than individuals working independently and have assessed JAD as a very successful development technique.

Many end users today are developing their own PC-based systems with technical assistance from IS personnel.
(Source: © 2000 PhotoDisc.)

end-user systems development

any systems development project in which the primary effort is undertaken by a combination of business managers and users

computer-aided software engineering (CASE)

tools that automate many of the tasks required in a systems development effort and enforce adherence to the systems development life cycle (SDLC)

The End-User Systems Development Life Cycle

Today, the term **end-user systems development** describes any systems development project in which the primary effort is undertaken by a combination of business managers and users. Rather than ignoring these initiatives, astute IS professionals encourage them by offering guidance and support. Technical assistance, communication of standards, and the sharing of "best practices" throughout the organization are just some of the ways IS professionals work with motivated managers and employees undertaking their own systems development. In this way, end-user-developed systems can be structured as complementary to, rather than in conflict with, existing and emerging information systems. In addition, this open communication among IS professionals, managers of the affected business area, and users allows the IS professionals to identify specific initiatives so that additional organizational resources, beyond the prerogative of the initiating business manager or user, are provided for its development.

Use of Computer-Aided Software Engineering (CASE) Tools

Computer-aided software engineering (CASE) tools automate many of the tasks required in a systems development effort. They also enforce adherence to the systems development life cycle (SDLC), thus ensuring a high degree of thoroughness and standardization throughout the entire systems development process.

CASE packages that focus on activities associated with the early stages of systems development are known as *upper-CASE tools*. These packages provide automated tools to assist with systems investigation, analysis, and design activities. Other CASE packages, called *lower-CASE tools*, focus on the later implementation stage of systems development and are capable of automatically generating structured program code. Some CASE tools provide links between upper- and lower-CASE packages, allowing lower-CASE packages to generate program code from upper-CASE package designs. These tools are called *integrated-CASE (I-CASE) tools*.

As with any team, coordinating the efforts of members of a systems development team can be a problem. So, many CASE tools allow more than one person to work on the same system at the same time via a multiuser interface, which coordinates and integrates the work performed by all members of the same design team. With this facility, a person working on one aspect of systems development can automatically share his or her results with someone working on another aspect of the same system. Advantages and disadvantages of CASE tools are listed in Table 8.2.

Once the general systems development approach has been selected and the tools and techniques determined, the team can begin to investigate the needs of the organization or department and consider the problems a new system can solve.

TABLE 8.2

Advantages and
Disadvantages of CASE Tools

Advantages	Disadvantages
Produce systems with a longer effective operational life	Produce initial systems that are more expensive to build and maintain
Produce systems that more closely meet user needs and requirements	Require more extensive and accurate definition of user needs and requirements
Produce systems with excellent documentation	May be difficult to customize
Produce systems that need less systems support	Require training of maintenance staff
Produce more flexible systems	May be difficult to use with existing systems

SYSTEMS INVESTIGATION

As discussed earlier, systems investigation is the first phase in the traditional SDLC of a new or modified business information system. The purpose is to identify potential problems and opportunities and consider them in light of the goals of the company. In general, systems investigation attempts to uncover answers to the following questions:

1. What primary problems might a new or enhanced system solve?
2. What opportunities might a new or enhanced system provide?
3. What new hardware, software, databases, telecommunications, personnel, or procedures will improve an existing system or are required in a new system?
4. What are the potential costs (variable and fixed)?
5. What are the associated risks?

Initiating Systems Investigation

Because systems development requests can require considerable time and effort to implement, many organizations have adopted a formal procedure for initiating systems development, beginning with systems investigation. The systems request form is a document that is filled out by someone who wants the IS department to initiate systems investigation. This form typically includes the following information:

- Problems in or opportunities for the system
- Objectives of systems investigation
- Overview of the proposed system
- Expected costs and benefits of the proposed system

The information in the systems request form helps to rationalize and prioritize the activities of the IS department. Based on the overall IS plan, the organization's needs and goals, and the estimated value and priority of the proposed projects, managers make decisions regarding the initiation of each systems investigation for such projects.

Feasibility Analysis

feasibility analysis

assessment of the technical, operational, schedule, economic, and legal feasibility of a project

A key step of the systems investigation phase is **feasibility analysis**, which assesses technical, operational, schedule, economic, and legal feasibility. *Technical feasibility* is concerned with whether the hardware, software, and other system components can be acquired or developed to solve the problem. *Operational feasibility* is a measure of whether the project can be put into action or operation. It can include logistical and motivational (acceptance of change) considerations. Motivational considerations are very important because new systems affect people and data flows and may have unintended consequences. As a result, power and politics may come into play, and some people may resist the new system. Operational feasibility can also include legal and ethical considerations. *Schedule feasibility* determines whether the project can be completed in a reasonable amount of time—a process that involves balancing the time and resource requirements of the project with other projects. *Economic feasibility* determines whether the project makes financial sense and whether predicted benefits offset the cost and time needed to obtain them. Economic feasibility can involve cash flow analysis such as that done in net present value calculations. *Legal feasibility* determines whether there are laws or regulations that may prevent or limit a systems development project. For example, an Internet site that allowed users to share music without paying musicians or music producers was sued. Legal feasibility involves an analysis of existing and future laws to determine the likelihood of legal action against the systems development project and the possible consequences.

The Systems Investigation Report

systems investigation report

a summary of the results of the systems investigation and the process of feasibility analysis; recommends a course of action

The primary outcome of systems investigation is a **systems investigation report**. This report summarizes the results of systems investigation and the process of feasibility analysis and recommends a course of action: continue on into systems analysis, modify the project in some manner, or drop it. A typical table of contents for the systems investigation report is shown in Figure 8.7.

SYSTEMS ANALYSIS

After a project has been approved for further study, the next step is to answer the question "What must the information system do to solve the problem?" The process needs to go beyond mere computerization of existing systems. The entire system, and the business process with which it is associated, should be evaluated. Often, a firm can make great gains if it restructures both business activities and the related information system simultaneously. The overall emphasis of analysis is gathering data on the existing system, determining the requirements for the new system, considering alternatives within these constraints, and investigating the feasibility of the solutions. The primary outcome

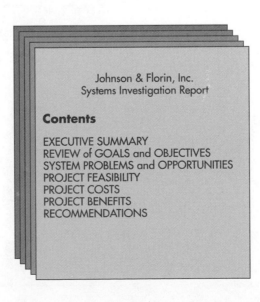

Johnson & Florin, Inc.
Systems Investigation Report

Contents

EXECUTIVE SUMMARY
REVIEW of GOALS and OBJECTIVES
SYSTEM PROBLEMS and OPPORTUNITIES
PROJECT FEASIBILITY
PROJECT COSTS
PROJECT BENEFITS
RECOMMENDATIONS

FIGURE 8.7

A Typical Table of Contents for a Systems Investigation Report of systems analysis is a prioritized list of systems requirements. Systems analysis begins with data collection and analysis.

Data Collection

The purpose of data collection is to seek additional information about the problems or needs identified in the systems investigation report. During this process, the strengths and weaknesses of the existing system are emphasized.

Identifying Sources of Data

Data collection begins by identifying and locating the various sources of data, including both internal and external sources (Figure 8.8).

Performing Data Collection

Once data sources have been identified, data collection begins. Figure 8.9 shows the steps involved. Data collection may require a number of tools and techniques, such as interviews, direct observation, and questionnaires. The choice depends on the analysis team's experience and the type and location of the organization's business processes. In a structured interview, the questions are written in advance. In an unstructured interview, the questions are not written in advance; the interviewer relies on experience in asking the best questions to uncover the problems of the existing system. With direct observation, one or more members of the analysis team directly observe the existing system in action. In today's far-flung organizations when many data sources are spread over a wide geographic area, questionnaires may be the best approach. Like interviews, questionnaires can be either structured or unstructured. In most cases, a pilot study is conducted to fine-tune the questionnaire. A follow-up questionnaire can also capture the opinions of those who do not respond to the original questionnaire.

Data Analysis

The data collected in its raw form is usually not adequate to determine the effectiveness and efficiency of the existing system or the requirements for the new system. The next step is to manipulate the collected data so that it is usable for the development team participating in systems analysis. This manipulation

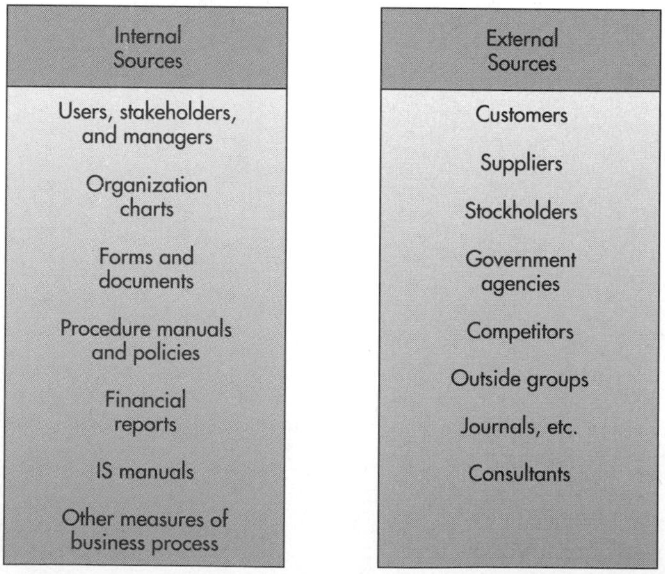

FIGURE 8.8

Internal and External Sources of Data for Systems Analysis

Internal Sources	External Sources
Users, stakeholders, and managers	Customers
Organization charts	Suppliers
Forms and documents	Stockholders
Procedure manuals and policies	Government agencies
Financial reports	Competitors
IS manuals	Outside groups
Other measures of business process	Journals, etc.
	Consultants

FIGURE 8.9

The Steps in Data Collection

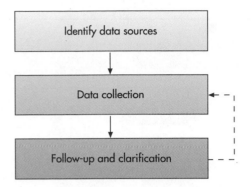

data analysis

manipulation of the collected data so that it is usable for the development team members who are participating in systems analysis

is called **data analysis**. Data and activity modeling, using entity-relationship diagrams and data-flow diagrams, are useful during data analysis to show data flows and the relationships among various objects, associations, and activities. As discussed previously, CASE tools can also be used.

Data Modeling

Data modeling, first introduced in Chapter 3, is a commonly accepted approach to modeling organizational objects and associations that employ both text and graphics. The exact way data modeling is employed, however, is governed by the specific systems development methodology.

Data modeling is most often accomplished through the use of entity-relationship (ER) diagrams. Recall from Chapter 3 that an entity is a generalized representation of an object type—such as a class of people (employee), events (sales), things (desks), or places (Philadelphia)—and that entities possess certain attributes. Objects can be related to other objects in numerous ways. An entity-relationship diagram, such as the one shown in Figure 8.10a, describes a number of objects and the ways they are associated. An ER diagram is not capable in and of itself of fully describing a business problem or solution, because it lacks descriptions of the related activities. It is, however, a good place to start, since it describes object types and attributes about which data may need to be collected for processing.

Activity Modeling

To fully describe a business problem or solution, it is necessary to describe the related objects, associations, and activities. Activities in this sense are events or items that are necessary to fulfill the business relationship or that can be associated with the business relationship in a meaningful way.

data-flow diagram (DFD)

a model of objects, associations, and activities that describes how data can flow between and around various objects

Activity modeling is often accomplished through the use of data-flow diagrams. A **data-flow diagram (DFD)** models objects, associations, and activities by describing how data can flow between and around various objects. DFDs work on the premise that for every activity there is some communication, transference, or flow that can be described as a data element. DFDs describe what activities are occurring to fulfill a business relationship or accomplish a business task, not how these activities are to be performed. That is, DFDs show the logical sequence of associations and activities, not the physical processes. A system modeled with a DFD could operate manually or could be computer based; if computer based, the system could operate with a variety of technologies. See Figure 8.10b.

Comparing entity-relationship diagrams with data-flow diagrams provides insight into the concept of top-down design. Figures 8.10a and b show an entity-relationship diagram and a data-flow diagram for the same business relationship—namely, a member of a golf course playing golf. Figure 8.10c provides a brief description of the business relationship for clarification. After data analysis, the development team can begin requirements analysis.

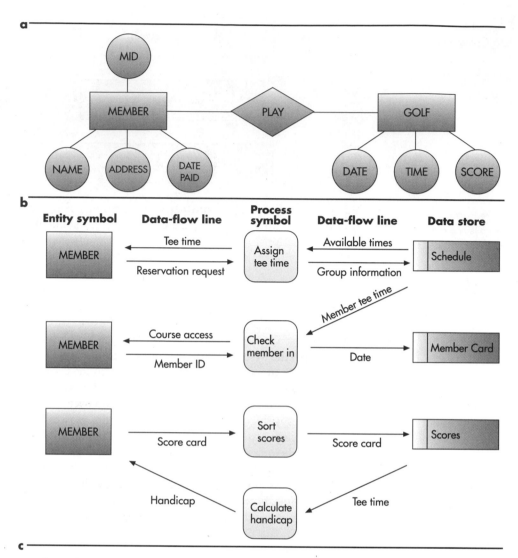

To play golf at the course, you must first pay a fee to become a member of the golf club. Members are issued member cards and are assigned member ID numbers. To reserve a tee time (a time to play golf), a member calls the club house at the golf course and arranges an available time slot with the reception clerk. The reception clerk reserves the tee time by writing the member's name and number of players in the group on the course schedule. When a member arrives at the course, he or she checks in at the reception desk where the reception clerk checks the course schedule and notes the date on the member's card. After a round of golf has been completed, the members leave their score card with the reception clerk. Member scores are tracked and member handicaps are updated on a monthly basis.

FIGURE 8.10

Data and Activity Modeling

(a) An entity-relationship diagram. (b) A data-flow diagram. (c) A semantic description of the business process.
(Source: G. Lawrence Sanders, *Data Modeling* [Danvers, MA: Boyd & Fraser Publishing, 1995].)

requirements analysis

determination of user, stakeholder, and organizational needs

Requirements Analysis

The overall purpose of **requirements analysis** is to determine user, stakeholder, and organizational needs. For an accounts payable application, for example, the stakeholders could include suppliers and members of the purchasing department. Questions that should be asked during requirements analysis include the following:

1. Are these stakeholders satisfied with the current accounts payable application?
2. What improvements could be made to satisfy suppliers and help the purchasing department?

Asking Directly

One the most basic techniques used in requirements analysis is asking directly. **Asking directly** is an approach that asks users, stakeholders, and other managers about what they want and expect from the new or modified system. This approach works best for stable systems in which stakeholders and users clearly understand the system's functions. Unfortunately, many individuals do not know exactly or are unable to adequately articulate what they want or need. The role of the systems analyst during the analysis phase is to critically and creatively evaluate needs and define them clearly so that the systems can best meet them.

Critical Success Factors

Another approach uses critical success factors (CSFs). Managers and decision makers are asked to list only those factors that are critical to the success of their area of the organization. A CSF for a production manager might be adequate raw materials from suppliers; a CSF for a sales representative could be a list of customers currently buying a certain type of product. Starting from these CSFs, the system inputs, outputs, performance, and other specific requirements can be determined.

The IS Plan

As we have seen, the IS plan translates strategic and organizational goals into systems development initiatives. The IS planning process often generates strategic planning documents that can be used to define system requirements. Working from these documents ensures that requirements analysis will address the goals set by top-level managers and decision makers (Figure 8.11). There are unique benefits to applying the IS plan to define systems requirements. Because the IS plan takes a long-range approach to using information technology within the organization, the requirements for a system analyzed in terms of the IS plan are more likely to be compatible with future systems development initiatives.

Requirements Analysis Tools

A number of tools can be used to document requirements analysis. Again, CASE tools are often employed. As requirements are developed and agreed upon, entity-relationship diagrams, data-flow diagrams, and other types of documentation will be stored in the CASE repository. These requirements might also be used later as a reference during the rest of systems development or for a different systems development project.

The Systems Analysis Report

Systems analysis concludes with a formal systems analysis report. It should cover the following elements:

1. The strengths and weaknesses of the existing system from a stakeholder's perspective
2. The user/stakeholder requirements for the new system (also called the *functional requirements*)
3. The organizational requirements for the new system
4. A description of what the new information system should do to solve the problem

FIGURE 8.11

Converting Organizational Goals into Systems Requirements

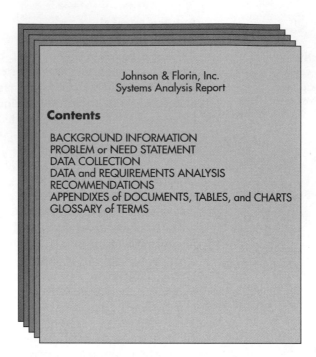

Johnson & Florin, Inc.
Systems Analysis Report

Contents

BACKGROUND INFORMATION
PROBLEM or NEED STATEMENT
DATA COLLECTION
DATA and REQUIREMENTS ANALYSIS
RECOMMENDATIONS
APPENDIXES of DOCUMENTS, TABLES, and CHARTS
GLOSSARY of TERMS

FIGURE 8.12

A Typical Table of Contents for a Report on an Existing System

Suppose analysis reveals that a marketing manager thinks a weakness of the existing system is its inability to provide accurate reports on product availability. These requirements and a preliminary list of the corporate objectives for the new system will be in the systems analysis report. Particular attention is placed on areas of the existing system that could be improved to meet user requirements. The table of contents for a typical report is shown in Figure 8.12. After the formal report is prepared, the team can move from systems analysis to the next phase—systems design.

SYSTEMS DESIGN

The purpose of systems design is to answer the question "How will the information system solve a problem?" The primary result of the systems design phase is a technical design that details system outputs, inputs, and user interfaces; specifies hardware, software, databases, telecommunications, personnel, and procedures; and shows how these components are related. The new system should overcome shortcomings of the existing system and help the organization achieve its goals. Of course, the system must also meet certain guidelines, including user and stakeholder requirements and the objectives defined during previous development phases.

Design can range in scope from multicorporate to individual projects (see Table 8.3). Nearly all companies are continually involved in designing systems for individuals, workgroups, and the enterprise. Increasingly, companies undertake multicorporate design, where two or more companies form a partnership or an alliance to design a new system. Design is also increasingly focusing on Internet applications.

TABLE 8.3

The Scope of Design

Design can range in scope from multicorporate to individual projects.

Multicorporate design projects	Increasingly
Enterprise design projects	Narrow
Workgroup design projects	Scope
Individual design projects	

logical design

description of the functional
requirements of a system

physical design

specification of the characteristics
of the system components necessary
to put the logical design into action

Information systems must be designed along two dimensions: logical and physical. The **logical design** refers to what the system will do. It describes the functional requirements of a system. That is, it conceptualizes what the system will do to solve the problems identified through earlier analysis. Without this step, the technical aspects of the system (such as which hardware devices should be acquired) often obscure the solution. Logical design involves planning the purpose of each system element, independent of hardware and software considerations. The **physical design** refers to how the tasks are accomplished, including how the components work together and what each component does. Physical design specifies the characteristics of the system components necessary to put the logical design into action. In this phase, the characteristics of each of the following components must be specified:

Hardware design. All computer equipment, including input, processing, and output devices, must be specified by performance characteristics. For example, if the logical design specified that the database must hold large amounts of historical data, then the system storage devices must have large capacity.

Software design. All software must be specified by capabilities. For example, if the logical design specifies that dozens of users must be able to update the database concurrently, then the physical design must specify a database management system that allows this to occur. In some cases, software can be purchased; in others it is developed internally. Logical design specifications for data inputs, processing requirements, and program outputs are also considered during the physical design of the software. For example, the ability to access data stored on certain disk files that the program will use is specified.

Database design. The type, structure, and function of the databases must be specified. The relationships between data elements established in the logical design must be mirrored in the physical design as well. These relationships include such things as access paths and file structure organization. Fortunately, many excellent database management systems exist to assist with this activity.

Telecommunications design. The necessary characteristics of the communications software, media, and devices must be specified. For example, if the logical design specifies that all members of a department must be able to share data and run common software, then the local area network configuration and the communications software that are specified in the physical design must possess this capability.

Personnel design. This step involves specifying the background and experience of individuals most likely to meet the job descriptions specified in the logical design.

Procedures and control design. How each application is to run, as well as what is to be done to minimize the potential for crime and fraud, must be specified. These specifications include auditing, backup, and output distribution methods.

Organizations need to protect their valuable data and equipment, so you may encounter some rules and procedures that you need to follow concerning the information system you use. Most IS departments establish tight systems controls to maintain data security and help prevent computer misuse, crime, and fraud by employees and others. Most IS departments have a set of general operating rules. Some information systems departments are *closed shops,* in which only authorized operators can run the computers. Other IS departments are *open shops,* in which other people, such as programmers and systems analysts, are also authorized to run the computers. Other rules specify the conduct of the IS department. All of these rules are *deterrence controls,* which help prevent problems before they occur. Making a computer more secure and less vulnerable to a break-in is another example. Good control techniques should also help an organization contain and recover from problems when they do occur. Containment tries to minimize the impact of a problem while it is occurring, and recovery control involves responding to a problem that has already occurred.

Many companies use ID badges to prevent unauthorized access to sensitive areas in the information systems facility.
(Source: Courtesy of Sensomatic Electronics Corporation.)

Generating Systems Design Alternatives

When additional hardware and software are not required, alternative designs are often generated without input from outside vendors. If the new system is complex, the original development team may want to involve other personnel in generating alternative designs. If new hardware and software need to be acquired from a vendor, a formal request for proposal (RFP) should be made.

request for proposal (RFP)

a document that specifies in detail required resources such as hardware and software

The **request for proposal (RFP)** is one of the most important documents generated during systems development. It often results in a formal bid that is used to determine who gets a contract for new or modified systems. The RFP specifies in detail the required resources such as hardware and software. The Justice Ministry of New Zealand, for example, developed an RFP to get proposals on its disaster recovery facility for its large law enforcement system.[8] The system never had disaster recovery provisions previously. Only two companies were asked to submit proposals because, according to the CIO, "These companies have been assessed as the suppliers most capable of addressing the complex requirements for a disaster recovery service." The disaster facility will be based outside Auckland, New Zealand.

In some cases, separate RFPs are developed for different needs. For example, a company might develop separate RFPs for hardware, software, and database systems. The RFP also communicates these needs to one or more vendors, and it provides a way to evaluate whether the vendor has delivered what was expected. The table of contents for a typical RFP is shown in Figure 8.13.

Financial Options

When it comes to acquiring computer systems, three choices are available: purchase, lease, or rent. Cost objectives and constraints set for the system play a significant role in the alternative chosen, as do the advantages and disadvantages of each. Table 8.4 summarizes the advantages and disadvantages of these financial options.

Johnson & Florin, Inc.
Systems Investigation Report

Contents

COVER PAGE (with company name and contact person)
BRIEF DESCRIPTION of the COMPANY
OVERVIEW of the EXISTING COMPUTER SYSTEM
SUMMARY of COMPUTER-RELATED NEEDS and/or PROBLEMS
OBJECTIVES of the PROJECT
DESCRIPTION of WHAT IS NEEDED
HARDWARE REQUIREMENTS
PERSONNEL REQUIREMENTS
COMMUNICATIONS REQUIREMENTS
PROCEDURES to BE DEVELOPED
TRAINING REQUIREMENTS
MAINTENANCE REQUIREMENTS
EVALUATION PROCEDURES (how vendors will be judged)
PROPOSAL FORMAT (how vendors should respond)
IMPORTANT DATES (when tasks are to be completed)
SUMMARY

TABLE 8.4

Advantages and
Disadvantages of Acquisition
Options

Renting (Short-Term Option)	
Advantages	**Disadvantages**
No risk of obsolescence	No ownership of equipment
No long-term financial investment	High monthly costs
No initial investment of funds	Restrictive rental agreements
Maintenance usually included	

Leasing (Longer-Term Option)	
Advantages	**Disadvantages**
No risk of obsolescence	High cost of canceling lease
No long-term financial investment	Longer time commitment than renting
No initial investment of funds	No ownership of equipment
Less expensive than renting	

Purchasing	
Advantages	**Disadvantages**
Total control over equipment	High initial investment
Can sell equipment at any time	Additional cost of maintenance
Can depreciate equipment	Possibility of obsolescence
Low cost if owned for a number of years	Other expenses, including taxes and insurance

Evaluating and Selecting a System Design

The next step in systems design is to evaluate the various alternatives and select the one that will offer the best solution for organizational goals. Performance, cost, control, and complexity objectives must be considered and balanced during design evaluation. Depending on their weight, any one of these objectives may result in the selection of one design over another. For example, financial concerns might make a company choose rental over equipment purchase. Specific performance objectives—say, that the new system must perform on-line data processing—may result in a complex network design for which control procedures must be established. Evaluating and selecting the best design involves a balance of system objectives that will best support organizational goals.

Normally, evaluation and selection involve both a preliminary and a final evaluation before a design is selected. A preliminary evaluation begins after all proposals have been submitted. The purpose of this evaluation is to dismiss unwanted proposals. Several vendors can usually be eliminated by investigating their proposals and comparing them with the original criteria. The final evaluation begins with a detailed investigation of the proposals offered by the remaining vendors. After any presentations and demonstrations have been given, the organization makes the final selection.

Freezing Design Specifications

Near the end of the design stage, an organization prohibits any further changes in the design of the system. The design specifications are then said to be frozen. Freezing systems design specifications means that the user agrees in writing that the design is acceptable. (Figure 8.14). Most consulting companies insist on this formal step to avoid cost overruns and missed user expectations.

FIGURE 8.14

Freezing Design Specifications

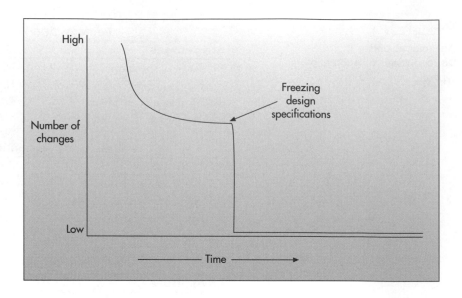

The Design Report

System specifications are the final results of systems design. They include a technical description that details system outputs, inputs, and user interfaces, as well as all hardware, software, databases, telecommunications, personnel, and procedure components and the way these components are related. The specifications are contained in a **design report**, which is the primary result of systems design. The design report reflects the decisions made for system design and prepares the way for systems implementation. The contents of the design report are summarized in Figure 8.15.

When developing a system, it is important to understand and thoroughly complete the systems development activities covered in this chapter. These phases provide the blueprints and groundwork for the rest of systems development. The activities of the next phases will be easier, faster, and more accurate and will result in a more efficient, effective system if the design is complete and well thought out.

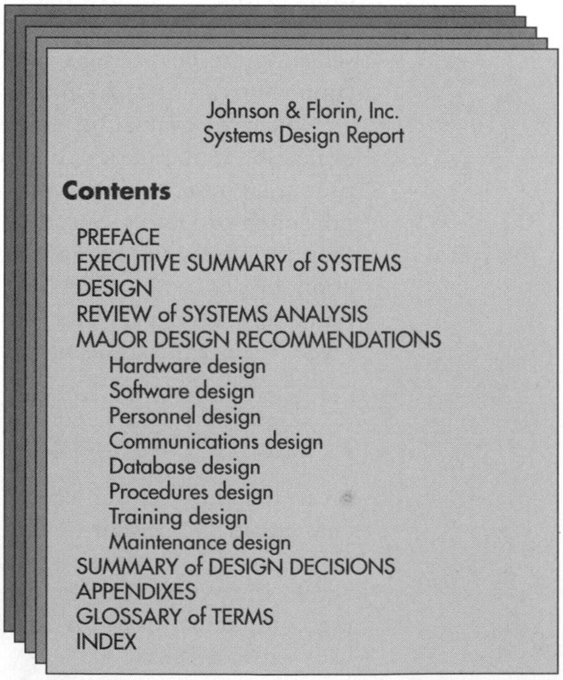

design report

the primary result of systems design, reflecting the decisions made for system design and preparing the way for systems implementation

Johnson & Florin, Inc.
Systems Design Report

Contents

PREFACE
EXECUTIVE SUMMARY of SYSTEMS DESIGN
REVIEW of SYSTEMS ANALYSIS
MAJOR DESIGN RECOMMENDATIONS
    Hardware design
    Software design
    Personnel design
    Communications design
    Database design
    Procedures design
    Training design
    Maintenance design
SUMMARY of DESIGN DECISIONS
APPENDIXES
GLOSSARY of TERMS
INDEX

FIGURE 8.15

A Typical Table of Contents for a Systems Design Report

SYSTEMS IMPLEMENTATION

After the information system has been designed, a number of tasks must be completed before the system is installed and ready to operate. This process, called systems implementation, includes hardware acquisition, software acquisition or development, user preparation, hiring and training of personnel, site and data preparation, installation, testing, start-up, and user acceptance. The typical sequence of these activities is shown in Figure 8.16.

Choices and trade-offs are made at each step of systems implementation, which involve analyzing the benefits in terms of performance, cost, control, and complexity. Unfortunately, many organizations do not take full advantage of these steps or carefully analyze the trade-offs and so never realize the full potential of new or modified systems. The hassles and carelessness must be avoided if organizations are to maximize the return on their information systems investment.

Acquiring Hardware from an Information Systems Vendor

To obtain the components for an information system, organizations can purchase, lease, or rent computer hardware and other resources. ABB, for example,

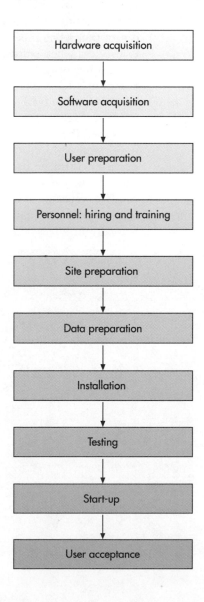

FIGURE 8.16

Typical Steps in Systems
Implementation

signed a $250 million lease deal with IBM for about 75,000 desktop PCs, 25,000 laptop computers, and 9,000 PC servers.[9] The PC lease is expected to save ABB $35 million over the next three years. Some vendors, such as Hewlett-Packard, offer "pay-as-you-go" options.[10] The approach, which is being used for Internet applications, eliminates the need for a company to pay up front for a system. It also allows a company to change or upgrade its information system to meet changing business needs.

Types of hardware vendors include general computer manufacturers (e.g., IBM and Hewlett-Packard), small computer manufacturers (e.g., Dell and Gateway), peripheral equipment manufacturers (e.g., Epson, Canon, and Piiceon), computer dealers and distributors (e.g., Canadian Communication Products, Dell, and Gateway) and leasing companies (e.g., National Computer Leasing, ECONOCOM-US, and Paramount Computer Rentals).

Acquiring Software: Make or Buy?

make-or-buy decision

the decision regarding whether to obtain the necessary software from internal or external sources

As with hardware, software can be acquired several ways. As previously mentioned, it can be purchased from external developers or developed in-house. This decision is often called the **make-or-buy decision**. Some of the reasons a company might purchase or lease externally developed software include lower costs, less risk regarding the features and performance of the package, and ease of installation. The cost of the software package is known, and there is little doubt that it will meet the company's needs. The amount of development effort is also less when software is purchased, compared with in-house development.

Another option is to make or develop software internally. This requires the company's IS personnel to be responsible for all aspects of software development. Some advantages inherent with in-house-developed software include meeting user and organizational requirements and having more features and increased flexibility in terms of customization and changes. Software programs developed within a company also have greater potential for providing a competitive advantage because they are not easily duplicated by competitors in the short term. In addition, sometimes software from other development efforts can be reused, reducing the time it takes to deliver in-house software. BankAmerica, for example, reuses previously developed software to deliver new software in 90 days or less.[11] In some cases, a company develops its own software internally then decides to sell it to other companies.[12] General Electric, for example, developed a better system for order processing. Now it is looking to sell the software to other companies with similar problems. If successful, this new business for General Electric will help pay for the software development process. It may even turn into a profitable business.

In some cases, companies use a blend of external and internal software development. That is, off-the-shelf or proprietary software programs are modified or customized by in-house personnel. The advantages and disadvantages of these approaches were discussed in Chapter 2.

Computer dealers, such as CompUSA, manufacture build-to-order computer systems and sell computers and supplies from other vendors.
(Source: Courtesy of CompUSA, Inc.)

Acquiring Database and Telecommunications Systems

Acquiring or upgrading database systems can be one of the most important steps of a systems development effort. Because databases are a blend of hardware and software, many of the approaches discussed earlier for acquiring hardware and software also apply to database systems. For example, an upgraded inventory control system may require database capabilities, including more hard disk storage or a new DBMS. If so, additional storage hardware will have to be acquired from an information systems vendor. New or upgraded software might also be purchased or developed in-house.

Telecommunications is one of the fastest-growing applications for today's businesses and individuals. Like database systems, telecommunications systems require a blend of hardware and software. For personal computer systems, the primary piece of hardware is a modem. For client/server and mainframe systems, the hardware can include multiplexers, concentrators, communications processors, and a variety of network equipment. Com-munications software will also have to be acquired from a software company or developed in-house. Again, the earlier discussion on acquiring hardware and software also applies to the acquisition of telecommunications hardware and software.

User Preparation

User preparation is the process of readying managers, decision makers, employees, other users, and stakeholders for the new systems. With the growing trend to employee empowerment, system developers need to provide users with proper training to make sure they use the information system correctly, efficiently, and effectively. User preparation can include active participation, marketing, training, documentation, and support. Top-management support is absolutely essential to ensure that sufficient time and resources are allocated to user preparation for a successful system start-up.

Informing and preparing users for new or modified systems can be done in a variety of ways. User preparation actually begins with user participation in system analysis. Some organizations also actively market new systems to users via brochures, newsletters, and seminars to promote them the way they would a new product or service.

site preparation

preparation of the location of the new system

Providing users with proper training can help ensure that the information system is used correctly, efficiently, and effectively.
(Source: Eyewire.)

IS Personnel: Hiring and Training

Depending on the size of the new system, an organization may have to hire and, in some cases, train new IS personnel. An information systems manager, systems analysts, computer programmers, data entry operators, and similar personnel may be needed for the new system.

As with users, the eventual success of any system depends on how it is used by the personnel within the organization. Training programs should be conducted for the IS personnel who will be using the computer system. These programs are similar to those for the users, although they may be more detailed in the technical aspects of the systems. Effective training will help IS personnel use the new system to perform their jobs and support other users in the organization.

Site Preparation

The location of the new system needs to be prepared in a process called **site preparation**. For a small system, site preparation can be as simple as rearranging the furniture in an office to make room for a computer. With a larger system, this process is not so easy because it may require special wiring and air-conditioning. One or two rooms may have to be completely renovated, and additional furniture may have to be purchased. A special floor may have to be built, under which the cables connecting the various computer components are placed, and a new security system may be needed to protect the equipment. For larger systems, additional power circuits may also be required.

Data Preparation

data preparation (data conversion)

conversion of manual files into computer files

If the organization is computerizing its work processes, all manual files must be converted to computer files in a process called **data preparation**, or **data conversion**. All permanent data must be placed on a permanent storage device, such as magnetic tape or disk. Usually the organization hires temporary, part-time data-entry operators or a service company to convert the manual data. Once the data has been converted, the temporary workers are no longer needed. A computerized database system or other software will then be used to maintain and update the computer files.

Installation

installation

the process of physically placing the computer equipment on the site and making it operational

Installation is the process of physically placing the computer equipment on the site and making it operational. Although normally the manufacturer is responsible for installing computer equipment, someone from the organization (usually the IS manager) should oversee the process, making sure that all equipment is installed at the proper location. After the system is installed, the manufacturer performs several tests to ensure that the equipment is operating as it should.

Testing

Good testing procedures are essential to make sure that the new or modified information system operates as intended. Inadequate testing can result in mistakes and problems. A popular tax preparation company, for example, implemented a Web-based tax preparation system, but people could see one another's tax returns.[13] The president of the tax preparation company called it "our worst-case scenario." In another case, the London Stock Exchange experienced a crash that lasted almost eight hours.[14] Some believe that the London Stock Exchange can salvage its reputation, but others disagree. The head of a trading firm said that this is "the nail in the coffin for the London Stock Exchange in the way it exists now. The ramifications are going to be huge." Better testing might have prevented these types of problems.

Several forms of testing should be used, including testing each of the individual programs (unit testing), testing the entire system of programs together (system testing), testing the application with a large amount of data (volume testing), and testing all related systems together (integration testing), as well as conducting any tests required by the user (acceptance testing).

start-up

the process of making the final tested information system fully operational

direct conversion

stopping the old system and starting the new system on a given date (also called *plunge* or *direct cutover*)

phase-in approach

slowly replacing components of the old system with those of the new one; this process is repeated for each application until the new system is running every application and performing as expected (also called *piecemeal approach*)

Start-Up

Start-up begins with the final tested information system. When start-up is finished, the system is fully operational. Various start-up approaches are available (Figure 8.17). **Direct conversion** (also called *plunge* or *direct cutover*) involves stopping the old system and starting the new system on a given date. Direct conversion is usually the least desirable approach because of the potential for problems and errors when the old system is shut off and the new system is turned on at the same instant. The **phase-in approach** is a popular technique preferred by many organizations. In this approach, sometimes called a *piecemeal approach,* components of the new system are slowly phased in while components of the old one are slowly phased out. When everyone is confident that the new system is performing as expected, the old system is completely phased out. This gradual replacement is repeated for each application until the new system is running every application. **Pilot start-up** involves running the new system for one group of users rather than all users. For example, a manufacturing company with a number of retail outlets throughout the country could use the pilot start-up approach and install a new inventory control system at one of the retail outlets. When this pilot retail outlet runs without problems, the new inventory control system can be implemented at other retail outlets. **Parallel start-up** involves running both the old and new systems for a period of time. The output of the new system is compared closely with the output of the old system, and any differences are reconciled. When users are comfortable that the new system is working correctly, the old system is eliminated.

User Acceptance

Most mainframe computer manufacturers use a formal **user acceptance document**—a formal agreement signed by the user that states that a phase of the installation or the complete system is approved. This is a legal document that usually removes or reduces the information systems vendor's liability for problems that occur after the user acceptance document has been signed. Because this document is so important, many companies get legal assistance

pilot start-up

running the new system for one group of users rather than all users

parallel start-up

running both the old and new systems for a period of time and comparing the output of the new system closely with the output of the old system; any differences are reconciled; when users are comfortable that the new system is working correctly, the old system is eliminated

user acceptance document

a formal agreement signed by the user that states that a phase of the installation or the complete system is approved

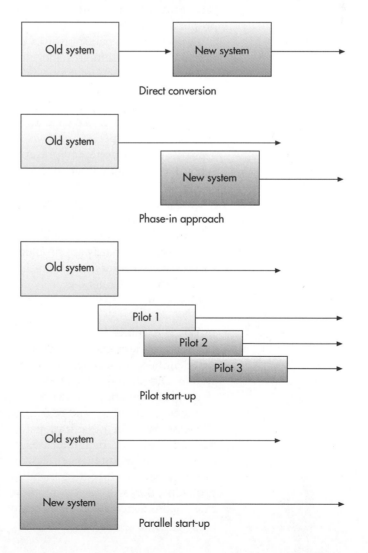

FIGURE 8.17

Start-Up Approaches

before they sign the acceptance document. Stakeholders may also be involved in acceptance to make sure that the benefits to them are realized. As seen in the "Ethical and Societal Issues" box, some systems development efforts may not be desirable or acceptable to everyone involved.

After a system has been selected, tested, and installed, the organization begins to use it for its day-to-day business functions. The focus of the organization then shifts to maintaining and reviewing the system to ensure that it is performing as expected.

SYSTEMS MAINTENANCE AND REVIEW

Since information systems are so critical to business functions today, organizations spend considerable time and money to maintain them and review their performance. As with the other phases of the systems development life cycle, both users and IS personnel are involved in maintenance and review—but to differing degrees.

Systems Maintenance

The cost of systems maintenance is staggering. For older programs, the total cost of maintenance can be up to five times greater than the total cost of development. In other words, a program that originally cost $25,000 to develop may cost $125,000 to maintain over its lifetime. Because it is so costly, the organization must follow designated steps to be sure that it carries out maintenance properly.

Systems maintenance involves checking, changing, and enhancing the system to make it more useful in achieving user and organizational goals. In some cases, an organization encounters major problems that involve recycling the entire systems development process. In other situations, minor modifications are sufficient.

Once a program is written, it is likely to need ongoing maintenance. To some extent, a program is like a car that needs oil changes, tune-ups, and repairs at certain times. Experience shows that frequent, minor maintenance to a program, if properly done, can prevent major system failures later. Some of the reasons for program maintenance are the following:

- Changes in business processes
- New requests from stakeholders, users, and managers
- Bugs or errors in the program
- Technical and hardware problems
- Corporate mergers and acquisitions
- Government regulations
- Change in the operating system or hardware on which the application runs

When it comes to making necessary changes, most companies modify their existing programs instead of developing new ones. That is, as new systems needs are identified, most often the burden of fulfilling the needs falls on the existing system. Old programs are repeatedly modified to meet ever-changing needs. Over time, these modifications tend to interfere with the system's overall structure, reducing its efficiency and making further modifications more burdensome.

Depending on organizational policies, the people who perform systems maintenance vary. In some cases, the team that designs and builds the system also performs maintenance. This ongoing responsibility gives the designers and programmers an incentive to build systems well from the outset: if there are problems, they will have to fix them. In other cases, organizations have a separate maintenance team. This team is responsible for modifying, fixing, and

ETHICAL AND SOCIETAL ISSUES
Monitoring Debit Card Usage

William Scheurer, the president of a large company, was frustrated. His 15-year-old daughter was always running out of money, which required him to make frequent trips to an ATM. Not only was it frustrating—it was also time-consuming. Being president of a $400 million credit card service company, Scheurer knew there had to be something he could do for himself and the millions of parents who faced similar problems. Could an innovative systems development project give a solution?

His answer was to implement a special debit card, called PocketCard, just for teenagers. A special account would be set up. Then the parents could deposit or withdraw funds as needed and warranted. The most interesting part of the system was its connection to the Internet and on-line computer systems. With this on-line capability, parents could monitor their teenagers' spending, on a second-by-second basis, as card charges worked their way through a Visa network. If a parent saw his or her kid purchasing alcohol at a liquor store or if the parent saw a charge at a forbidden nightclub, all the remaining money could be transferred out of the account, and the teenager could be disciplined when he or she arrived back home. The new teenager accounts could generate revenues for the card issuer by charging to activate the account, requiring a maintenance fee, and charging for each transaction. The systems development effort could cost from $5 million to $15 million. In addition to teenagers, the new debit card could be used with baby-sitters, secretaries, or even employees. With the new card, expenses could be monitored through the Internet or by making a simple phone call.

Although the system sounds like a blessing for parents, it poses ethical and societal concerns. If used with secretaries or employees, would the new card be an invasion of their privacy? Would the use of such a card in a corporate environment cause more problems than it would solve? These questions have yet to be resolved, but the novel approach is generating attention. Other companies, including SpendCash and Cobalt Card, are also getting into the debit-card business. SpendCash, for example, will be sold in $10, $20, $50, and $100 increments at retail and convenience stores. The company will post a 2.25 percent charge for each transaction.

Discussion Questions

1. What are the advantages and disadvantages of this new debit card?
2. Do you think this systems development effort will be acceptable to all involved?

Critical Thinking Questions

3. Assume that you are the systems development manager for the debit card project. How would you modify the teenager debit card system to make the card appropriate for use with employees, secretaries, and managers of a company?
4. Are there any privacy protections that you could build into the system during the systems implementation stage?

Sources: Adapted from Chana Schoenberger, "Big Brother (and Sister)," *Forbes*, January 24, 2000, p. 142; "Only Time Will Tell," *ITS World*, January 2000, p. 1086; and "Online Music Discounter to Promote PocketCard," *Card Fax*, February 25, 2000, p. 2.

updating existing software. Because experience and skills are important in maintenance, some organizations use a specialized maintenance team or department. Java and the object-oriented programming languages hold the promise of reducing the program maintenance effort.

Systems Review

Systems review, the final step of systems development, is the process of analyzing systems to make sure they are operating as intended. This analysis often compares the performance and benefits of the system as it was designed with the actual performance and benefits of the system in operation. Cost, performance, control, and complexity factors investigated during design are revisited after the system has been operating. Problems and opportunities uncovered during systems review will trigger systems development and begin the process anew. For example, as the number of users of an interactive system increases, it is not unusual for system response time to increase. If the lag is too great, it may be necessary to redesign some of the system, modify databases, or increase the power of the computer hardware.

Systems review can be performed by internal employees, external consultants, or both. When the problems or opportunities are industrywide, people from several firms may get together, sometimes at an IS conference or in a private meeting involving several firms. For some situations, state or federal governments may be involved. For example, after a rash of Internet stoppages, former President Clinton called a meeting with top Internet firms, the attorney general, the commerce secretary, and the national security advisor.[15] The purpose was to review the stoppages and develop a plan to help prevent this type of hacking from happening in the future.

There are two types of review procedures: event driven and time driven (Table 8.5). An **event-driven review** is triggered by a problem or opportunity such as an error, a corporate merger, or a new market for products. In some cases, companies wait until a large problem or opportunity occurs before a change is made, ignoring minor problems. In contrast, other companies use a continuous-improvement approach to systems development. With this approach, an organization makes changes to a system even when small problems or opportunities occur. Although continuous improvement can keep the system current and responsive, doing the repeated design and implementation can be both time-consuming and expensive.

A **time-driven review** is performed after a specified amount of time. Many application programs are reviewed every six months to a year. With this approach, an existing system is monitored on a schedule. If problems or opportunities are uncovered, a new systems development cycle may be initiated. A payroll application, for example, may be reviewed once a year to make sure it is still operating as expected. If it is not, changes are made.

Many companies use both approaches. A billing application might be reviewed once a year for errors, inefficiencies, and opportunities to reduce operating costs—a time-driven approach. In addition, the billing application might be redone if there is a corporate merger, if one or more new managers require different information or reports, or if federal laws on bill collecting and privacy change—an event-driven approach. If a systems review reveals significant problems or opportunities, systems development is started again from the beginning.

event-driven review

a review triggered by a problem or opportunity such as an error, a corporate merger, or a new market for products

time-driven review

a review performed after a specified amount of time

TABLE 8.5

Examples of Review Types

Event Driven	Time Driven
A problem with an existing system	Monthly review
A merger	Yearly review
A new accounting system	Review every few years
An executive decision that an upgraded Internet site is needed to stay competitive	Five-year review

INFORMATION SYSTEMS IN ACTION
Developing Distance Learning Courses for Flexibility

John Vargo, *University of Canterbury*

Systems development efforts are often a balance between the benefits of a new or enhanced system and the costs of designing and implementing the system. In addition, many organizations have restricted budgets and increasing demands for systems development projects. This was the case with the University of Canterbury in Christchurch, New Zealand. Students have to work longer hours to keep their finances afloat and still try to attend fixed lecture schedules. The university, on the other hand, has limited lecture space during a period of growing student demand. So the university decided to develop a digital lecture system to solve these and other problems, including scheduling clashes between popular courses that make it impossible for students to attend lectures for both courses. The new system for digital lectures consist of digital video files captured from a live lecture, indexed for ease of use, and delivered asynchronously through a Web browser interface. These digital lectures can also be linked to other Web-based resources.

The university developed a pilot program involving two large introductory courses (with 1,100 students combined) that included delivery via CDs through the university library and trial delivery using a streaming media server over the campus LAN. The impact on the campus network is potentially huge when you consider the bandwidth demands of such streamed media. Using industry standard compression (a means for reducing the number of bytes of information required to represent a transmitted message) can still result in bandwidth demands that could bring the campus network to a grinding halt! Resolving this problem requires complementary solutions including further compression, the use of multicast technology, appropriate physical location of the media server, and use of switching hubs. Future Internet-based "home" delivery will use MPEG video, limited by bandwidth considerations, as well as MP3 compressed audio. Like many other systems development projects, this one at the University of Canterbury shows that multiple IS solutions are often needed to solve a problem or satisfy an important organizational need.

● SUMMARY

PRINCIPLE • Effective systems development requires a team effort, and it starts with careful planning.

The systems development team consists of stakeholders, users, managers, systems development specialists, and various support personnel. The development team is responsible for determining the objectives of the information system and delivering to the organization a system that meets its objectives.

Stakeholders are individuals who, either themselves or through the area of the organization they represent, ultimately benefit from the systems development project. Users are the employees, managers, customers, or suppliers who will interact with the system regularly. Managers on development teams typically represent stakeholders or may be stakeholders themselves. In addition, managers are most capable of initiating and maintaining change.

A systems analyst is a professional who specializes in analyzing and designing business systems. The programmer is responsible for modifying or developing programs to satisfy user requirements. Other support personnel on the development team include technical specialists, either IS department employees or outside consultants. Depending on the magnitude of the systems development project and the number of IS systems development specialists on the team, the team may also include one or more IS managers.

● ● ●

Information systems planning refers to the translation of strategic and organizational goals into systems development initiatives. Benefits of IS planning include a long-range view of information technology use and better use of information systems resources. Planning requires developing overall IS objectives; identifying IS projects; setting priorities and selecting projects; analyzing resource requirements; setting schedules, milestones, and deadlines; and developing the IS planning document.

PRINCIPLE • Systems development often uses a predetermined methodology to select, implement, and monitor projects.

The five phases of the traditional SDLC are investigation, analysis, design, implementation, and maintenance and review. Systems investigation involves identifying potential problems and opportunities and considering them in light of organizational goals. Systems analysis seeks a general understanding of the solution required to solve the problem; the existing system is studied in detail and weaknesses are identified. Systems design involves creating new or modified system requirements. Systems implementation encompasses programming, testing, training, conversion, and operation of the system. Systems maintenance and review entails monitoring the system and performing enhancements or repairs.

Prototyping is an iterative approach that involves defining the problem, building the initial version, having users utilize and evaluate the initial version, providing feedback, and incorporating suggestions into the second version. Prototypes can be fully operational or nonoperational, depending on how critical the system under development is and how much time and money the organization has to spend on prototyping.

Rapid application development (RAD) uses tools and techniques designed to speed application development. Its use reduces paper-based documentation, automates program source code generation, and facilitates user participation in development activities. RAD makes extensive use of the joint application development (JAD) process to gather data and perform requirements analysis. JAD involves group meetings in which users, stakeholders, and IS professionals work together to analyze existing systems, propose possible solutions, and define the requirements for a new or modified system.

End-user systems development supports projects where the primary effort is undertaken by a combination of business managers and users. End-user systems development is becoming increasingly important as more users develop systems for their personal computers.

PRINCIPLE • **Systems development starts with investigation and analysis of existing systems.**

In most organizations, a systems request form initiates the investigation process. The systems investigation is designed to assess the feasibility of implementing solutions for business problems. An investigation team follows up on the request and performs a feasibility analysis that addresses technical, economic, operational, schedule, and legal feasibility. As a final step in the investigation process, a systems investigation report should be prepared to document relevant findings.

If the project under investigation is feasible, major goals are set for the system's development, including performance, cost, managerial goals, and procedural goals. A systems development methodology must be selected. As a final step in the investigation process, a systems investigation report should be prepared to document relevant findings.

. . .

Systems analysis is the examination of existing systems, which begins once approval for further study is received from management. Additional study of a selected system allows those involved to further understand the systems' weaknesses and potential improvement areas. An analysis team is assembled to collect and analyze data on the existing system.

Data collection methods include observation,

interviews, and questionnaires. Data analysis manipulates the collected data to provide information. The overall purpose of requirements analysis is to determine user and organizational needs.

Data modeling is used to model organizational objects and associations using text and graphical diagrams. It is most often accomplished through the use of entity-relationship (ER) diagrams. Activity modeling is often accomplished through the use of data-flow diagrams (DFDs), which model objects, associations, and activities by describing how data can flow between and around various objects. DFDs use symbols for data flows, processing, entities, and data stores.

PRINCIPLE • **Designing new systems or modifying existing ones should always be aimed at helping an organization achieve its goals.**

The purpose of systems design is to prepare the detailed design needs for a new system or modifications to the existing system. Logical systems design refers to the way the various components of an information system will work together. Physical systems design refers to the specification of the actual physical components. The physical design must specify characteristics for hardware and software design, database and telecommunications, and personnel and procedures design.

. . .

Key steps taken during the design phase include generating design alternatives, evaluating and selecting a system design, freezing design specifications, and developing a design report.

. . .

If new hardware or software will be purchased from a vendor, a formal request for proposal (RFP) is needed. The RFP outlines the company's needs; in response, the vendor provides a written reply. In addition to responding to the company's stated needs, the vendor provides data on its operations. This data might include the vendor's reliability and stability, the type of postsale service offered, the vendor's ability to perform repairs and fix problems, vendor training, and the vendor's reputation.

RFPs from various vendors are reviewed and narrowed down to the few most likely candidates. Financial options, including leasing, purchasing, and renting, are considered. After the vendor is chosen, contract negotiations can begin.

PRINCIPLE • **The primary emphasis of systems implementation is to make sure that the right information is delivered to the right person in the right format at the right time.**

The purpose of systems implementation is to install

the system and make everything, including users, ready for its operation. Systems implementation includes hardware acquisition, software acquisition or development, user preparation, hiring and training of personnel, site and data preparation, installation, testing, start-up, and user acceptance. Hardware acquisition requires purchasing, leasing, or renting computer resources from a vendor.

Implementation must also address database and telecommunications systems, user preparation, and personnel requirements. User preparation involves readying managers, employees, and other users for the new system. New IS personnel may need to be hired, and users must be well trained in the system's functions. The physical site of the system must be prepared, and any existing data to be used in the new system will require conversion to the new format. Hardware is installed during the implementation step, and testing is done. Testing includes program (unit) testing, systems testing, volume testing, integration testing, and acceptance testing.

Start-up begins with the final tested information system. When start-up is finished, the system is fully operational. There are a number of different start-up approaches. Direct conversion (also called *plunge* or *direct cutover*) involves stopping the old system and starting the new system on a given date. With the phase-in approach, sometimes called a *piecemeal approach*, components of the new system are slowly phased in while components of the old one are slowly phased out. When everyone is confident that the new system is performing as expected, the old system is completely phased out. Pilot start-up involves running the new system for one group of users rather than for all users. Parallel start-up involves running both the old and new systems for a period of time. The output of the new system is compared closely with the output of the old system, and any differences are reconciled.

When users are comfortable that the new system is working correctly, the old system is eliminated. The final step of implementation is user acceptance.

· · ·

Software can be purchased from external vendors or developed in-house—a decision termed the *make-or-buy decision*. A purchased software package usually has a lower cost, less risk regarding the features and performance, and easy installation. The amount of development effort is also less when software is purchased. Developing software can result in a system that more closely meets the business needs and has increased flexibility in terms of customization and changes. Developing software also has greater potential for providing a competitive advantage.

PRINCIPLE • Maintenance and review add to the useful life of a system but can consume large amounts of resources.

Systems maintenance involves checking, changing, and enhancing the system to make it more useful in obtaining user and organizational goals. Maintenance is critical for the continued smooth operation of the system. Some major causes of maintenance are new requests from stakeholders and managers, enhancement requests from users, bugs or errors, technical or hardware problems, newly added equipment, changes in organization structure, and government regulations.

· · ·

Systems review is the process of analyzing systems to make sure that they are operating as intended. It involves monitoring systems to be sure they are operating as designed and measuring how well the system is supporting the mission and goals of the organization. An event-driven review is triggered by a problem or opportunity. A time-driven review is started after a specified amount of time.

● REVIEW QUESTIONS

1. What is an information system stakeholder?
2. What is the goal of information systems planning? What steps are involved in IS planning?
3. Identify each of the four systems development life cycles and summarize their strengths and weaknesses.
4. What are the steps of the systems development life cycle?
5. What is the purpose of systems investigation?
6. List the different types of feasibility.
7. What is the purpose of systems analysis?
8. How does the JAD technique support the RAD systems development life cycle?
9. What is the purpose of systems design?
10. What is an RFP? What is typically included in one? How is it used?
11. What activities go on during the user preparation phase of system implementation?
12. What are the major steps of systems implementation?
13. What are the financial options of acquiring hardware?
14. What are some of the reasons for program maintenance? Explain the three types of maintenance.

● DISCUSSION QUESTIONS

1. Why is it important for business managers to have a basic understanding of the systems development process?

2. Briefly describe the role of a system user in the systems investigation and systems analysis stages of a project.

3. For what types of systems development projects might prototyping be especially useful? What are the characteristics of a system developed with a prototyping technique?

4. Imagine that your firm has never developed an information systems plan. What sort of issues between the business functions and IS organization might exist?

5. Assume that you are responsible for a new payroll program. What steps would you take to ensure a high-quality payroll system?

6. How important are communications skills to IS personnel? Consider this statement: "IS personnel need a combination of skills—one-third technical skills, one-third business skills, and one-third communications skills." Do you think this is true? How would this affect the training of IS personnel?

7. Imagine that you are a highly paid consultant who has been retained to evaluate an organization's systems development processes. With whom would you meet? How would you make your assessment?

8. You are a senior manager of a functional area in which a critical system is being developed. How can you safeguard this project from mushrooming out of control?

9. Assume that you are the owner of a company that is about to start marketing and selling bicycles over the Internet. Describe your top three objectives in developing a new Web site for this systems development project.

10. Identify some of the advantages and disadvantages of purchasing versus leasing hardware.

11. Identify the various forms of testing. Why are there so many different types of tests?

12. What is the goal of conducting a systems review? What factors need to be considered during systems review?

13. How would you go about evaluating a software vendor?

● PROBLEM-SOLVING EXERCISES

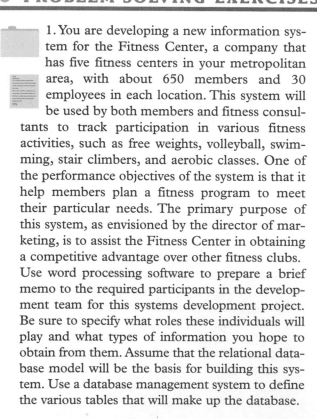

1. You are developing a new information system for the Fitness Center, a company that has five fitness centers in your metropolitan area, with about 650 members and 30 employees in each location. This system will be used by both members and fitness consultants to track participation in various fitness activities, such as free weights, volleyball, swimming, stair climbers, and aerobic classes. One of the performance objectives of the system is that it help members plan a fitness program to meet their particular needs. The primary purpose of this system, as envisioned by the director of marketing, is to assist the Fitness Center in obtaining a competitive advantage over other fitness clubs. Use word processing software to prepare a brief memo to the required participants in the development team for this systems development project. Be sure to specify what roles these individuals will play and what types of information you hope to obtain from them. Assume that the relational database model will be the basis for building this system. Use a database management system to define the various tables that will make up the database.

2. A project team has estimated the costs associated with the development and maintenance of a new system. One approach requires a more complete design and will result in a slightly higher design and implementation cost but a lower maintenance cost over the life of the system. The second approach cuts the design effort, saving some dollars but with a likely increase in maintenance cost.

a. Enter the following data in the spreadsheet. Print the result.

The Benefits of Good Design

	Good Design	Poor Design
Design Costs	$14,000	$10,000
Implementation Cost	$42,000	$35,000
Annual Maintenance Cost	$32,000	$40,000

b. Create a stacked bar graph that shows the total cost, including design, implementation, and maintenance costs. Be sure that the chart has a title and that the costs are labeled on the chart.

c. Use your word processing software to write a paragraph that recommends which approach to take and why.

● TEAM ACTIVITIES

1. System development is more of an art and less of a science, with a wide variety of approaches in how companies perform this activity. You and the members of your team are to interview members of an information systems organization's development group. Prepare a list of questions to determine whether they follow the approach outlined in this chapter. During the course of your interview, when you find discrepancies between the approach they follow and the process suggested in the text, find out why there is a difference. Also, learn what tools and techniques are most frequently employed. Prepare a short report on your findings.

2. Your team has been hired to determine the requirements and layout of the Web pages for a new company that sells fishing equipment over the Internet. Using the approaches discussed in this chapter, develop a rough sketch of at least five Web pages that you would recommend. Make sure to show the important features and the hyperlinks for each page.

3. Assume that your team is working for a medium-size consulting firm. You have been hired to develop an on-line tax preparation program to complete federal 1040 forms.
   a. Develop a project proposal for this effort including an estimate of the time required and stating potential project benefits.
   b. Develop a mock-up report showing what will be displayed to the senior consultants and partners if they want a project status report.
   c. Use a database management system to document the key forms to be included with the 1040, including descriptions and how the forms are to be linked to the 1040.

● WEB EXERCISES

1. A number of companies were discussed in this chapter. Locate the Web site of one of these companies. What are the goods and services that this company produces? After visiting the company Web site, describe how systems development could be used to improve the goods or services it produces. You may be asked to develop a report or send an e-mail message to your instructor about what you found.

2. Locate a company on the Internet that sells products, such as books or clothes. Write a report describing the strengths and weaknesses of the Web pages you encountered. In your opinion, what are the most important steps of the systems development process that could be used to improve the Internet site?

3. Accounting programs can be developed or purchased. Using the Internet and a search engine, such as Yahoo!, find one or more companies that sell accounting software, such as payroll, billing, order processing, or inventory control software. Describe the software package and the company that sells it. Describe the features of the software package. What are the advantages of purchasing this software package versus developing the software?

● CASES

 IT Projects at Coca-Cola

There are many reasons companies initiate a systems development project. In some cases, a merger can cause a major systems development effort. In other cases, companies start systems development to take advantage of a new marketing opportunity. For some companies, a downturn in revenues and profits can cause a big shakeup, which often involves new systems development initiatives.

A 21-year-old student in Atlanta used to drink Diet Coke all day long. She has changed, though, saying, "As I've gotten older, I've realized that drinking five Diet Cokes a day isn't good for you." And she is not alone. More people are switching from colas and caffeine to water and noncarbonated beverages. What may be a good health move for some has meant a drop in sales for cola companies, including Coca-Cola. After decades of steady growth, Coca-Cola recently experienced flat sales. Net operating profits fell a staggering 20 percent, and Coca-Cola's stock price took a dip. But things are changing at the traditional cola company under the new leadership of Douglas Daft. According to Steve Jones, chief marketing officer at Coca-Cola, "For us to achieve the growth rate that people are expecting, we have to become more diversified. We have to move beyond Coke and the carbs." In addition to moving into new products, Coca-Cola is moving into new IS projects through an aggressive systems development effort.

In addition to slashing 6,000 jobs, Coca-Cola is also aggressively launching new systems development initiatives. One involves an automated inventory system. The inventory systems development project involves the use of handheld computers on trucks for route drivers and sophisticated software to transfer data to computers to track inventory more accurately. The company is also developing systems to strengthen its relationship with local franchises. Coca-Cola sells concentrate to local bottlers, who add sugar or artificial sweeteners and water. The local bottlers then package and distribute the soft drink. According to chief operating officer Daft, "We've spent years building the brands, infrastructure, and technology needed to be successful at the local level."

Discussion Questions

1. Why did Coca-Cola initiate new systems development projects?
2. What was Coca-Cola trying to accomplish with its new inventory system?

Critical Thinking Questions

3. If you were a manager of a cola company, how big would a downturn in sales and profits have to be for you to cut costs and initiate systems development projects?
4. Would you suggest that Coca-Cola heavily invest in an Internet site? If so, what type of Internet site would you recommend? Would it be oriented toward customers or local bottlers and franchises?

Sources: Adapted from Stacy Collett, "IT Projects Part of Coca-Cola Realignment," *Computerworld*, January 31, 2000, p. 4; Dean Foust et al., "Coke Is No Longer It," *Business Week*, February 28, 2000, p. 148; and "Debunking Coke," *The Economist*, February 12, 2000.

Mergers Drive Systems Development

The executives of Glaxo Wellcome and SmithKline Beecham are busy making plans for a huge $76 billion merger. The new company would be the largest pharmaceutical company in the world, with 7.3 percent of the total pharmaceutical market and more than 100,000 employees.

While the executives of these two giant companies are planning the merger, the systems development staffs of both companies are scrambling to merge the IS departments and operations. Unfortunately, the companies are using different enterprise resource planning packages.

Glaxo Wellcome is currently using the SAP R/3 enterprise resource planning package. SmithKline Beecham, however, is using an ERP package from J.D. Edwards. Getting the systems to work together is as important as getting the two companies together. "The pharmaceutical business is an incredibly competitive market, and everybody is facing the same huge urgency to improve speed to market and globalize operations. You can't do that with fragmented systems," says Steve Shaha, an analyst for the Gartner Group. Shaha believes that the new company, to be called Glaxo SmithKline if the merger goes through, will need to plot a new IS strategy and have it operational in two to three years.

The two ERP systems now being used, however, will complicate any effort to integrate the information systems of the two companies. "Technically, this will not be a slam dunk," says Shaha. To make matters even more difficult, both companies have to make

sure that any new or modified systems meet federal food and drug regulations. According to research, it is not uncommon for these governmental compliance requirements to take up to 40 percent of an IS department's total budget. A fragmented system could cost even more in compliance costs. "That's a scary number," says Roddy Martin, an industry analysis.

Discussion Questions

1. Why is the merger of Glaxo and SmithKline going to have a significant impact on the IS department of the merged company?
2. Would you recommend using a system from one of the existing companies or would you recommend that the merged companies explore new alternatives for integrating their technological infrastructure?

Critical Thinking Questions

3. Comment on the following: It appears that the decision to merge was made before any detailed considerations about integrating the two different information systems.
4. What impact do you think the merger will have on staffing for the IS area of the merged company? If you were a middle-level project manager in the IS department for one of the companies, would you be worried about your job security or job duties in the new company?

Sources: Adapted from Craig Stedman, "Market Pressures Will Make IT a Priority in Drug Merger," *Computerworld*, January 24, 2000, p. 2; Carol Sliwa, "Drug Giants' Merger to Bring Systems Integration Hurdles," *Computerworld*, January 3, 2000, p. 12; and Amy Barrett, "Addicted to Mergers?" *Business Week*, December 6, 1999, p. 84.

3 Developing a Wireless Net to Improve Customer Service

As a result of the deregulation of the electric power business in many states, power companies are striving to find ways to improve their business and customer service. Failure to take these actions could result in a substantial loss of business. In 1999, commercial electric customers in Illinois were free to choose their power company. In 2000, home users were free to choose their power company. According to Roger Koester, supervisor of energy delivery technology for Illinois Power Company, "A few years ago, there weren't that many competitors. But since then, the competition has just exploded." Illinois Power, realizing the potential impact of increased competition from deregulation, decided to embark on a systems development project to build a wireless communications system to improve customer service.

Illinois Power is a subsidiary of Illinova Corporation, a $2.4 billion company. Illinois Power provides both gas and electric service to almost one million customers, primarily in central Illinois. Traditionally, Illinois Power received calls from customers requesting repairs or service over the phone at 26 field offices. This information was then placed on service and repair orders and given to dispatchers, who coordinated the work field crews. Although the system worked, it was slow and inefficient. After years of planning and millions of dollars invested in the systems development effort, Illinois Power was ready to roll out its new wireless system.

With the new wireless system, service and repair orders are captured at a centralized computer system. The work orders are then sent via wireless transmission directly to laptop computers of repair and service crews in the field. "The driver gets in his truck and instead of searching through paper, his day's work is loaded on the laptop," said Koester. In addition, drivers do not need to travel to a central office to get their work. They can wake up in the morning, get into their trucks, and download the day's service and repair work. Eliminating trips to a central office has saved the company thousands of dollars in gas, while making the crews more efficient. Each truck is also equipped with a global positioning system (GPS) to allow a central dispatcher to locate the truck closest to an emergency repair. With the implementation of the new system, Illinois Power hopes to increase profits and market share.

Discussion Questions

1. Why did Illinois Power decide to invest in a new wireless system?
2. What are the benefits of the new system?

Critical Thinking Questions

3. From a customer's perspective, what could Illinois Power do to further increase customer service and satisfaction for home users?
4. Some believe that the power business has little room for increasing profits. What other systems development efforts might help Illinois Power increase its market share and profitability in its commercial power market?

Sources: Adapted from Matt Hamblen, "Wireless Net Helps Utility Improve Customer Service," *Computerworld*, January 24, 2000, p. 36; Dennis Berman, "Keeping the Home Fires Burning," *Business Week*, September 20, 1999, p. 6; and "EPA Bares Its Teeth," *Power Economics*, January 31, 2000, p. 17.

● NOTES

Sources for the opening vignette on p. 283: Adapted from Lisa Levenson, "AT&T Offers Service for Online Software Renters," *The Rocky Mountain News*, January 31, 2000, p. 21B; Steve Ulfelder, "Evaluate the ASP Phenomenon," *Computerworld*, January 2, 2000, p. S22; and Kavita Kaur, "Application Service Providers: Rent an App," *Computers Today*, February 29, 2000, p. 43.

1. Matthew Gallagher, "Oil Trading Headed for Cyberspace," *The Wall Street Journal*, March 6, 2000, p. 13B.
2. Julekha Dash, "Skills Shortage," *Computerworld*, March 27, 2000, p. 10.
3. Deborah Radcliff, "Aligning Marriott," *Computerworld*, April 10, 2000, p. 58.
4. Mike Simons, "Anatomy of an IT Disaster," *Computer Weekly*, February 3, 2000, p. 1.
5. Varun Grover et al., "The Role of Organizational and Information Technology Antecedents in Reengineering Initiation Behavior," *Decision Sciences*, Summer 1999, p. 749.
6. Laurie Toupin, "Computer Productivity Tools," *Design News*, January 3, 2000, p. 100.
7. "Techniques: MFC for Non-Believers," *EXE*, February 1, 2000, pp. 24–33.
8. Amanda Wells, "Police System Tender Issued," *NZ Infotech Weekly*, January 17, 2000, p. 1.
9. Jaikumar Vijayan, "ABB, IBM Sign $250M Agreement," *Computerworld*, April 10, 2000, p. 6.
10. Jaikumar Vijayan, "HP Net Hosting Has Pay-as-You-Go Model," *Computerworld*, March 20, 2000, p. 8.
11. Thomas Hoffman, "Bank's Reuse Project Survives Mergers," *Computerworld*, January 25, 1999, p. 39.
12. Spikumar Rao, "General Electric, Software Vendor," *Forbes*, January 24, 2000, p. 144.
13. Darnell Little, "H&R Block Gets All Snarled Up in the Web," *Business Week*, April 17, 2000, p. 200.
14. "Computer Snag Halts London Market 8 Hours," *The Wall Street Journal*, April 6, 2000, p. A14.
15. "Clinton to Hold Internet Security Summit," *The Wall Street Journal*, February 11, 2000, p. A3.

CHAPTER 9

Security, Privacy, and Ethical Issues in Information Systems and the Internet

There are 13-year-old kids without degrees breaking into systems from their bedrooms. Security should be the subject of every Master and Ph.D. [information systems] student's thesis.

— John V. Daley, telecom specialist at the Federal Aviation Administration and part-time community-college professor

Principles	Learning Objectives
Policies and procedures must be established to avoid computer waste and mistakes.	• *Describe some examples of waste and mistakes in information systems, their causes, and possible solutions.* • *Identify policies and procedures useful in eliminating waste and mistakes.*
Computer crime is a serious and rapidly growing area of concern requiring management attention.	• *Explain the types and effects of computer crime.* • *Identify specific measures to prevent computer crime.* • *Discuss the principles and limits of an individual's right to privacy.*
Jobs, equipment, and working conditions must be designed to avoid negative health effects.	• *List the important effects of computers on the work environment.* • *Identify specific actions that must be taken to ensure the health and safety of employees.*

Los Alamos National Laboratory

*Compromise of Nuclear
Weapon Secrets*

The United States agreed in 1992 to end nuclear weapons testing. The country had already developed such sophisticated computer models that weapons could be tested entirely on simulation software using data from decades of actual bomb testing. In August 1999, U.S. Energy Secretary Bill Richardson confirmed that these classified nuclear weapon models and data were transferred by a Chinese lab employee at the Los Alamos National Laboratory to an unclassified computer system. Where they went from there is not known; however, the incident has been at the center of allegations that China pilfered U.S. nuclear secrets.

Although the software and data may not enable someone to make an exact copy of a U.S. weapon, they clearly would be enormously valuable to another country's nuclear weapons program. That's especially so for China, which also signed the test ban treaty and so must rely on computer simulations. We may never know whether the secret files were given to China, or what, if anything, China has done with them. But the possibilities are scary indeed.

This incident reinforces three basic computer security principles. First, computer security is often neglected because systems and attacks against them are often invisible—out of sight, out of mind. Second, the vast majority of computer crimes are committed by insiders unaffected by many of the standard protective measures, such as firewalls. A disgruntled employee is a bigger threat than a hacker because he or she already has access and knowledge of how a system works, whereas a hacker usually comes in blind. Third, the effort invested in protecting computer assets should reflect both the likelihood that such attempts will be made and the impact on an organization if those attempts succeed. At Los Alamos, risk times cost equals infinity. Yet security efforts apparently fell short.

It is now clear that the Los Alamos monitoring systems were not as stringent as they needed to be. Moving files from a classified to an unclassified system should have triggered an immediate alert. The secretary of energy charged that his department wasn't getting the help it needed from Congress to develop such systems and that the fiscal year 2000 budget in Congress killed needed security funding. By Congress's denying $35 million in funds for cybersecurity upgrades, it was impossible to provide real-time cyberintrusion detection and protection for all DOE sites.

As you read this chapter, consider the following:

- What motivates an individual to commit a computer crime?

- What can be done to detect and avoid computer crime?

TABLE 9.1

Social Issues in Information
Systems

● Computer waste and mistakes	● Health concerns
● Computer crime	● Ethical issues
● Privacy	● Patent and copyright violations

Use of computer-based information systems in business has led to increased profits, superior goods and services, and a higher quality of life. In fact, today's businesspeople would have difficulty imagining work without them. Yet, the Information Age has also brought some potential problems for workers, companies, and society in general (see Table 9.1).

In this chapter we discuss detailed problems as a reminder of the social and ethical considerations underlying the use of computer-based information systems. Everyone who uses an IS has a responsibility to see that it is used correctly. Managers and users at all levels play a major role in helping organizations achieve the positive benefits of IS and in minimizing or eliminating the negative consequences.

Many of the problems presented in this chapter should cause you to think back to some of the systems design and control issues we have already discussed. They should also help you look forward to how these issues and your choices might affect your future career in business.

COMPUTER WASTE AND MISTAKES

Computer-related waste and mistakes are major contributors to unnecessarily high costs and lost profits. Computer waste involves the inappropriate use of computer technology and resources. Computer-related mistakes refer to errors, failures, and other computer problems that make computer output incorrect or not useful, caused mostly by human error. In this section we explore the damage that can be done as a result of computer waste and mistakes.

Computer Waste

The U.S. government is the largest single user of information systems in the world. It should come as no surprise then that it is also perhaps the largest misuser. The government is not unique in this regard—the same type of waste and misuse also exists in the private sector. Some companies discard old software and even complete computer systems when they still have value. Others waste corporate resources to build and maintain complex systems never used to their fullest extent. A less dramatic, yet still relevant, example of waste is the amount of company time and money employees may waste playing computer games, sending unimportant e-mail, or accessing the Internet. People also receive hundreds of unwanted e-mail messages and faxes advertising products and services, which wastes not only time but paper and computer resources. Waste typically has one common cause: improper management of information systems and resources.

Computer-Related Mistakes

Despite many people's distrust, computers themselves rarely make mistakes. Even the most sophisticated hardware cannot produce meaningful output if users do not follow proper procedures. Mistakes can also be caused by unclear expectations and a lack of feedback. Or a program might contain errors. In other cases, a data-entry clerk might enter the wrong data. Unless errors are prevented or caught early, the speed of computers can intensify mistakes. Here are some recent examples:

- The wireless communications venture Globalstar Telecommunications lost 12 of its satellites when the Ukrainian-built Zenit-2 rocket crashed just minutes after liftoff from the Baikonur Cosmodrome. NPO Yuzhnoye, the company that manufactured the rocket, said that two computer glitches occurred in rapid succession and as a result the computer sent an order to cut the engines. The satellites were valued at $185 million.
- A Utah retiree was ordered to pay back nearly $14,000 in excess retirement benefits he received. A retirement fund employee testified in the trial that a portion of the retiree's benefits were calculated by hand and another portion by computer, and the two figures were accidentally combined, resulting in double payments. The company overpaid in two ways—it paid a lump-sum distribution that was $14,000 too high, and it overpaid the retiree's monthly benefits.
- Students in Chippewa Falls, Wisconsin, received free school lunches as a result of a computer glitch. During tests of new software, duplicate student identification numbers were issued. On the first day of the school year, students were sent home with notices saying that they were eligible for free lunches that week.

Preventing Computer-Related Waste and Mistakes

To remain profitable in a competitive environment, organizations must use all resources wisely. Thus, preventing computer-related waste and mistakes like those just described should be a goal. Employees and managers alike should strive to minimize waste and mistakes by defining and monitoring effective policies and procedures.

Establishing Policies and Procedures

The first step to prevent computer-related waste is to establish policies and procedures regarding efficient acquisition, use, and disposal of systems and devices. Most companies have implemented stringent policies on the acquisition of computer systems and equipment, such as formal justifications for computer equipment purchases, definition of standard computing platforms, and the use of preferred vendors for all acquisitions.

To control and prevent potential problems, companies have developed policies and procedures that cover the following:

- Acquisition and use of computers, with a goal of avoiding waste and mistakes
- Training programs for individuals and workgroups
- Manuals and documents on how computer systems are to be maintained and used
- Approval of certain systems and applications before they are implemented to ensure compatibility and cost-effectiveness
- Filing documentation and descriptions of certain applications with a central office, including all cell formulas for spreadsheets and a description of all data elements and relationships in a database system; such standardization can ease access and use for all personnel

Once companies have planned and developed policies and procedures, they must consider how best to implement them.

Implementing Policies and Procedures

Most companies develop policies and procedures to minimize waste and mistakes with advice from an internal auditing group or an external auditing firm.

TABLE 9.2

Useful Policies to Eliminate
Waste and Mistakes

- Changes to critical tables should be tightly controlled, with all changes authorized by responsible owners and documented.
- A user manual should be available that covers operating procedures and that documents the management and control of the application.
- Each system report should indicate its general content in its title and specify the time period it covers.
- The system should have controls to prevent invalid and unreasonable data entry.
- Controls should exist to ensure that data input is valid, applicable, and posted in the right time period.
- Users should implement proper procedures to ensure correct input data.

The policies often focus on data editing to ensure accuracy and completeness, and assigning clear responsibility for data accuracy within each information system. Table 9.2 lists some useful policies to minimize waste and mistakes.

Training is another key aspect of implementation. Many users are not properly trained to use or develop applications, and their mistakes can be very costly. More and more people use computers in their daily work, so it is important that they understand how to use them and follow policies and procedures. Companies converting to ERP systems typically invest weeks of training for key users of the system's modules.

Monitoring Policies and Procedures

To ensure that users are following established procedures, organizations monitor routine practices and take corrective action if necessary. Many organizations use internal audits to measure actual results against established goals, such as percentage of end-user reports produced on time, percentage of data input errors rejected, and number of input transactions entered per eight-hour shift.

Reviewing Policies and Procedures

The final step in mistake prevention and detection is to review existing policies and procedures and determine whether they are adequate. People should ask the following questions:

- Do current policies cover existing practices adequately? Were any problems or opportunities uncovered during monitoring?
- Does the organization plan any new activities in the future? If so, does it need new policies or procedures on who will handle them and what must be done?
- Are contingencies and disasters covered?

This review and planning often allows companies to avert disasters since they are alerted to upcoming crises in information systems that could have a profound effect on many business activities.

In addition to policies and procedures to prevent and detect unintended errors and mistakes, organizations need to implement security measures and legal protections to deter and uncover computer crime.

COMPUTER CRIME

Even the best IS policies may not be able to predict or prevent computer crime. A computer's ability to process millions of pieces of data in less than a second can help a thief steal data worth millions of dollars. Instead of facing the physical dangers of robbing a bank or retail store with a gun, a computer criminal with the right equipment and know-how can steal large amounts of money from the privacy of a home. Computer crime often defies detection, the amount stolen or diverted can be substantial, and the crime is "clean" and nonviolent.

Here is a summary of recent computer crimes from 1999 to 2000. In March 1999, the Melissa virus caused an estimated $80 million in damage when it swept around the world, paralyzing e-mail systems. In December 1999, 300,000 credit card numbers were snatched from on-line music retailer CD Universe. In February 2000, hackers were able to crash the Web sites of several large e-commerce companies. In March, hackers-for-hire pleaded guilty to breaking into phone giants AT&T, GTE, and Sprint, among others, for calling card numbers that eventually made their way to organized crime gangs in Italy. According to the FBI, the phone companies were hit for an estimated $2 million.[1] In April, the U.S. energy secretary confirmed that classified nuclear weapon computer software at Los Alamos National Laboratory in New Mexico was transferred by a lab employee to an unclassified computer system.[2] In June, an embarrassing loss of computer disks containing additional classified nuclear information at the Los Alamos National Laboratory was uncovered. The computer disks reportedly contained information on how to disarm Russian and American nuclear devices.[3] In September, the FTC filed a case against individuals in Portugal and Australia who engaged in "pagejacking" and "mousetrapping" when they captured unauthorized copies of U.S.-based Web sites (including those of PaineWebber and *The Harvard Law Review*) and produced look-alike versions that were indexed by major search engines. The defendants diverted unsuspecting consumers to a sequence of porno sites that they couldn't exit.[4]

Although no one really knows how pervasive cybercrime is, most agree that it is growing rapidly. At least 60 percent of all attacks go undetected, according to security experts. What's more, of the attacks that are exposed, only an estimated 15 percent are reported to law enforcement agencies. Why? Companies don't want the bad press. When Russian organized crime used hackers to break into Citibank to steal $10 million—all but $400,000 was recovered—competitors used the news in marketing campaigns against the bank. Such publicity makes the job even tougher for law enforcement. Most companies that have been electronically attacked won't talk to the press. A big concern is loss of public trust and image—not to mention the fear of encouraging copycat hackers.[5]

The increase in computer security breaches is raising concern that the American workforce may not have enough troops to battle cybersnoops. Too few colleges offer security courses. Only about a half-dozen U.S. academic institutions have graduate programs in computer security—and that number hasn't changed much in ten years. Even more unsettling is that no quick solution is in sight. Computer assurance still isn't a recognized discipline at most colleges.[6]

Highlights of the annual Computer Crime and Security Survey are shown in Table 9.3. The survey is based on responses from 643 companies and government agencies. The Computer Security Institute, with the San Francisco Federal Bureau of Investigation (FBI) Computer Intrusion Squad, conducts this survey to raise awareness and determine the scope of computer crime in the United States.

TABLE 9.3

Summary of Key Data from 2000 Computer Crime and Security Survey

Source: Data from "2000 CSI-FBI Survey Results," http://www.gocsi.com/prelea_000321.htm, accessed August 6, 2000.

Incident	2000 Results
Companies reporting serious computer breaches	70%
Companies that acknowledge suffering financial losses from computer security breaches	74%
Companies that are able to quantify the financial losses from computer security breaches	42%
Companies with Web sites that had detected unauthorized access or misuse of Web site	19%
Companies reporting virus contamination	85%

All told, the FBI estimates that computer losses exceed $10 billion a year. Law enforcement officials estimate that up to 60 percent of break-ins are from employees. For example, an entertainment company that was suspicious about an employee called in a digital detective from PricewaterhouseCoopers in Los Angeles. The employee was under financial pressure and had installed a program called Back Orifice on three of the company's servers, allowing him to take over those machines, read passwords, and access all the company's financial data. Fortunately, the employee was terminated before any damage could be done.[7]

Today, computer criminals are a new breed—bolder and more creative. With the increased use of the Internet, computer crime is also becoming global. Attacks from overseas, particularly eastern European countries, are on the rise. Nabbing bad guys overseas is a particularly thorny issue, though. In 1998, intruders broke into Aye.Net, a small Internet service provider, and knocked it off the Net for four days. The director of systems engineering discovered the hackers and found messages in Russian. He reported it to the FBI, but no one has been able to track down the hackers.[8]

Regardless of its nonviolent image, computer crime is still a crime. Part of what makes computer crime unique and hard to combat is its dual nature—it can be both the tool used to commit a crime and the object of that crime.

The Computer as a Tool to Commit Crime

A computer can be used to gain access to valuable information and to steal millions of dollars. Many individuals who commit computer-related crime claim they do it for the challenge, not for the money. Credit card fraud, in which a criminal accesses someone else's account with stolen credit card numbers, is a major concern for today's banks and financial institutions. Criminals need two capabilities to commit most computer crimes: knowledge of how to gain access to the computer system and how to manipulate the system to produce the desired result. Dallas FBI agent Mike Morris estimates that in at least a third of the computer crime cases he's investigated in five years, an individual has been talked out of a critical computer password, a practice called **social engineering**. Or, the attackers simply go through the garbage—**dumpster diving**—for important pieces of information that can help crack the computers or persuade someone at the company to give them more access.[9] In addition, security experts estimate that nearly 2,000 Web sites offer the digital tools—for free—that will let people snoop, crash computers, hijack control of a machine, or retrieve a copy of every keystroke.[10]

social engineering

the practice of talking an individual out of a critical computer password

dumpster diving

searching through the garbage for important pieces of information that can help crack an organization's computers or be used to persuade someone at the company to give criminals access to the computers

Also, with today's sophisticated desktop publishing programs and high-quality printers, crimes involving counterfeit money, bank checks, traveler's checks, and stock and bond certificates are on the rise. As a result, the U.S. Treasury Department redesigned and printed new currency that is much more difficult to counterfeit.

The Computer as the Object of Crime

A computer can also be the object of a crime. Tens of millions of dollars of computer time and resources are stolen every year. These crimes fall into several categories: illegal access and use, data alteration and destruction, information and equipment theft, software and Internet piracy, computer scams, and international computer crime.

Illegal Access and Use

Crimes involving illegal system access and use of computer services are a concern to both government and business. A 28-year-old computer expert allegedly tied up thousands of US West computers in an attempt to solve a classic math problem, racking up ten years of computer processing time. The alleged hacking was discovered after company officials noticed that computers were taking up to five minutes to retrieve telephone numbers, instead of only three to five seconds. At one point, customer calls had to be rerouted to other states, and the delays threatened to close down the Phoenix Service Delivery Center.

Since the outset of information technology, computers have been plagued by criminal hackers. A **hacker** is a person who enjoys computer technology and spends time learning and using computer systems. A **criminal hacker**, also called a **cracker**, is a computer-savvy person who attempts to gain unauthorized or illegal access to computer systems. In many cases, criminal hackers are looking for fun and excitement—the challenge of beating the system. In others, they steal passwords, files and programs, or even money.

Catching and convicting criminal hackers remains difficult because their methods are often hard to determine. Even if the method behind the crime is known, tracking down the criminals can take a lot of time. It took years for the FBI to arrest one criminal hacker for the alleged "theft" of almost 20,000 credit card numbers that had been sent over the Internet. Table 9.4 provides some guidelines to follow in the event of a computer security incident.

Data Alteration and Destruction

Data and information are valuable corporate assets, and intentionally altering or destroying data is as much a crime as destroying tangible goods. Common programs used are viruses and worms, which are software programs that, when loaded into a computer system, will destroy, interrupt, or cause errors in processing. There are more than 53,000 known computer viruses today, with more than 6,000 new viruses and worms being discovered each year. A **virus** is a program that attaches itself to other programs. A **worm** functions as an independent program, replicating its own program files until it interrupts network and computer operations. In other cases, a virus or a worm can destroy important data and programs. If backups are inadequate, the data and programs may never fully function again.

A personal computer can get a virus from an infected disk, an application, or e-mail attachments. A virus or worm that attacks a network or client/server system is usually more severe because it can affect hundreds or thousands of personal computers and other devices attached to the network. Workplace

---

hacker

a person who enjoys computer technology and spends time learning and using computer systems

criminal hacker (cracker)

a computer-savvy person who attempts to gain unauthorized or illegal access to computer systems

virus

a program that attaches itself to other programs

worm

an independent program that replicates its own program files until it interrupts the operation of networks and computer systems

---

- Follow your site's policies and procedures for a computer security incident. (They are documented, aren't they?)
- Contact the incident response group responsible for your site as soon as possible.
- Inform others, following the appropriate chain of command.
- Further communications about the incident should be guarded to ensure intruders do not intercept information.
- Document all follow-up actions (phone calls made, files modified, system jobs that were stopped, etc.).
- Make backups of damaged or altered files.
- Designate one person to secure potential evidence.
- Make copies of possible intruder files (malicious code, log files, etc.) and store them off-line.
- Evidence, such as tape backups and printouts, should be secured in a locked cabinet, with access limited to one person.
- Get the National Computer Emergency Response Team involved if necessary.
- If you are unsure of what actions to take, seek additional help and guidance before removing files or halting system processes.

TABLE 9.4

How to Respond to a Security Incident

logic bomb

an application or system virus designed to "explode" or execute at a specified time and date

Trojan horse

a program that appears to be useful but actually masks a destructive program

password sniffer

a small program hidden in a network or a computer system that records identification numbers and passwords

TABLE 9.5

Sources of Information about Viruses

(Source: Data from Leslie Goff, "Resources," *Computerworld*, May 25, 1998)

computer virus infections are increasing rapidly due to the increased spread of viruses in e-mail attachments. The number of infections per 1,000 PCs was 21.45 in 1997; in 1998 it had grown to 31.85, according to the International Computer Security Association. The primary ways to avoid viruses and worms are to install virus scanning software on all systems, update it routinely, and avoid using disks or files from unknown or unreliable sources. People should also avoid opening files from people they know—unless they are expecting them. Many worms are sent as e-mails to people in the initial victim's address book so that it appears to be a file from someone you know.

Another type of program that can destroy a system is a **logic bomb**, an application or system virus designed to "explode" or execute at a specified time and date. Logic bombs are often disguised as a **Trojan horse**, a program that appears to be useful but actually masks the destructive program.

On April 1, 1999, a 31-year-old New Jersey programmer was arrested by federal and state officials and charged with creating and disseminating the Melissa virus, which began spreading across the Internet on March 26, 1999. Melissa is a virus with an unusual payload. When a user opens an infected document, the virus attempts to e-mail a copy of the document to 50 other people, using Microsoft Outlook. In a plea bargain with prosecutors, the programmer pleaded guilty to one charge of computer theft. The Melissa program caused more than $80 million in damage—most related to the time systems administrators spent to clear the virus off affected computers. Based on the agreement with the man's attorneys, prosecutors recommended a sentence of ten years in prison, the maximum allowable in such crimes, and a fine of $150,000.[11]

Hoax, or false, viruses are another problem. Criminal hackers sometimes warn the public of a new and devastating virus that doesn't exist. Companies can spend hundreds of hours warning employees and taking preventive action against a nonexistent virus. Security specialists recommend that organizations establish a formal paranoia policy to thwart panic among gullible end users. Such policies stress the importance of forwarding a potential problem to the help desk or the security team before alerting colleagues and higher-ups. Table 9.5 lists some of the most informative sites about viruses. Table 9.6 lists the top ten viruses according to McAfee, a provider of antivirus software. Note that several of those listed are hoaxes.

Information and Equipment Theft

To obtain illegal access to data, criminal hackers need identification numbers and passwords. Some try different identification numbers and passwords until they find ones that work. Using password sniffers is another approach. A **password sniffer** is a small program hidden in a network or a computer system that records identification numbers and passwords. In a few days, a password sniffer can record thousands of identification numbers and passwords. LOpht, founded in 1992 by a group of hackers who provide computer and network

Web Page	Address
CIAC Internet Hoaxes Page	http://hoaxbusters.ciac.org
Computer Virus Myths Homepage	http://www.vmyths.com
Trusecure Corp.	http://www.trusecure.com
The Truth About E-Mail Viruses	http://www.gerlitz.com/virushoax/
Dr. Solomon's Software	http://www.drsolomon.com
Symantec Corp.	http://www.symantec.com
Network Associates, Inc.	http://www.networkassociates.com

TABLE 9.6

Top 10 Viruses
(Data source: McAfee Web site.)

Virus Name	Date Discovered	Risk Assessment
VBS/Loveletter.a	5/4/00	High
IRC/Stages.worm	5/26/00	High
W32/Pretty.worm.unp	2/15/00	Medium
W32/Ska	1/27/99	Medium
W32/Pretty.Worm	5/26/99	Medium
APStrojan.qa	1/18/00	Medium
W32/FunLove.4099	11/9/99	Medium
VBS/Netlog.worm.a	2/2/00	Low
W97M/Ethan.a	1/21/99	Medium

security consulting, has developed a software product called AntiSniff that runs nonintrusive tests to determine whether a remote computer user is listening in on network communications.[12]

In addition to data, all types of computer systems and equipment have been stolen from offices. Computer theft is now second only to automobile theft, according to recent U.S. crime statistics. In some cases, the data and information stored in these systems are more valuable than the equipment. Without adequate protection and security measures, equipment can easily be stolen.

Software and Internet Piracy

Each time you use a word processing program or other software, you are using someone else's intellectual property. Like books and movies—other intellectual properties—software is protected by copyright laws. Often, people who would never think of plagiarizing another author's written work have no qualms about using and copying software programs they have not paid for. Such illegal duplicators are called pirates; the act of illegally duplicating software is called **software piracy**.

Internet piracy involves illegally gaining access to and using the Internet. Although not yet as prevalent as software piracy, Internet piracy is growing rapidly. Many companies on the Internet receive customer fees for their information, services, and even products. Some investment firms, for example, offer market analysis and investment information for a monthly or annual fee. Other companies offer sports information or provide research for a fee. Typically, Internet companies give customers identification numbers or passwords, and some customers illegally share them with others. Sometimes criminal hackers obtain these numbers illegally. When unauthorized people use such services, Internet firms lose valuable revenues.

software piracy

the act of illegally duplicating software

Internet piracy

illegally gaining access to and using the Internet

To fight computer crime, many companies use devices such as BookLock (shown here), which disables the disk drive and locks the computer to the desk. (Source: Kensington Notebook MicroSaver®)

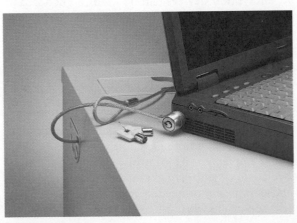

Computer Scams

People have lost hundreds of thousands of dollars on real estate, stock, and other business scams. Today, many scams involve computers. Internet scam artists offer get-rich-quick schemes involving real estate deals, bank fraud, telephone lotteries, penny stocks, and tax avoidance. In one Internet scam, a lottery investment of $129 was reported to guarantee a return that could range from $600 per week to $10,000 per week. The scam raised more than $3 million in less than a year. In most cases, only the scam artists get rich.

If you need advice about an Internet or on-line solicitation, or you want to report a possible scam, use the Online Reporting Form or Online Question & Suggestion Form features on the Web site for the National Fraud Information Center at http://fraud.org, or call the NFIC hotline at 1-800-876-7060.

International Computer Crime

Estimates of software piracy in the global marketplace indicate that more than one-quarter of software is pirated, adding up to more than $3 billion in lost revenue. In China alone, 96 percent of software is pirated, with lost revenue totaling more than $1 billion. Another issue in international computer crime relates to illegally obtaining and selling restricted information in countries with less stringent laws. To avoid these problems, many countries require that computer equipment and software be registered with appropriate authorities before it can be brought into the country. With cash and funds being transferred electronically, some are concerned that international drug dealers and criminals are using information systems to launder funds. Computer terrorism is another aspect of international computer crime.

Preventing Computer-Related Crime

Because of increased computer use today, greater emphasis is placed on preventing and detecting computer crime. Although more than 45 states have passed computer crime bills, some believe that they are not effective because companies do not always detect and pursue computer crime, security is inadequate, and convicted criminals are not severely punished. But all over the United States, private users, companies, employees, and public officials are trying to curb computer crime—recently with some success.

Crime Prevention by State and Federal Agencies

State and federal agencies have begun aggressive attacks on computer criminals, including criminal hackers of all ages. In 1986, Congress enacted the Computer Fraud and Abuse Act, which mandates punishment based on the victim's dollar loss. The Department of Defense also supports the Computer Emergency Response Team (CERT), which responds to network security breaches and monitors systems for emerging threats. Law enforcement agencies are also increasing their efforts to stop criminal hackers, and many states are now passing new, comprehensive computer crime bills. Recent court cases and police reports show that lawmakers are ready to introduce newer and tougher computer crime legislation. Several states have passed laws to outlaw spam, the practice of sending large amounts of unsolicited e-mail to overwhelm users' e-mail boxes or the e-mail servers on a network.

Crime Prevention by Corporations

Companies take crime-fighting efforts seriously. Many businesses have designed procedures and specialized hardware and software to protect their corporate data and systems. For example, encryption devices can be used to encode data and information to help prevent unauthorized use. **Biometrics**, another way to protect important data and information systems, involves the measurement of a living trait, whether physical or behavioral. Biometric techniques compare a person's unique characteristics against a stored set to detect differences between them. Biometric systems can scan fingerprints, faces, handprints, and retinal images to prevent unauthorized access to important data and computer resources. Fingerprints hit the middle ground between price and usability. Iris and retina scans are more accurate, but they involve more expensive equipment.

Crime-fighting procedures usually require additional controls on the information system. Before designing and implementing controls, organizations must consider the types of computer-related crime that might occur, the consequences of these crimes, and the cost and complexity of needed controls. In most cases, organizations conclude that the trade-off between crime and the additional cost

biometrics

the measurement of a living trait, whether physical or behavioral

Bioscrypt Enterprise is a fingerprint authentication device that provides security in the PC environment by using fingerprint information instead of passwords.
(Source: Courtesy of Mytec Technologies, Inc.)

and complexity weighs in favor of better system controls. Some companies actually hire former criminals to thwart other criminals. Table 9.7 provides a set of useful guidelines to protect your organization's computer from hackers.

Companies are also joining to fight crime. The Software Publishers Association (SPA), which was formed by a number of leading software companies, audits companies and checks for software licenses. Organizations that are illegally using software can be fined or sued. Depending on the violation, the fine can be hundreds of thousands of dollars. While the SPA is an effective deterrent in the United States, curbing software abuse in other countries is much more difficult.

Using Antivirus Programs

antivirus programs

programs or utilities that prevent viruses and recover from them if they infect a computer

To protect their computer systems and networks from viruses, companies and individuals use **antivirus programs** or utilities to prevent viruses and recover from them if they infect a computer. These programs range in cost from free (shareware) to a few hundred dollars. Antivirus programs are developed for different operating systems. Norton Antivirus for Windows 95/98 and NT workstations, Norton Antivirus for Macintosh, Quarterdeck Utility Pack for Windows 95/98, Dr. Solomon's Anti-Virus Toolkit for Windows 95/98, and Network Associate's Virex for Macintosh are just a few examples of antivirus programs. Proper use of antivirus software requires the following steps:

1. *Install a virus scanner and run it often.* Many of these programs automatically check for viruses each time you boot up your computer or insert a diskette, and some even monitor all transmissions and copying operations.
2. *Update the virus scanner often.* Old programs may fail to detect new viruses.
3. *Scan all diskettes before copying or running programs from them.* Hiding on diskettes, viruses often move between systems. If you carry document or program files on diskettes between computers at school or work and your home system, always scan them.
4. *Install software only from a sealed package produced by a known software company.* Even software publishers can unknowingly distribute viruses on their program diskettes or software downloads. Most scan their own systems, but viruses may still remain.
5. *Follow careful downloading practices.* If you download software from the Internet or a bulletin board, check your computer for viruses immediately after completing the transmission.
6. *If you detect a virus, take immediate action.* Early detection often allows you to remove a virus before it does any serious damage.

TABLE 9.7

How to Protect Corporate Data From Hackers

- Install strong user authentication and encryption capabilities on your firewall.
- Install the latest security patches, which are often available at the vendor's Internet site.
- Disable guest accounts and null user accounts that let intruders access the network without a password.
- Do not provide overfriendly log-in procedures for remote users (e.g., an organization that used the word *welcome* on their initial log-on screen found they had difficulty prosecuting a hacker).
- Give an application (e-mail, file transfer protocol, and domain name server) its own dedicated server.
- Restrict physical access to the server and configure it so that breaking into one server won't compromise the whole network.
- Turn audit trails on.
- Consider installing caller ID.

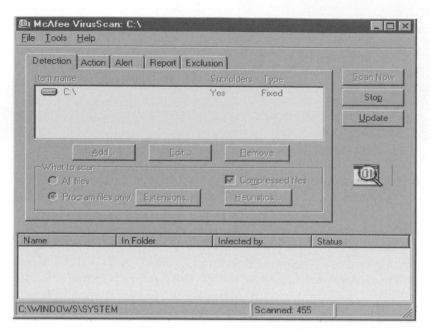

Despite careful precautions, viruses can still cause problems. They can elude virus-scanning software by lurking almost anywhere in a system. Future antivirus programs may incorporate "nature-based models" that check for unusual or unfamiliar computer code. The advantage of this type of virus program is the ability to detect new viruses that are not part of an antivirus database.

Internet Laws for Libel and Protection of Decency

The Telecommunications Act of 1996 included an act called the Communications Decency Act. One of the original provisions of this act is the ability of the government to jail or fine anyone up to $100,000 for sending indecent materials to a minor electronically.

Antivirus software should be used and updated often.

Many people and companies were very concerned about the free speech implications of this provision and turned their Web pages to black in protest. Subsequently, the courts limited the Communications Decency Act. In an unrelated case, the German government forced CompuServe to suspend about 200 newsgroups on its service that Germany claimed violated Section 184 of the German Criminal Code, which deals with the distribution of pornographic materials to minors.

To help parents control what their children see on the Internet, some companies are developing *filtering software* to help screen Internet content. Many of these screening programs also prevent children from sending personal information over e-mail or through chat groups. The two approaches used are filtering, which blocks certain Web sites, and rating, which places a rating on Web sites. Examples include Border Manager, Choice Net, Click & Browse Junior, Cybersitter, Cyber Patrol, Net Nanny, Specs for Kids, Surf Guard, and SurfWatch.

With the increased popularity of networks and the Internet, libel and decency become important legal issues. A publisher, such as a newspaper, can be sued for libel, which involves publishing a written statement that is damaging to a person's reputation. Generally, a bookstore cannot be held liable for statements made in a newspaper or other publications it sells. On-line services, such as CompuServe and America Online Time Warner, have control over who puts information on their service but may not have direct control over the content of what is published by others on their service. Can on-line services be sued for libel for content that someone else publishes on their service? Are on-line services more like a newspaper or a bookstore? This legal issue has not been completely resolved, but some court cases have been decided. The *Cubby, Inc. v. CompuServe* case ruled that CompuServe was more like a bookstore and not liable for content put on its service by others.

Net Nanny is a filtering software program that helps block unwanted Internet content from children and young adults.
(Source: Courtesy of Net Nanny Software International.)

Preventing Crime on the Internet

As mentioned in Chapter 4, Internet security can include firewalls and a number of methods to secure financial transactions. A firewall can include hardware and software that act as a barrier between an organization's information system and the outside world. A number of systems help safeguard financial transactions on the Internet.

To help prevent crime on the Internet, an organization can take the following steps:

1. Develop effective Internet and security policies for all employees.
2. Use a stand-alone firewall (hardware and software) with network monitoring capabilities.
3. Monitor managers and employees to make sure they are using the Internet for business purposes only.
4. Use Internet security specialists to perform audits of all Internet and network activities.

Even with these precautions, computers and networks can never be completely protected against crime. Although firewalls provide good control to prevent crime from the outside, procedures and protection measures are needed for personnel. Passwords, identification numbers, and tighter control of employees and managers also help prevent Internet-related crime. Read the "E-Commerce" box to gain an appreciation of the need to provide safeguards to protect Web sites.

PRIVACY

Another important social issue in information systems involves privacy. In 1890, U.S. Supreme Court Justice Louis Brandeis stated that the "right to be left alone" is one of the most "comprehensive of rights and the most valued by civilized man." Basically, the issue of privacy deals with this right to be left alone or to be withdrawn from public view. With information systems, privacy deals with the collection and use or misuse of data. Data is constantly being collected and stored on each of us. This data is often distributed over easily accessed networks and without our knowledge or consent. Concerns of privacy regarding this data must be addressed.

The right to privacy is an especially challenging problem today. More data and information is produced and used than ever before. A difficult question to answer is, "Who owns this information and knowledge?" If a public or private organization spends time and resources in obtaining data on you, does the organization own the data and can it use the data any way it desires? Government legislation answers these questions to some extent for federal agencies, but the questions remain unanswered for private organizations.

Privacy Issues

The issue of privacy is important because data on an individual can be collected, stored, and used without that person's knowledge or consent. When someone is born, takes certain high school exams, starts working, enrolls in a college course, applies for a driver's license, purchases a car, serves in the military, gets married, buys insurance, gets a library card, applies for a charge card or loan, buys a house, or merely purchases certain products, data is collected and stored somewhere in computer databases.

Privacy and the Federal Government
The federal government is perhaps the largest collector of data. Close to four billion records exist on individuals, collected by about 100 federal agencies. Other data collectors include state and local governments and profit and nonprofit organizations of all types and sizes.

In recent years, events have led to a concern for privacy of on-line data. Cookies were planted on the hard drives of visitors to the Web site of the White House drug enforcement office.[13] Echelon, a global surveillance network run by the National Security Agency and allied intelligence bureaus, monitors every

E-COMMERCE

Business Entrepreneurs Carve New Path

Scour was founded in December 1997 by five UCLA computer science students. Its core service is a search engine that specializes in finding multimedia on the Internet. Unlike conventional search engines that index text found on Web pages, Scour's unique search engine is used to find audio, video, still images, and animation, including music videos, movie trailers, and full-length movies. The search engine takes you directly to the individual multimedia files that you're looking for. (Centerspan Communications recently acquired all the assets of Scour and plans to relaunch Scour as a secure and legal distribution channel for audio, video, images, etc.).

The sites linked from the Scour Web site are not under Scour's control, and Scour does not assume any responsibility or liability for communications or materials available at such linked sites. Scour does offer a free application, the Scour Media Agent, to let you download files. If you want a copy of "Old Time Rock and Roll" by Bob Seger, Scour Exchange will connect you to a user with that track on his hard drive, and zap the song your way.

Scour wants to be viewed as a cooperative partner of the entertainment industry. It works with dozens of content owners to provide licensed premium content such as music downloads, music videos, radio stations, movie trailers, short films, animations, and much more. Scour states that it has worked with the leading copyright experts to design its service to conform to all applicable laws, including the Digital Millennium Copyright Act of 1998 (a summary of which may be found on the U.S. Copyright Office Web site at http://lcweb.loc.gov/copyright/legislation/dmca.pdf). This law was specifically designed to protect copyrights on-line. In accordance with this law, Scour provides a system for copyright owners to notify them of any sites, linked to by the Scour search engine, that have posted material without the proper permission of the copyright owner. Also, Scour must respond quickly to all claims of copyright infringement. Scour's Web site contains procedures for submitting a claim of copyright infringement.

Most entertainment executives and artists in the movie and music business consider Scour illegal because it facilitates copyright infringement. As a result, the Motion Picture Association of America (MPAA), the Recording Industry Association of America (RIAA), the National Music Publishers Association (NMPA), together with every major record label and movie studio, filed a lawsuit against Scour seeking to shut down Scour's search engine products and services. The law firm representing these associations is asking for damages of $150,000 for each illegally downloaded file. Scour has three million unique monthly visitors, so even assuming that 1 percent of them downloaded an illegally copied file, Scour would be forced to pay more than $4.5 billion!

Scour's owners believe that they are fighting a battle that will affect the future of technology and the Internet. A bad result for Scour would have repercussions for almost every business on the Internet. This is why Scour has assembled a powerhouse legal team to defend itself and its right to develop new technologies for the Internet.

Discussion Questions

1. Read the Digital Millennium Copyright Act of 1998 and summarize the key points that apply to Scour.
2. Do you believe that Scour complies with the intent and spirit of this act? Why or why not?

Critical Thinking Questions

3. How might the results of this suit affect other Internet companies?
4. How might Scour modify its basic services to avoid future lawsuits?

Sources: Amy Kover, "Napster Who? Scour Raises Hollywood Ire, Investment," *Fortune*, September 15, 2000, p. 52; and the Scour Technology Freedom Center at the Scour Web site, http://www.scour.com, accessed September 19, 2000.

electronic communication in the world—cell phone calls, satellite transmissions, e-mail messages.[14] In Maryland, a banker accessed medical records to find people diagnosed with cancer. Once he identified them, the bank called in their loans.[15]

As a result, the Federal Trade Commission is considering action, and about 70 laws are pending before Congress regarding Internet privacy. The European Union has already passed a data-protection directive that requires firms transporting data across national boundaries to have certain privacy procedures in place. This directive affects virtually any company doing business in Europe, and it is driving much of the attention concerning privacy in the United States.

Most companies and computer vendors are wary of having the federal government dictate Internet privacy standards. A group called the Online Privacy

Alliance is developing a voluntary code of conduct. It is backed by companies such as AT&T, IBM, Dun & Bradstreet, AOL Time Warner, Walt Disney, the Lexis-Nexis division of Reed Elsevier, Microsoft, and Netscape. The alliance's guidelines will call on companies to notify users when they are collecting data at Web sites to gain consent for all uses of that data, to provide for the enforcement of privacy policies, and to have a clear process in place for receiving and addressing user complaints. The alliance's policy can be found at http://www.privacyalliance.org.

Privacy at Work

The right to privacy at work is also an important issue. Some experts believe that a collision is looming between workers who want to maintain their privacy and companies that demand to know more. Recently, companies that have been monitoring their employees have been in the spotlight. Workers may be closely monitored via computer technology that is tied directly into workstations; specialized computer programs can track every keystroke a user makes. The system knows not only what workers are doing at the keyboard but also when they are not using the computer. So, these systems can estimate how many breaks a person takes. Obviously, many workers consider this type of supervision dehumanizing.

E-Mail Privacy Issues

E-mail also raises some interesting issues about work privacy. Federal law permits employers to monitor e-mail sent and received by employees. Also, e-mail messages that have been erased from hard disks may be retrieved and used in lawsuits because the laws demand that companies produce all relevant business documents. On the other hand, the use of e-mail among public officials may violate "open meeting" laws. These laws, which apply to many local, state, and federal agencies, prevent public officials from meeting in private about matters that affect the state or local area.

The legal community has debated long and hard on the applicability of privileges—specifically, the attorney-client privilege—to communications made electronically, including electronic mail. The state of New York recently amended its Civil Practice Law and Rules and clarified the matter: A privileged communication does not lose its privileged character if it is communicated or transmitted electronically. More than a dozen states either have put a similar law into effect or are debating this matter. In the past few decades, significant laws have been passed regarding an individual's right to privacy. Others relate to business privacy rights and the fair use of data and information. Table 9.8 lists additional laws related to privacy.

Individual Efforts to Protect Privacy

Although many state and federal laws deal with privacy, laws cannot completely protect individuals. And not all companies have privacy policies. As a result, people are taking control of their privacy. Some of the steps that individuals can take to protect personal privacy include the following:

- *Find out what is stored about you in existing databases.* Call the major credit bureaus to get a copy of your credit report for $8 (you can obtain one free if you have been denied credit in the last 60 days). The major companies are Equilan (800-392-1122), Trans Union (312-258-1717), and Equifax (800-685-1111). You can also submit a Freedom of Information Act request to a federal agency that you suspect may have information stored on you.
- *Be careful when you share information about yourself.* Don't share information unless it is absolutely necessary. Every time you give information about yourself through an 800, 888, or 900 call, your privacy is at risk. You can ask your doctor, bank, or financial institution not to share information about you with others without your written consent.

Law	Provisions
Fair Credit Reporting Act of 1970 (FCRA)	Regulates operations of credit-reporting bureaus, including how they collect, store, and use credit information
Tax Reform Act of 1976	Restricts collection and use of certain information by the Internal Revenue Service
Electronic Funds Transfer Act of 1979	Outlines the responsibilities of companies that use electronic funds transfer systems, including consumer rights and liability for bank debit cards
Right to Financial Privacy Act of 1978	Restricts government access to certain records held by financial institutions
Freedom of Information Act of 1970	Guarantees access for individuals to personal data collected about them and about government activities in federal agency files
Education Privacy Act	Restricts collection and use of data by federally funded educational institutions, including specifications for the type of data collected, access by parents and students to the data, and limitations on disclosure
Computer Matching and Privacy Act of 1988	Regulates cross-references between federal agencies' computer files (e.g., to verify eligibility for federal programs)
Video Privacy Act of 1988	Prevents retail stores from disclosing video rental records without a court order
Telephone Consumer Protection Act of 1991	Limits telemarketers' practices
Cable Act of 1992	Regulates companies and organizations that provide wireless communications services, including cellular phones
Computer Abuse Amendments Act of 1994	Prohibits transmissions of harmful computer programs and code, including viruses
Children's Online Privacy Protection Act of 1998	Sets standards for sites that collect information from children. The goal is to prohibit unfair or deceptive acts or practices in connection with the collection, use, or disclosure of personally identifiable information from and about children on the Internet

TABLE 9.8

Federal Privacy Laws and their Provisions

- *Be proactive to protect your privacy.* You can get an unlisted phone number and ask the phone company to block caller ID systems from reading your phone number. If you change your address, don't fill out a change-of-address form with the U.S. Postal Service; you can notify the people and companies you want to have your new address. Be careful about sending personal e-mail messages over a corporate e-mail system. You can also avoid junk mail and telemarketing calls by contacting the Direct Marketing Association at P.O. Box 3861, New York, NY 10163.

Navigating these privacy concerns will be a challenge for organizations in the Information Age. Read the "Ethical & Societal Issues" box to learn more about the monitoring of employee e-mail.

Privacy of Hardware and Software Consumers

Recent actions by major hardware and software vendors have shown that consumers should be on guard to protect their privacy. Privacy groups protested in 1999 after Intel announced that it would release Pentium III chips with a processor serial number. The groups said the number could be used to track people's habits as they surfed the Internet. Intel said the number would be used to help information systems managers keep track of their computers, but it eventually agreed to work with personal computer makers to turn the feature off during manufacturing. Consumer groups worried that the presence of the number still left unanswered questions about privacy protection. So Intel announced in April 2000 that it would stop stamping serial numbers in its processors, starting with its Willamette chip due out late 2000, but it will continue the number in its Pentium III chips.[16]

ETHICAL AND SOCIETAL ISSUES
Monitoring E-Mail—An Invasion of Privacy?

A two-month investigation by Dow Chemical, sparked by complaints from an employee, resulted in the firing of 50 workers for sending explicit pornographic images through the company's e-mail system. Another two hundred workers were disciplined by the chemical manufacturer for distributing, downloading, or saving pictures that were either pornographic or violent. Pharmaceutical giant Merck fired or disciplined an undisclosed number of employees for inappropriate use of e-mail. The *New York Times* fired 23 workers because they had allegedly distributed offensive jokes on the company's e-mail system.

Such incidents soon become public, raising protests of conflicts with employees' rights to free speech. In spite of these disputes, more and more companies are gaining greater control over employees' use of e-mail. Concerns about company e-mail abuse—and the decision about whether to monitor e-mail to avoid problems—are issues that businesses of all sizes wrestle with every day.

Companies' need to monitor e-mail reflects the realities of the modern workplace. Employers are concerned over how easily classified business information can be distributed via e-mail. They also worry about e-mail content, given what happened in the Microsoft antitrust trial, when e-mails thought to be private were dredged up in court and used against the software company. Employees sometimes write things in e-mails they would never say in public, including offensive remarks that can leave a company defenseless in a lawsuit. In fact, courts have ruled that companies are liable for harassment charges if they institute monitoring of e-mail but don't uncover messages that led to the charges.

In most instances, employers have the legal right to monitor employee e-mail, as long as they notify workers that they might do so. State and federal courts have ruled in numerous lawsuits that employees can't always demand privacy in the workplace.

To monitor e-mail, companies install basic monitoring software to scan incoming and outgoing mail for words and key phrases that managers have compiled in lists. The software can also identify e-mail viruses. Flagged messages can be reviewed for possible policy violations. More sophisticated systems have advanced features that help avoid false alarms. But e-mail monitoring is costly and labor-intensive.

The key to gaining employee support—or at least tolerance—of e-mail monitoring is to educate employees about why it's necessary and to provide carefully written e-mail usage policies. Written policies also help protect against potential legal problems. All workers need to be informed about the policies they're subject to on the job. Company managers should fully explain the details of their policies and the reasons for them. And employees need to understand that it's the company's e-mail—not theirs.

Discussion Questions

1. What are the key elements of an employee policy statement that explains the need for e-mail monitoring?
2. What steps must you take to successfully implement this policy?

Critical Thinking Questions

3. Obtain and read a copy of your company's or school's e-mail policy. Do you think the policy is clear and complete in addressing all the issues mentioned in the box? Who is responsible for enforcing this policy?
4. What steps would you take if you received "hate mail" or pornographic or violent pictures through your company's or school's e-mail system?

Sources: Adapted from Thomas York, "Invasion of Privacy? E-Mail Monitoring Is on the Rise," *Information Week*, February 21, 2000, pp. 142–146; Jennifer Disabatino, "Congress Weighs E-Mail, Net Monitoring Legislature," *Computerworld*, July 31, 2000, p. 30; Todd R. Weiss, "Dow Chemical Fires 50 Workers for Sending Pornographic E-Mail," *Computerworld*, July 28, 2000, accessed at http://www.computerworld.com/cwi/story/0,1199,NAV47_STO47689,00.html; and Christopher Lindquist, "You've Got Dirty Mail," *Computerworld*, March 13, 2000, accessed at http://www.computerworld.com/cwi/story/0,1199,NAV47_STO42841,00.html.

Microsoft acknowledged that Windows 98 and other Microsoft applications, such as Word and Excel, automatically record the author of electronic documents and the computer on which the documents were created. Documents created using Microsoft's popular Word and Excel programs in tandem with its Windows 98 operating system inserted into documents a hidden 32-digit number that was unique to the computer. This number, called a Globally Unique Identifier (GUID), is unique to each personal computer and is based on the hardware components of the system, including the Ethernet card, if it's present. Since many offices and home businesses use Ethernet cards for local area networking and high-speed Internet access, the GUID can trace a document back to its originator, even if the author strives to maintain anonymity. The computer's GUID was sent to Microsoft when the owner ran Win98's registration wizard. So Microsoft knows your personal computer's ID and can trace documents you create.[17]

Responding to repeated customer requests and significant negative press, Microsoft released two software patches that removed the Globally Unique

Identifier (GUID) from Windows 98 systems and removed the GUID from Office 97 documents and spreadsheets.[18]

Privacy and the Internet

Some people assume that there is no privacy on the Internet and that you use it at your own risk. Others believe that company Web sites should have strict privacy procedures and be accountable for privacy invasion. Regardless of your view, the potential for privacy invasion on the Internet is huge. People wanting to invade your privacy could be criminal hackers, marketing companies, or corporate bosses, and your personal and professional information could be seized without your knowledge or consent. E-mail is a prime target, as discussed previously. Sending an e-mail message is like having an open conversation in a large room—people can "listen" to your messages. When you visit a Web site on the Internet, information about you and your computer can be captured. When this information is combined with other information, companies can know what you read, what products you buy, and what your interests are. According to an executive of an Internet software monitoring company, "It's a marketing person's dream."

Most people who buy products on the Web say it's very important for a site to have a policy explaining what information is collected, how personal information is used, and what will and will not be done with it. Yet the Federal Trade Commission found that privacy policies were posted at only about 14 percent of 1,400 Web sites it recently surveyed. A *Business Week* check of the top 100 Web sites found that 43 percent displayed privacy policies. Of the notices posted, some were difficult to find and inconsistent in explaining how data is tracked and used. Registration information should be a particular concern to Internet users. If a site requests your name and address, you have every right to know why and what will be done with them. If you buy something and provide a shipping address, will it be sold to other retailers? Will your e-mail address be sold on a list of active Internet shoppers? If so, it's no different from the lists compiled by catalog retailers. You have the right to be taken off any mailing address.

Selling information to other companies can be so lucrative that many companies store and sell the data they collect on customers, employees, and others. But organizations need to determine when this information storage and use is fair and reasonable. Do individuals have a right to know about data stored about them and to decide what data is stored and used? Table 9.9 lists four issues that should be addressed: knowledge, control, notice, and consent.

TABLE 9.9

The Right to Know and the Ability to Decide

Fairness Issues	Database Storage	Database Usage
The right to know	Knowledge	Notice
The ability to decide	Control	Consent

Knowledge. Should individuals have knowledge of what data is stored on them? In some cases, individuals are informed that information on them is stored in a corporate database. In others, individuals do not know that their personal information is stored in corporate databases.

Control. Should individuals have the ability to correct errors in corporate database systems? This is possible with most organizations, although it can be difficult in some cases.

Notice. Should an organization that uses personal data for a purpose other than the original purpose notify individuals in advance? Most companies don't do this.

Consent. If information on individuals is to be used for other purposes, should these individuals be asked to give their consent before data on them is used? Many companies do not give individuals the ability to decide if information on them will be sold or used for other purposes.

HEALTH CONCERNS

Computer use may affect physical health. Strains, sprains, tendonitis, and other problems account for more than 60 percent of all occupational illnesses and about a third of workers' compensation claims, according to the Joyce Institute in Seattle. The cost to U.S. corporations for these health problems is as high as $27 billion annually. Claims relating to **repetitive motion disorder** (also called **repetitive stress injury**, or **RSI**), which can be caused by working with computer keyboards and other equipment, have increased greatly. The problems can include tendonitis, tennis elbow, the inability to hold objects, and sharp pain in the fingers. Also common is **carpal tunnel syndrome (CTS)**, which is the aggravation of the pathway for nerves that travel through the wrist (the carpal tunnel). CTS causes wrist pain, a feeling of tingling and numbness, and difficulty grasping and holding objects. It may be triggered by many factors, such as stress, lack of exercise, and the repetitive motion of typing on a computer keyboard. Decisions on workers' compensation related to repetitive stress syndrome have been decided both for and against employees.

Other work-related health hazards involve emissions from improperly maintained and used equipment. Some studies show that poorly maintained laser printers may release ozone into the air; others dispute the claim. Studies on the impact of emissions from display screens are also inconclusive. Some medical authorities believe that long-term exposure can cause cancer, but the issue remains unresolved at this time. Many organizations are developing conservative and cautious policies to be safe.

Most computer manufacturers publish technical information on radiative emissions from their screens, and many companies pay close attention to this information. San Francisco was one of the first cities to propose a video display terminal (VDT) bill. The bill requires companies with 15 or more employees who spend at least four hours a day working with computer screens to give 15-minute breaks every two hours. In addition, adjustable chairs and workstations are required if requested by employees.

Avoiding Health and Environmental Problems

Many computer-related health problems are minor and caused by a poorly designed work environment. The computer screen may be hard to read, with glare and poor contrast. Desks and chairs may also be uncomfortable. Keyboards and computer screens may be fixed in place or difficult to move. These unfavorable conditions are collectively referred to as *work stressors*. Although these problems may not concern casual users of computer systems, continued stressors such as repetitive motion, awkward posture, and eyestrain may cause more serious and long-term injuries. If nothing else, they can severely limit productivity and performance.

The study of designing and positioning computer equipment, called **ergonomics**, has suggested a number of approaches to reduce these health problems and achieve "no pain" computing. The slope of the keyboard, the positioning and design of display screens, and the placement and design of computer tables and chairs have been carefully studied. Flexibility is a major component of ergonomics and an important feature of computer devices. People require different positioning of equipment for best results. Some people, for example, want to have the keyboard in their lap; others prefer to place the keyboard on a solid table. Because of these individual differences, computer designers are attempting to develop systems that provide a great deal of flexibility.

repetitive motion disorder (repetitive stress injury; RSI)

an injury that can be caused by working with computer keyboards and other equipment

carpal tunnel syndrome (CTS)

the aggravation of the pathway for nerves that travel through the wrist (the carpal tunnel)

ergonomics

the study of designing and positioning computer equipment for employee health and safety

Prolonged use of keyboards and computer equipment may cause RSI and other problems. (Source: © Steve Kahn 1993.)

INFORMATION SYSTEMS IN ACTION
TRUSTe: Building a Reputation of Trust On-Line

Rebecca Lawson, Lansing Community College

The Internet and the World Wide Web have evolved tremendously since the mid-1990s. With technological and graphical improvements, e-commerce has become an additional tool for businesses of all sizes to advertise, promote and market their products. Software advances make it easy for anyone to create a professional-looking Web page or an on-line storefront. But this ease in creating a storefront and the open network architecture of the Internet make it very hard for consumers to be certain whom they are really dealing with in any given transaction.

E-commerce raises some distinct problems. How do customers select an on-line store that carries their desired product? Whom can they trust as a reputable business? How do customers protect themselves from fraud and privacy invasion while using the Internet for shopping?

TRUSTe is an organization trying to foster trust on the Internet. The basic principles governing its program are that all Internet users have a right to informed consent about a site's privacy policy. TRUSTe further states that no single privacy policy is adequate for all situations. The pilot program was launched in 1996 with 100 sites participating. By 1997, the program had reached businesses around the world. Many major sites have joined TRUSTe, including CNET, Yahoo! America Online, Microsoft, and Netscape. In addition, many major sites joined with TRUSTe to help launch the Privacy Partnership.

The Privacy Partnership is a consumer education campaign aimed at raising awareness of Internet privacy issues. The Privacy Partnership originally used banner advertising space donated by nearly 800 key sites, making it one of the largest Internet advertising campaigns. Now, members of the Privacy Partnership display the TRUSTe trustmark to show they are part of the program.

The TRUSTe trustmark is similar to the Good Housekeeping seal of approval. It conveys to consumers that they are dealing with a reputable on-line storefront, one that values them as customers by carefully guarding the personal information they provide to conduct a business transaction. As e-commerce develops further, more and more Web sites will proudly display the TRUSTe trustmark. This logo means that the site adheres to established privacy practices and is approved by the Federal Trade Commission, the U.S. Department of Commerce, and prominent organizations in the industry.

Each site displaying the TRUSTe trustmark adopts a privacy policy regarding the sharing of consumers' personal information gathered on-line. The site discloses its user information practices and gives consumers control over the collection and use of that information. The site also secures and protects consumers' personal information against misuse or alteration. These privacy policy guidelines assure consumers that they are dealing with a storefront that can be trusted when they make purchases on-line.

Sources: "The TRUSTe Story," http://www.truste.org/about/about_truste.html, accessed November 2, 2000; and "The TRUSTe Program: How It Protects Your Privacy," http://www.truste.org/users/users_how.html, accessed November 2, 2000.

• SUMMARY

PRINCIPLE • **Policies and procedures must be established to avoid computer waste and mistakes.**

At the corporate level, computer waste and mistakes impose unnecessarily high costs for an information system and drag down profits. Waste often results from poor integration of IS components, leading to duplication of efforts and overcapacity. Inefficient procedures also waste IS resources, as do thoughtless disposal of useful resources and misuse of computer time for games and personal tasks. Inappropriate processing instructions, inaccurate data entry, mishandling of IS output, and poor systems design all cause computer mistakes.

Careful programming practices, thorough testing, flexible network interconnections, and rigorous backup procedures can help an information system prevent and recover from many kinds of mistakes. Companies should develop manuals and training programs to avoid waste and mistakes. Company policies should specify criteria for new resource purchases and user-developed processing tools to help guard against waste and mistakes.

PRINCIPLE • **Computer crime is a serious and rapidly growing area of concern requiring management attention.**

Some criminals use computers to execute their crimes. Other crimes target computer systems, including illegal access to computer systems by criminal hackers, alteration and destruction of data and programs by viruses, and simple theft of computer resources. A virus is a program that attaches itself to other programs. A worm functions as an independent program, replicating its own program files until it destroys other systems and programs or interrupts the operation of computer systems and networks. A logic bomb is designed to "explode" or execute at a specified time and date. Because of increased computer use, greater emphasis is placed on the prevention and detection of computer crime. Software and Internet piracy may represent the most common computer crime. Computer scams have cost individuals and companies thousands of dollars. Computer crime is also an international issue.

• • •

Preventing computer crime is done by state and federal agencies, corporations, and individuals. Security measures, such as using passwords, identification numbers, and data encryption, help to guard against illegal access, especially when supported by effective control procedures. Virus scanning software identifies and removes damaging computer programs. Law enforcement agencies armed with new legal tools enacted by Congress now actively pursue computer criminals.

• • •

Federal law serves as a nationwide moral guideline for privacy rights and activities by private organizations. Some states supplement federal protections and limit activities within their jurisdictions by private organizations. A business should develop a clear and thorough policy about privacy rights for customers, including database access. That policy should also address the rights of employees, including electronic monitoring systems and e-mail. Fairness in information use for privacy rights emphasizes knowledge, control, notice, and consent for people profiled in databases. Individuals should have knowledge of the data that is stored about them and have the ability to correct errors in corporate database systems. If information on individuals is to be used for other purposes, these individuals should be asked to give their consent beforehand. Each individual has the right to know and the ability to decide.

• • •

PRINCIPLE • JOBS, EQUIPMENT, AND WORKING CONDITIONS MUST BE DESIGNED TO AVOID NEGATIVE HEALTH EFFECTS.

Computers have changed the makeup of the workforce and even eliminated some jobs, but they have also expanded and enriched employment opportunities in many ways. Some critics blame computer systems for emissions of ozone and electromagnetic radiation.

• • •

Computers and related devices affect employees' emotional and physical health, especially by causing repetitive stress injury (RSI). The study of designing and positioning computer equipment, called ergonomics, has suggested a number of approaches to reducing these health problems.

● REVIEW QUESTIONS

1. What can organizations do to prevent computer-related waste and mistakes?
2. Identify four specific actions that can be taken to reduce crime on the Internet.
3. How might the computer be the object of crime?
4. What are the major problems caused by criminal hackers?
5. What is the difference between a worm and a virus?
6. What is software piracy, and why is it so common?

7. What is ergonomics? How can it be applied to office workers?
8. What is Internet piracy, and how can companies avoid it?
9. What four issues should be addressed when considering the individual's right to privacy?
10. What specific actions can you take to avoid RSI?
11. What is the difference between CTS and RSI?
12. Under what conditions is the monitoring of e-mail not considered an invasion of privacy?

● DISCUSSION QUESTIONS

1. How can a criminal use dumpster diving to enhance his opportunity of success at social engineering?

2. You are surprised when you receive a check from the IRS for a tax refund for $10,000 more than you are owed. How could this have happened? What would you do?

3. Your marketing department has just opened a Web site and is requesting visitors to register at the site to enter a promotional contest where the chances of winning a prize are better than one in three. Visitors must provide the information necessary to contact them plus fill out a brief survey about their use of your company's products. What data privacy issues may arise?

4. What new laws do you think are necessary to improve the data privacy of Internet users? Is there a danger that some of these laws could be found unconstitutional? What portions of the Constitution would be most at risk of being violated?

5. Based on a number of recent workers' compensation cases, your employer has pledged to spend additional money for office furniture, equipment, and computers that are ergonomically designed so that employees avoid repetitive stress injuries. Will this solve the problem? Why or why not?

6. How could you use the Internet to help improve your health?

7. In the chapter opening quote John Daley says: "There are 13-year-old kids without degrees breaking into systems from their bedrooms. Security should be the subject of every Master and Ph.D. student's thesis." While this may be taking the issue too far, what actions would you recommend?

8. Using information presented in this chapter on federal privacy legislation, identify which federal law regulates the following areas and situations: cross-checking IRS and social security files to verify the accuracy of information; use of data from youngsters gathered over the Internet; credit bureaus processing home loans; customer liability for debit cards; individuals' right to access data contained in federal agency files; the IRS obtaining personal information; government obtaining financial records; and employers' access to university transcripts.

9. How can organizations ensure that their information systems are used ethically and morally? Should anyone audit the organization, departments, and employees to ensure that this is indeed the case?

● PROBLEM-SOLVING EXERCISES

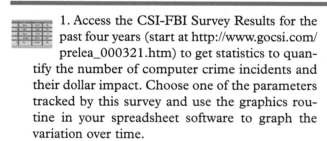 1. Access the CSI-FBI Survey Results for the past four years (start at http://www.gocsi.com/prelea_000321.htm) to get statistics to quantify the number of computer crime incidents and their dollar impact. Choose one of the parameters tracked by this survey and use the graphics routine in your spreadsheet software to graph the variation over time.

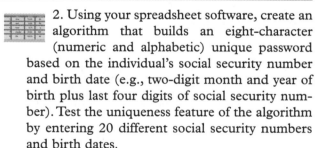 2. Using your spreadsheet software, create an algorithm that builds an eight-character (numeric and alphabetic) unique password based on the individual's social security number and birth date (e.g., two-digit month and year of birth plus last four digits of social security number). Test the uniqueness feature of the algorithm by entering 20 different social security numbers and birth dates.

● TEAM ACTIVITIES

1. Your team has been hired as consultants to improve the security at the Los Alamos National Laboratory. You have been given a budget of $10 million and a time limit of 90 days. How would you identify opportunities for improvement? What actions would your team take?

2. Your team is assigned the task of developing policies for the acquisition of computers and office equipment to provide the best ergonomic office environ-

ment. Develop an initial draft of your policies. How would you select computer systems and office equipment to ensure good ergonomic features?

3. Have each member of your team access ten different Web sites and summarize his or her findings in terms of the existence of data privacy policy statements—did the site have such a policy, was it easy to find, was it complete and easy to understand?

● SOME WEB EXERCISES

1. Visit the Web site of McAfee, Symantec, or some other provider of computer security software. Develop a "top ten" list of the current viruses that are rated as having the highest risk assessment. Create a simple chart that identifies the symptoms or impact of each of these viruses.
2. Several computer-related organizations, including AITP, ACM, IEEE, and CPSR, provide codes of ethics for IT professionals. Locate the Web pages for any two of these associations. Choose one of the codes and modify it to meet the needs of the general computer user.
3. Do Web research to identify any new federal acts that support ethics in the area of information technology. In your search, did you identify any such acts that have been overturned by the courts and ruled unconstitutional? Write a brief summary of your findings.

● CASES

 ### Predatory Hiring Practices

Executive employee defections to rival software vendors are commonplace. Senior managers are often enticed with stock and salary incentives. When one company loses a number of key employees to another over a short period of time, however, lawsuits are likely.

SAP America brought two lawsuits against Siebel Systems over the defections of more than 25 SAP employees to the rival business-management software vendor. SAP alleged that Siebel had engaged in "predatory hiring practices" in a systematic effort to injure SAP's business and impede its ability to compete with Siebel in the customer relationship management software marketplace. The parties eventually mutually agreed to settle the litigation, and the lawsuits were dismissed.

Microsoft and Borland International (now Inprise) clashed over the same issue. Borland sued Microsoft, alleging that the software giant had systematically tried to raid its staff to gain a competitive advantage. This suit was also settled out of court.

Discussion Questions

1. How is the company that loses employees affected when a number of key executives leave for a rival firm?
2. Are key executives who are wooed away from their employer by a rival firm in violation of any personal code of ethics? Is your answer true under any set of circumstances?

Critical Thinking Questions

3. Every firm competes to employ the best people available. This need for skill and experience often means hiring people from rival firms. How would you distinguish between predatory hiring practices and "normal" hiring practices?
4. What can a firm do to protect itself from losing key executives to rival firms?

Sources: Adapted from Jack McCarthy, "SAP, Siebel Settle Recruitment Lawsuit," *Computerworld*, March 20, 2000, accessed at http://www.computerworld.com/cwi/story/ 0,1199,NAV47_STO41954,00.html; and Jack McCarthy, "SAP Files Suit against Siebel," *Computerworld*, November 15, 1999, accessed at http://www.computerworld.com/cwi/story/ 0,1199,NAV47_STO37591,00.html.

 ### The Children's Online Privacy Protection Act of 1998

The Children's Online Privacy Protection Act of 1998 (COPPA) sets standards for sites that collect information from children. The goal of the legislation is to prohibit unfair or deceptive practices in the collection, use, or disclosure of personal information from and about children on the Internet. Under this rule, operators of Web sites directed to children must post prominent links on their Web sites to a notice of how they collect, use, and disclose personal information from the children; must notify parents that they wish to collect information from their children and obtain parental consent before collecting, using, or disclosing such information; cannot condition a child's participation in on-line activities on the provision of more personal information than is reasonably necessary to participate in the activity; must give parents the opportunity to review their children's information, have it deleted from the operator's database, and prohibit further collection from the child; and must establish procedures to protect the confidentiality, security, and integrity of personal information they collect from children.

In June 2000, a federal appeals court ruled that COPPA is unconstitutional. The law, which would have made it a federal crime to use the World Wide Web to communicate "for commercial purposes" material considered "harmful to minors," included penalties of up to $150,000 for each day of violation and up to six months in prison. The decision upholds a lower court ruling from February 1999.

Passed by Congress in 1998, COPPA was quickly challenged in court by the American Booksellers Foundation for Free Expression (ABFFE), Powell's Books, and A Different Light Bookstores, which joined 16 other companies and civil liberties groups in a lawsuit filed by the American Civil Liberties Union. The suit contended that the law would unlawfully restrict the ability of adults to obtain a wide range of constitutionally protected material on the Internet, including novels, poetry, art and photography books, and works on health and sex education.

In the COPPA ruling, the appeals court declared that the current definition of "harmful to minors" cannot be applied to cyberspace without censoring a wide variety of constitutionally protected materials. In a bricks-and-mortar store, when someone is accused of selling a work that is harmful to minors, a jury must decide whether the material is harmful by applying "contemporary community standards." However, the court noted that on the World Wide Web content might be viewed by people in many communities simultaneously. "Because of the peculiar geography-free nature of cyberspace," the court wrote in the 34-page decision, "a 'community standards' test would essentially require every Web communication to abide by the most restrictive community's standards," in essence forcing booksellers and others to eliminate any material that might be viewed as harmful in the most conservative community in the country. This, noted the court, "in and of itself, imposes an impermissible burden on constitutionally protected First Amendment speech."

Even though COPPA had been ruled unconstitutional, in July 2000, the Federal Trade Commission (FTC) announced that it was sending e-mail messages to "scores of Web sites" that target children to alert companies that they could face legal action as early as September if they did not comply with COPPA. FTC staffers recently surfed the Internet to determine whether Web sites were in compliance with the act. They then sent e-mails to alleged offenders, according to an agency spokeswoman. Violators could face civil penalties of $11,000 per violation, and the commission said it has a number of private investigations under way.

Many sites have been slow to conform to COPPA. Organizations tried to become COPPA-compliant by the April 2000 deadline because they expected the FTC to crack down on them. But when the FTC failed to take action, those sites stopped efforts to conform because they thought the law would not be enforced.

Discussion Questions

1. What is the contemporary community standards test? Can it be applied to material on the Internet? Why or why not?

2. What actions has the FTC taken to enforce COPPA? Do these actions seem reasonable to you? Why or why not?

Critical Thinking Questions

3. Do you agree with the June 2000 federal appeals court ruling that COPPA is unconstitutional? Why or why not? What can you find out about the most current status of COPPA?

4. Why have so many sites been slow to meet the COPPA requirements? Is their lack of action justifiable? Why or why not?

Sources: Adapted from Linda Rosencrance, "FTC Warns Sites to Comply with Children's Privacy Law," *Computerworld,* July 24, 2000, accessed at http://www.computerworld.com/cwi/story/0,1199,NAV47_STO47439,00.html; and Dan Cullen, "Appeals Court Upholds First Amendment in Cyberspace," Industry Newsroom, American Booksellers Association, June 30, 2000, accessed at http://www.bookweb.org/home/news/btw/3444.html.

 Taxing Internet Sales

According to sales-tax law, a catalog company or other remote retail company is required to collect sales tax only from customers who reside in states where a company has a physical presence—referred to in legal terms as a *nexus*. Through recent court decisions, the concept of nexus has

been interpreted to include a company's headquarters, distribution centers, retail stores, and other substantial operations. Businesses, consumers, and governments were willing to accept earlier tax laws regarding remote purchases because mail-order sales—and the resulting loss in tax revenue—were not great enough to prompt action. With the advent of e-commerce and

Internet sales, the dollar value of remote purchases has increased dramatically.

Estimates vary widely on just how much revenue states lose from not taxing Internet sales—the estimated loss for 1998 is between $120 million and $4 billion. Yet e-commerce clearly offers a large and rapidly growing potential for increased tax revenue. Government agencies want to tap this source of income and eliminate the tax advantage remote sellers have over local businesses. They propose assessing sales tax based on where the purchaser lives, rather than the seller. Making this change in the tax code would eliminate consumers' tax incentive to favor one type of retailer. But with roughly 6,500 taxing jurisdictions in the United States created by separate state, county, and city laws, such an approach would be a nightmare for retailers—they would have to calculate different taxes for each area, burying them in a mountain of tax forms. Adopting a uniform sales tax across all jurisdictions is highly unlikely because local governments use taxes to address unique social, economic, and political issues.

The debate on Internet taxation began in earnest in October 1998, when Congress passed the Internet Tax Freedom Act. This act placed a three-year moratorium on federal, state, and local tax authorities' creating new taxes specially aimed at Internet transactions. It basically buys time for legislatures and the e-commerce industry to figure out the best approach for dealing with e-commerce sales taxes. An Advisory Commission on Electronic Commerce was created by Congress and charged with producing recommendations on e-commerce and tax policy. The commission completed its work in April 2000. In May 2000 the U.S. House of Representatives endorsed the conclusions of the commission by passing legislation to extend the moratorium for five years, prohibiting multiple or discriminatory taxes on e-commerce and eliminating taxes on Internet access fees. A three-year moratorium had been imposed in 1998 by the Internet Tax Freedom Act, which also grandfathered states taxing Internet access on or before October 1, 1998, by allowing them to continue doing so. The new legislation, if signed into law, would extend the moratorium to October 2006 and would eliminate the grandfather provision.

The United States is not alone in its struggle to define e-commerce taxation rules. Many other countries are also struggling to integrate this new form of retailing into their existing tax systems.

Discussion Questions

1. What is the current status of legislation with regard to taxing Internet purchases? Do research and write a short paragraph summarizing your findings.
2. Why do you think the Advisory Commission on Electronic Commerce decided to continue the moratorium on Internet taxes for five more years?

Critical Thinking Questions

3. Are you in favor of taxing Internet purchases or do you oppose such taxes? Why?
4. What do you think will happen after the expiration of the five-year moratorium?

Sources: Adapted from Charles Waltner, "Internet Develops Its Own Tax Code," *Information Week*, December 6, 1999, pp. 110–116; and Advisory Commission on Electronic Commerce Press Room, "Advisory Commission on Electronic Commerce Concludes Business," May 4, 2000, http://www.ecommercecommission.org/releases/acec0525.htm, accessed August 2, 2000.

● NOTES

Sources for the opening vignette on p. 319: Adapted from Ann Harrison, "Audit Trails Might Have Fingered Los Alamos Insiders," *Computerworld*, May 24, 1999, http://www.computerworld.com/cwi/story/0,1199,NAV47_STO35791,00.html; Gary H. Anthes, "Computer Security Bombs at Los Alamos," *Computerworld*, May 10, 1999, http://www.computerworld.com/cwi/story/0,1199,NAV47_STO35589,00.html; and Patrick Thibodeau, "Energy Dept. Seeks More IT Security Funds," *Computerworld*, September 29, 1999, http://www.computerworld.com/cwi/story/0,1199,NAV47_STO29046,00.html.

1. Ira Sage, "CyberCrime," *Business Week*, February 21, 2000, pp. 37–42.
2. Ann Harrison, "Audit Trails Might Have Fingered Los Alamos Insiders," *Computerworld*, May 24, 1999, accessed at http://www.computerworld.com/cwi/story/0,1199,NAV47_STO35791,00.html.

3. Jaikumar Vijayan, "Government Investigates Loss of Disks at Los Alamos," *Computerworld*, June 15, 2000 accessed at http://www.computerworld.com/cwi/story/0,1199,NAV47_STO45843,00.html.
4. Ira Sage, "CyberCrime," *Business Week*, February 21, 2000, pp. 37–42.
5. Ira Sage, "CyberCrime," *Business Week*, February 21, 2000, pp. 37–42.
6. Nicole St. Pierre, "Who's Going to Train the Cyber Security Pros?" *Business Week*, February 16, 2000, p. 32.
7. Ira Sage, "CyberCrime," *Business Week*, February 21, 2000, pp. 37–42.
8. Ira Sage, "CyberCrime," *Business Week*, February 21, 2000, pp. 37–42.
9. Ira Sage, "CyberCrime," *Business Week*, February 21, 2000, pp. 37–42.

10. Ira Sage, "CyberCrime", *Business Week,* February 21, 2000, pp. 37–42.

11. Eric Luening, "Smith Pleads Guilty to Melissa Charges," CNET News.com, December 9, 1999, accessed at http://news.cnet.com/news/0-1005-200-1489249.html?tag= st.ne.1002.thed.1005-200-.

12. Ann Harrison, "Security Think Tank Releases Sniffer Tool," *Computerworld,* August 9, 1999, p. 28.

13. Robert Scheer, "Nowhere to Hide," Yahoo! Internet Life, October 2000, pp. 100–102.

14. Jeff Howe, "Global Evasedroppers," Yahoo! Internet Life, October 2000, p. 103.

15. "Defend Your Data," ACLU Web site, http://www.aclu.org/ privacy/, accessed September 20, 2000.

16. Jack McCarthy, "Intel to Phase Out Processor Serial Numbers," *Computerworld,* April 28, 2000, accessed at http://www.computerworld.com/cwi/story/0,1199,NAV47_ STO44013,00.html.

17. Dave Murphy, "Microsoft Documents Secretly Track Author," March 7, 1999, International Association of Information Technology Trainers Web site, http://itrain.org/ itinfo/1999/it990307.html, accessed August 9, 2000.

18. Dave Murphy, "Microsoft Patches Win98 GUID Privacy Bug," March 19, 1999, International Association of Information Technology Trainers Web site, http://itrain.org/ itinfo/1999/it990319.html, accessed August 9, 2000.

GLOSSARY

accounting MIS an information system that provides aggregate information on accounts payable, accounts receivable, payroll, and many other applications

ad hoc DSS a DSS concerned with situations or decisions that come up only a few times during the life of the organization

antivirus programs programs or utilities that prevent viruses and recover from them if they infect a computer

application program interface (API) interface that allows applications to make use of the operating system

application service provider (ASP) a company that provides both end user support and the computers on which to run the software from the user's facilities

application software programs that help users solve particular computing problems

arithmetic/logic unit (ALU) portion of the CPU that performs mathematical calculations and makes logical comparisons

ARPANET a project started by the U.S. Department of Defense (DOD) in 1969 as both an experiment in reliable networking and a means to link DOD and military research contractors, including a large number of universities doing military-funded research

artificial intelligence (AI) a field in which the computer system takes on the characteristics of human intelligence

artificial intelligence systems people, procedures, hardware, software, data, and knowledge needed to develop computer systems and machines that demonstrate characteristics of intelligence

asking directly an approach to gathering data that asks users, stakeholders, and other managers about what they want and expect from the new or modified system

attribute a characteristic of an entity

backbone one of the Internet's high-speed, long-distance communications links

backward chaining the process of starting with conclusions and working backward to the supporting facts

batch processing system method of computerized processing in which business transactions are accumulated over a period of time and prepared for processing as a single unit or batch

best practices the most efficient and effective ways to complete a business process

biometrics the measurement of a living trait, whether physical or behavioral

brainstorming decision-making approach that often involves members offering ideas "off the top of their heads"

business intelligence the process of getting enough of the right information in a timely manner and usable form and analyzing it so that it can have a positive impact on business strategy, tactics, or operations

business resumption planning the process of anticipating and providing for disasters

business-to-business (B2B) e-commerce a form of e-commerce in which the participants are organizations

business-to-consumer (B2C) e-commerce a form of e-commerce in which customers deal directly with the organization, avoiding any intermediaries

byte (B) eight bits together that represent a single character of data

carpal tunnel syndrome (CTS) the aggravation of the pathway for nerves that travel through the wrist (the carpal tunnel)

catalog management software software that automates the process of creating a real-time interactive catalog and delivering customized content to a user's screen

central processing unit (CPU) the part of the computer that consists of two primary elements: the arithmetic/logic unit and the control unit

centralized processing processing alternative in which all processing occurs in a single location or facility

certification a process for testing skills and knowledge that results in an endorsement by the certifying authority that an individual is capable of performing a particular job

character basic building block of information, consisting of uppercase letters, lowercase letters, numeric digits, or special symbols

chat room a facility that enables two or more people to engage in interactive "conversations" over the Internet

choice stage the third stage of decision making, which requires selecting a course of action

client/server an architecture in which multiple computer platforms are dedicated to special functions such as database management, printing, communications, and program execution

clock speed a series of electronic pulses, produced at a predetermined rate, that affect machine cycle time

command-based user interface a user interface that requires that text commands be given to the computer to perform basic activities

common carriers long-distance telephone companies

communications protocols rules and standards that ensure communications among computers of different types and from different manufacturers

communications software software that provides a number of important functions in a network

compact disk read-only memory (CD-ROM) a common form of optical disk on which data, once it has been recorded, cannot be modified

competitive advantage a significant and (ideally) long-term benefit to a company over its competition

competitive intelligence a continuous process involving the legal and ethical collection of information, analysis, and controlled dissemination of information to decision makers

compiler a language translator that converts a complete program into a machine language to produce a program that the computer can process in its entirety

computer network the communications media, devices, and software needed to connect two or more computer systems and/or devices

computer programs sequences of instructions for the computer

computer system platform the combination of a particular hardware configuration and systems software package

computer-aided software engineering (CASE) tools that automate many of the tasks required in a systems development effort and enforce adherence to the systems development life cycle

computer-based information system (CBIS) a single set of hardware, software, databases, telecommunications, people, and procedures that are configured to collect, manipulate, store, and process data into information

concurrency control a method of dealing with a situation in which two or more people need to access the same record in a database at the same time

content streaming a method for transferring multimedia files over the Internet so that the data stream of voice and pictures plays more or less continuously, without a break, or very few of them; enables users to browse large files in real time

contract software software developed for a particular company

control unit part of the CPU that sequentially accesses program instructions, decodes them, and coordinates the flow of data in and out of the ALU, primary storage, and even secondary storage and various output devices

cookie a text file that an Internet company can place on the hard disk of a computer system

coprocessor part of the computer that speeds processing by executing specific types of instructions while the CPU works on another processing activity

counterintelligence the steps an organization takes to protect information sought by "hostile" intelligence gatherers

criminal hacker (cracker) a computer-savvy person who attempts to gain unauthorized or illegal access to computer systems

cryptography the process of converting a message into a secret code

and changing the encoded message back to regular text

cybermall a single Web site that offers many products and services at one Internet location

data raw facts, such as an employee's name and number of hours worked in a week, inventory part numbers, or sales orders

data analysis manipulation of the collected data so that it is usable for the development team members who are participating in systems analysis

data collection the process of capturing and gathering all data necessary to complete transactions

data correction the process of reentering miskeyed or misscanned data that was found during data editing

data definition language (DDL) a collection of instructions and commands used to define and describe data and data relationships in a specific database

data dictionary a detailed description of all the data used in the database

data editing the process of checking data for validity and completeness

data integrity the degree to which the data in any one file is accurate

data item the specific value of an attribute

data manipulation the process of performing calculations and other data transformations related to business transactions

data manipulation language (DML) the commands that are used to manipulate the data in a database

data mart a subset of a data warehouse

data mining an information analysis tool that involves the automated discovery of patterns and relationships in a data warehouse

data model a diagram of data entities and their relationships

data preparation (data conversion) conversion of manual files into computer files

data redundancy duplication of data in separate files

data storage the process of updating one or more databases with new transactions

data warehouse a database that collects business information from many sources in the enterprise, covering all aspects of the company's processes, products, and customers

data-flow diagram (DFD) a model of objects, associations, and activities that describes how data can flow between and around various objects

database an organized collection of facts and information

database approach to data management an approach whereby a pool of related data is shared by multiple application programs

database management system (DBMS) a group of programs that manipulate the database and provide an interface between the database and the user of the database and other application programs

decentralized processing processing alternative in which processing devices are placed at various remote locations

decision room a room that supports decision making, with the decision makers in the same building, combining face-to-face verbal interaction with technology to make the meeting more effective and efficient

decision support system (DSS) an organized collection of people, procedures, software, databases, and devices used to support problem-specific decision making

decision-making phase the first part of problem solving, including three stages: intelligence, design, and choice

dedicated line a communications line that provides a constant connection between two points; no switching or dialing is needed, and the two devices are always connected

delphi approach a decision-making approach in which group decision makers are geographically dispersed; this approach encourages diversity among group members and fosters creativity and original thinking in decision making

demand reports reports developed to give certain information at a manager's request

design report the primary result of systems design, reflecting the decisions made for system design and preparing the way for systems implementation

design stage the second stage of decision making, during which alternative solutions to the problem are developed

dialogue manager user interface that allows decision makers to easily access and manipulate the DSS and use common business terms and phrases

digital certificate an attachment to an e-mail message or data embedded in a Web page that verifies the identity of a sender or a Web site

digital computer camera input device used with a PC to record and store images and video in digital form

digital signature an encryption technique used to verify the identity of a message sender for the processing of on-line financial transactions

digital subscriber line (DSL) a communications line that uses existing phone wires going into today's homes and businesses to provide transmission speeds exceeding 500 Kbps at a cost of $20 or more per month

digital video disk (DVD) storage format used to store digital video or computer data

direct access retrieval method in which data can be retrieved without the need to read and discard other data

direct access storage device (DASD) device used for direct access of secondary storage data

direct conversion stopping the old system and starting the new system on a given date (also called *plunge* or *direct cutover*)

distance learning the use of telecommunications to extend the classroom

distributed processing processing alternative in which computers are placed at remote locations but connected to each other via telecommunications devices

documentation text that describes the program functions to help the user operate the computer system

domain the allowable values for data attributes

domain expert the individual or group whose expertise or knowledge is captured for use in the expert system

drill down reports reports providing increasingly detailed data about a situation

dumpster diving searching through the garbage for important pieces of information that can help crack an organization's computers or be used to persuade someone at the company to give criminals access to the computers

e-commerce any business transaction executed electronically between parties such as companies (business-to-business), companies and consumers (business-to-consumer), business and the public sector, and consumers and the public sector

e-commerce software software that supports catalog management, product configuration, shopping cart facilities, and e-commerce transaction processing

e-commerce transaction processing software software that provides the basic connection between participants in the e-commerce economy, enabling communications between trading partners regardless of their technical infrastructure

electronic bill presentment a method of billing in which the biller posts an image of your statement on the Internet and alerts you by e-mail that your bill has arrived

electronic cash an amount of money that is computerized, stored, and used as cash for e-commerce transactions

electronic data interchange (EDI) an intercompany, application-to-application communication of data in standard format, permitting the recipient to perform the functions of a standard business transaction

electronic exchange an electronic forum where manufacturers, suppliers, and competitors buy and sell goods, trade market information, and run back-office operations

electronic retailing (e-tailing) the direct sale from business to consumer through electronic storefronts, typically designed around an electronic catalog and shopping cart model

electronic shopping cart a model used by many e-commerce sites to track the items selected for purchase, allowing shoppers to view what is in their cart, add new items to it, and remove items from it

electronic wallet a computerized stored value that holds credit card information, electronic cash, owner identification, and address information

encryption the conversion of a message into a secret code

end-user systems development any systems development project in which the primary effort is undertaken by a combination of business managers and users

enterprise data modeling data modeling done at the level of the entire enterprise

entity generalized class of people, places, or things for which data is collected, stored, and maintained

entity-relationship (ER) diagrams a data model that uses basic graphical symbols to show the organization of and relationships between data

ergonomics the study of designing and positioning computer equipment for employee health and safety

event-driven review a review triggered by a problem or opportunity such as an error, a corporate merger, or a new market for products

exception reports reports automatically produced when a situation is unusual or requires management action

execution time (E-time) the time it takes to execute an instruction and store the results

executive support system (ESS), or **executive information system (EIS)** specialized DSS that includes all hardware, software, data, procedures, and people used to assist senior-level executives within the organization

expandable storage devices storage that uses removable disk cartridges to provide additional storage capacity

expert system a system that gives a computer the ability to make suggestions and act as an expert in a particular field

expert system shell a collection of software packages and tools used to develop expert systems

explanation facility component of an expert system that allows a user or decision maker to understand how the expert system arrived at certain conclusions or results

Extensible Markup Language (XML) markup language for Web documents containing structured information, including words, pictures, and other elements

extranet a network that links selected resources of the intranet of a company with its customers, suppliers, or other business partners; based on Web technologies

feasibility analysis assessment of the technical, operational, schedule, economic, and legal feasibility of a project

feedback output that is used to make changes to input or processing activities

field typically a name, number, or combination of characters that describes an aspect of a business object or activity

file a collection of related records

file server an architecture in which the application and database reside on the one host computer, called the file server

file transfer protocol (FTP) a protocol that describes a file transfer process between a host and a remote computer and allows users to copy files from one computer to another

financial MIS an information system that provides financial information to all financial managers within an organization

financial model model that provides cash flow, internal rate of return, and other investment analysis

firewall a device that sits between your internal network and the outside Internet and limits access into and out of your network based on your organization's access policy

five-force model a widely accepted model that identifies five key factors that can lead to attainment of competitive advantage: rivalry among existing competitors, the threat of new market entrants, the threat of substitute products and services, the bargaining power of buyers, and the bargaining power of suppliers

forward chaining the process of starting with the facts and working forward to the conclusions

fuzzy logic a special research area in computer science that allows shades of gray and does not require conditions to be black/white, yes/no, or true/false

geographic information system (GIS) a computer system capable of assembling, storing, manipulating, and displaying geographic information (i.e., data identified according to its location)

graphical modeling program software package that assists decision makers in designing, developing, and using graphic displays of data and information

graphical user interface (GUI) an interface that uses icons and menus displayed on screen to send commands to the computer system

group consensus approach decision-making approach that forces members in the group to reach a unanimous decision

group decision support system (GDSS) software application that consists of most elements in a DSS, plus software needed to provide effective support in group decision making

hacker a person who enjoys computer technology and spends time learning and using computer systems

hardware computer equipment used to perform input, processing, and output activities

heuristics commonly accepted guidelines or procedures that usually find a good solution

hierarchical database model a data model in which data is organized in a top-down, or inverted tree, structure

hierarchy of data bits, characters, fields, records, files, and databases

highly structured problems problems that are straightforward and require known facts and relationships

home page a cover page for a Web site that has titles, graphics, and text

HTML tags codes that let the Web browser know how to format text: as a heading, as a list, or as body text and whether images, sound, and other elements should be inserted

human resource MIS an information system that is concerned with activities related to employees and potential employees of an organization

hypermedia tools that connect the data on Web pages, allowing users to access topics in whatever order they wish

hypertext markup language (HTML) the standard page description language for Web pages

if-then statements rules that suggest certain conclusions

implementation stage the stage of problem solving during which the solution is put into effect

inference engine part of the expert system that seeks information and relationships from the knowledge base and provides answers, predictions, and suggestions the way a human expert would

information a collection of facts organized in such a way that they have additional value beyond the value of the facts themselves

information center a support function that provides users with assistance, training, application development, documentation, equipment selection and setup, standards, technical assistance, and troubleshooting

information system (IS) a set of interrelated components that collect, manipulate, and disseminate data and information and provide a feedback mechanism to meet an objective

information systems planning the translation of strategic and organizational goals into systems development initiatives

input the activity of gathering and capturing raw data

installation the process of physically placing the computer equipment on the site and making it operational

instant messaging a method that allows two or more individuals to communicate on-line using the Internet

institutional DSS a DSS that handles situations or decisions that occur more than once, usually several times a year

instruction time (I-time) the time it takes to perform the fetch-instruction and decode-instruction steps of the instruction phase

intelligence stage the first stage of decision making, during which potential problems and opportunities are identified and defined

intelligent behavior the ability to learn from experience and apply knowledge acquired from experience, handle complex situations, solve problems when important information is missing, determine what is important, and react quickly and correctly to a new situation

international network a network that links systems between countries

Internet the world's largest computer network, actually consisting of thousands of interconnected networks, all freely exchanging information

Internet piracy illegally gaining access to and using the Internet

Internet protocol (IP) a communication standard that enables traffic to be routed from one network to another as needed

Internet service provider (ISP) any company that provides individuals and organizations with access to the Internet

interpreter a language translator that translates one program statement at a time into machine code

intranet an internal corporate network built using Internet and World Wide Web standards and products; used by the employees of the organization to access corporate information

Java an object-oriented programming language based on C++ that allows small programs (applets) to be embedded within an HTML document

joining data manipulation that combines two or more tables

joint application development (JAD) a process for data collection and requirements analysis

key a field or set of fields in a record that is used to identify the record

key-indicator report summary of the previous day's critical activities; typically available at the beginning of each workday

knowledge acquisition facility part of the expert system that provides convenient and efficient means of capturing and storing all components of the knowledge base

knowledge base a component of an expert system that stores all relevant information, data, rules, cases, and relationships used by the expert system

knowledge engineer an individual who has training or experience in the design, development, implementation, and maintenance of an expert system

knowledge management the process of capturing a company's collective expertise wherever it resides—in computers, on paper, in people's heads—and distributing it wherever it can help produce the biggest payoff

knowledge user the individual or group who uses and benefits from the expert system

language translator systems software that converts a programmer's source code into its equivalent in machine language

learning system a combination of software and hardware that allows the computer to change how it functions or reacts to situations based on feedback it receives

linking data manipulation that combines two or more tables using common data attributes to form a new table with only the unique data attributes

local area network (LAN) a network that connects computer systems and devices within the same geographic area

logic bomb an application or system virus designed to execute at a specified time and date

logical design description of the functional requirements of a system

machine cycle the instruction phase followed by the execution phase

magnetic disk common secondary storage medium; bits are represented by magnetized areas

magnetic tape common secondary storage medium; Mylar film coated with iron oxide, with portions of the tape magnetized to represent bits

magneto-optical disk a hybrid between a magnetic disk and an optical disk

mainframe computer large, powerful computer often shared by hundreds of concurrent users connected to the machine via terminals

make-or-buy decision the decision regarding whether to obtain the necessary software from internal or external sources

market segmentation the identification of specific markets and targeting them with advertising messages

marketing MIS an information system that supports managerial activities in product development, distribution, pricing decisions, promotional effectiveness, and sales forecasting

management information system (MIS) an organized collection of people, procedures, software, databases, and devices used to provide routine information to managers and decision makers

meta tag a special HTML tag, not visible on the displayed Web page, that contains keywords representing your site's content, which search engines use to build indexes pointing to your Web site

metadata the data that describes the contents of a database

midrange computer formerly called minicomputer, a system about the size of a small three-drawer file cabinet that can accommodate several users at one time

model base part of a DSS that provides decision makers access to a variety of models and assists them in decision making

model management software software that coordinates the use of models in a DSS

monitoring stage the final stage of the problem-solving process, during which decision makers evaluate the implementation

Moore's Law a hypothesis that states that transistor densities on a single chip will double every 18 months

multiprocessing simultaneous execution of two or more instructions

multitasking capability that allows a user to run more than one application at the same time

natural language processing processing that allows the computer to understand and react to statements and commands made in a "natural" language, such as English

network computer a cheaper-to-buy and cheaper-to-run version of the personal computer that is used primarily for accessing networks and the Internet

network management software software that enables a manager on a networked desktop to monitor the use of individual computers and shared hardware (such as printers), scan for viruses, and ensure compliance with software licenses

network model an expansion of the hierarchical database model with an owner-member relationship in which a member may have many owners

network operating system (NOS) systems software that controls the computer systems and devices on a network and allows them to communicate with each other

networks connected computers and computer equipment in a building, around the country, or around the world to enable electronic communications

neural network a computer system that can simulate the functioning of a human brain

nominal group technique decision-making approach that encourages feedback from individual group members; the final decision is made by voting, similar to the way public officials are elected

nonprogrammed decisions decisions that deal with unusual or exceptional situations

object code machine language code

object-relational database management system (ORDBMS) a DBMS capable of manipulating audio, video, and graphical data

off-the-shelf software existing software program

on-line analytical processing (OLAP) software that allows users to explore data from a number of different perspectives

on-line transaction processing (OLTP) computerized processing in which each transaction is processed immediately, without the delay of accumulating transactions into a batch

open database connectivity (ODBC) standards that ensure that software written to comply with these standards can be used with any ODBC-compliant database

operating system (OS) a set of computer programs that controls the computer hardware and acts as an interface with application programs

optical disk a rigid disk of plastic onto which data is recorded by special lasers that physically burn pits in the disk

optimization model a process to find the best solution, usually the one that will best help the organization meet its goals

order processing systems systems that process order entry, sales configuration, shipment planning, shipment execution, inventory control, invoicing, customer interaction, and routing and scheduling

organization a formal collection of people and other resources established to accomplish a set of goals

output production of useful information, usually in the form of documents and reports

parallel processing a form of multiprocessing that speeds processing by linking several processors to operate at the same time, or in parallel

parallel start-up running both the old and new systems for a period of time and comparing the output of the new system closely with the output of the old system; any differences are reconciled; when users are comfortable that the new system is working correctly, the old system is eliminated

password sniffer a small program hidden in a network or a computer system that records identification numbers and passwords

perceptive system a system that approximates the way a human sees, hears, and feels objects

personal computer (PC) relatively small, inexpensive computer system, sometimes called a microcomputer

personal productivity software software that enables users to improve their personal effectiveness, increasing the amount of work they can do and its quality

phase-in approach slowly replacing components of the old system with those of the new one; this process is repeated for each application until the new system is running every application and performing as expected (also called *piecemeal approach*)

physical design specification of the characteristics of the system components necessary to put the logical design into action

pilot start-up running the new system for one group of users rather than all users

pixel a dot of color on a photo image or a point of light on a display screen

planned data redundancy a way of organizing data in which the logical database design is altered so that certain data entities are combined, summary totals are carried in the data records rather than calculated from elemental data, and some data attributes are repeated in more than one data entity to improve database performance

point-of-sale (POS) device terminal used in retail operations to enter sales information into the computer system

point-to-point protocol (PPP) a communications protocol that transmits packets over telephone lines

primary key a field or set of fields that uniquely identifies the record

problem solving a process that goes beyond decision making to include the implementation stage

procedures the strategies, policies, methods, and rules for using a CBIS

processing converting or transforming data into useful outputs

product configuration software software used by buyers to build the product they need on-line

productivity a measure of the output achieved divided by the input required

programmed decisions decisions made using a rule, procedure, or quantitative method

programmer a specialist responsible for modifying or developing programs to satisfy user requirements

programming languages coding schemes used to write both systems and application software

project management model model used to coordinate large projects and identify critical activities and tasks that could delay or jeopardize an entire project if they are not completed on time and cost-effectively

projecting data manipulation that chooses columns in a table

proprietary software a one-of-a-kind program for a specific application

prototyping an iterative approach to the systems development process

public network services systems that give personal computer users access to vast databases and other services, usually for an initial fee plus usage fees

push technology automatic transmission of information over the Internet rather than making users search for it with their browsers

quality the ability of a product (including services) to meet or exceed customer expectations

random access memory (RAM) a form of memory in which instructions or data can be temporarily stored

rapid application development (RAD) a systems development approach that employs tools, techniques, and methodologies designed to speed application development

read-only memory (ROM) a nonvolatile form of memory

record a collection of related data fields

redundant array of independent/inexpensive disks (RAID) method of storing data that allows the system to create a "reconstruction map" so that if a hard drive fails, it can rebuild lost data

relational model a database model that describes data in which all data elements are placed in two-dimensional tables, called relations, that are the logical equivalent of files

repetitive motion disorder (repetitive stress injury; RSI) an injury that can be caused by working with computer keyboards and other equipment

request for proposal (RFP) a document that specifies in detail required resources such as hardware and software

requirements analysis determination of user, stakeholder, and organizational needs

return on investment (ROI) one measure of IS value that investigates the additional profits or benefits that are generated as a percentage of the investment in information systems technology

robotics mechanical or computer devices that perform tasks requiring a high degree of precision or that are tedious or hazardous for humans

rule a conditional statement that links given conditions to actions or outcomes

satisficing model a model that will find a good—but not necessarily the best—problem solution

scalability the ability of the computer to handle an increasing number of concurrent users smoothly

scheduled reports reports produced periodically, or on a schedule, such as daily, weekly, or monthly

schema a description of the entire database

search engine a Web search tool

secondary storage devices that store larger amounts of data, instructions, and information more permanently than allowed with main memory

selecting data manipulation that chooses rows according to certain criteria

semistructured or unstructured problems more complex problems in which the relationships among the data are not always clear, the data may be in a variety of formats, and the data is often difficult to manipulate or obtain

sequential access retrieval method in which data must be accessed in the order in which it is stored

sequential access storage device (SASD) device used to sequentially access secondary storage data

serial line Internet protocol (SLIP) a communications protocol that transmits packets over telephone lines

site preparation preparation of the location of the new system

smart card a credit card–sized device with an embedded microchip to provide electronic memory and processing capability

social engineering the practice of talking an individual out of a critical computer password

software the computer programs that govern the operation of the computer

software piracy the act of illegally duplicating software

software suite a collection of single-application software packages in a bundle

source code high-level program code written by the programmer

source data automation the process of capturing data at its source to record it accurately, in a timely fashion, with minimal manual effort, and in a form that can be directly entered into the computer rather than keying the data from a separate document

sphere of influence the scope of problems and opportunities addressed by a particular organization

stakeholders individuals who, either themselves or through the area of the organization they represent, ultimately benefit from the systems development project

start-up the process of making the final tested information system fully operational

statistical analysis model model that can provide summary statistics, trend projections, hypothesis testing, and more

storage area network (SAN) technology that uses computer servers, distributed storage devices, and networks to tie the storage system together

storefront brokers companies that act as intermediaries between your Web site and on-line merchants that have the products and retail expertise

strategic planning determining long-term objectives by analyzing the strengths and weaknesses of the organization, predicting future trends, and projecting the development of new product lines

subschema a file that contains a description of a subset of the database and identifies which users can view and modify the data items in the subset

supercomputers the most powerful computer systems, with the fastest processing speeds

superconductivity a property of certain metals that allows current to flow with minimal electrical resistance

supply chain management a key value chain composed of demand planning, supply planning, and demand fulfillment

switched line a communications line that uses switching equipment to allow one transmission device to be connected to other transmission devices

syntax a set of rules associated with a programming language

systems analysis the systems development phase involving the study of existing systems and work processes to identify strengths, weaknesses, and opportunities for improvement

systems analyst a professional who specializes in analyzing and designing business systems

systems design the systems development phase that defines how the information system will do what it must do to obtain the problem solution

systems development the activity of creating or modifying existing business systems

systems implementation the systems development phase that involves creating or acquiring various system components detailed in the systems design, assembling them, and placing the new or modified system into operation

systems investigation the systems development phase during which problems and opportunities are identified and considered in light of the goals of the business to gain a clear understanding of the problem to be solved or opportunity to be addressed

systems investigation report a summary of the results of the systems investigation and the process of feasibility analysis; recommends a course of action

systems maintenance and review the systems development phase that ensures that the system operates and modifies the system so that it continues to meet changing business needs; involves checking, changing, and enhancing the system to make it more useful in achieving user and organizational goals

systems software the set of programs designed to coordinate the activities and functions of the hardware and various programs throughout the computer system

technology acceptance model (TAM) a description of the factors that can lead to higher acceptance and use of technology in an organization, including perceived usefulness of the technology, ease of its use, quality of the information system, and the degree to which the organization supports the use of the information system

technology diffusion a measure of how widely technology is spread throughout an organization

technology infrastructure all the hardware, software, databases, telecommunications, people, and procedures that are configured to collect, manipulate, store, and process data into information

technology infusion the extent to which technology is deeply integrated into an area or department

telecommunications the electronic transmission of signals for communications; enables organizations to carry out their processes and tasks through effective computer networks

telecommunications medium anything that carries an electronic signal and interfaces between a sending device and a receiving device

telecommuting a work arrangement in which employees work away from the office using personal computers and networks to communicate via e-mail with other workers and to pick up and deliver results

Telnet a terminal emulation protocol that enables users to log on to other computers on the Internet to gain access to public files

terminal-to-host an architecture in which the application and database reside on one host computer, and the user interacts with the application and data using a "dumb" terminal

time-driven review review performed after a specified amount of time

time-sharing capability that allows more than one person to use a computer system at the same time

total cost of ownership (TCO) a measure of the total cost of owning computer equipment, including desktop computers, networks, and large computers

traditional approach to data management an approach whereby separate data files are created and stored for each application program

transaction any business-related exchange such as payments to employees, sales to customers, and payments to suppliers

transaction processing cycle the process of data collection, data editing, data correction, data manipulation, data storage, and document production

transaction processing system (TPS) an organized collection of people, procedures, software, databases, and devices used to record completed business transactions

transport control protocol (TCP) a protocol that includes rules that computers on a network use to establish and break connections

Trojan horse a program that appears to be useful but actually masks a destructive program

tunneling the process by which VPNs transfer information by encapsulating traffic in IP packets over the Internet

uniform resource locator (URL) an assigned address on the Internet for each computer

user acceptance document a formal agreement signed by the user that states that a phase of the installation or the complete system is approved

user interface element of the operating system that allows individuals to access and command the computer system

users individuals who will interact with the system regularly

value chain a series (chain) of activities that includes inbound logistics, warehouse and storage, production, finished product storage, outbound logistics, marketing and sales, and customer service

value-added carriers companies that have developed private telecommunications systems and offer their services for a fee

videoconferencing a telecommunication system that combines video and phone call capabilities with data or document conferencing

virtual private network (VPN) a network that transfers information by encapsulating traffic in IP packets and sending the packets over the Internet

virtual reality immersive virtual reality, which means the user becomes fully immersed in an artificial, three-dimensional world that is completely generated by a computer

virtual reality system system that enables one or more users to move and react in a computer-simulated environment

virtual workgroups teams of people located around the world working on common problems

virus a program that attaches itself to other programs

vision system the hardware and software that permit computers to capture, store, and manipulate visual images and pictures

voice mail technology that enables users to leave, receive, and store verbal messages for and from other people around the world

voice-over-IP (VOIP) technology that enables network managers to route phone calls and fax transmissions over the same network they use for data

voice-recognition device an input device that recognizes human speech

Web appliance a device that can connect to the Internet, typically through a phone line

Web browser software that creates a unique, hypermedia-based menu on your computer screen that provides a graphical interface to the Web

Web page construction software software that uses Web editors to produce both static and dynamic Web pages

Web site hosting companies companies that provide the tools and services required to set up a Web page and conduct e-commerce within a matter of days and with little up-front cost

Web site traffic data analysis software software that processes and analyzes data from the Web log file to provide useful information to improve Web site performance

wide area network (WAN) a network that ties together large geographic regions using microwave and satellite transmission or telephone lines

workstation computer that fits between high-end personal computers and low-end midrange computers in terms of cost and processing power

World Wide Web (WWW, or W3) an Internet service comprising tens of thousands of independently owned computers that work together as one

worm an independent program that replicates its own program files until it interrupts the operation of networks and computer systems

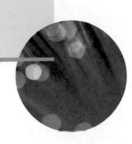

INDEX

A boldface page number indicates a key term and the location where its definition can be found.

A

ABB, hardware acquisition by, 303–304
Abilene, 142
Acceptance testing, 306
Access (Microsoft), 107, 108, 117
Accessible information, 6
Accounting MIS, 226
Accurate information, 6
Acquisitions
 ERP in, 207–208
 of hardware and software, 27
 options advantages and disadvantages, 301
 policies and procedures for, 321
ActionPlan products, 244
Active-matrix displays, 53
Activity modeling, for data analysis, 295, 296
Ad hoc DSS, **227**
Advertising
 DoubleClick and, 179
 ethics of, 162
Advisory Commission on Electronic Commerce, 343
AI. *See* Artificial intelligence systems (AI)
AI/ES. *See* Artificial intelligence/expert systems (AI/ES)
Aircraft maintenance training, virtual technology for, 272
Airlines, OLTP and, 191
Algorithms, for encryption, 162–163
Allina Health System, 96–97
 data warehouse at, 127
Alphanumeric data, 5
Altair, 82
ALU. *See* Arithmetic/logic unit (ALU)
Amazon.com, 153
American Airlines, SABRE system of, 21
American Telephone & Telegraph (AT&T), 134, 269
 computer crime against, 323
 as server, 145
 software rented by, 283

America Online (AOL), 145. *See also* On-line services
 merger with Time Warner, 20, 151
 public network services of, 150
Andreessen, Marc, 157
ANI. *See* Automatic number identification (ANI)
ANN. *See* Artificial neural network (ANN)
Anonymous input, in GDSS, 232
Antivirus programs, **329**–330
Aonix, 120
API. *See* Application program interface (API)
Apple Computer, 54, 65, 82
 operating systems of, 65
Applets, 155, 156
Application program interface (API), **62**
Application service provider (ASP), **69**, 72, 283
Application software, 10, **58**–59, 60, 61, 68–77
 customized package, 69
 enterprise, 76–77
 filing information about, 321
 off-the-shelf, 68–69
 personal, 69–74
 proprietary, 68
 workgroup, 74–76
Application-specific files, 94
Applix, iTM1 of, 116
Aqua Mac interface, 65
Arithmetic/logic unit (ALU), **41**
ARPANET, **142**
Artificial intelligence/expert systems (AI/ES), 191
Artificial intelligence systems (AI), **13**, **250**–256. *See also* Expert system (ES)
 applications of, 267–269
 conceptual model of, 254
 elements of, 15
 expert systems as, 253
 learning systems as, 256
 for monitoring and controlling networks, 279

natural intelligence compared with, 253
 natural language processing as, 255
 neural networks as, 256
 robotics as, 254–255
 vision systems as, 255
Artificial neural network (ANN), 273
Asking directly, **297**
Ask Jeeves search engine, 249
AskMe.com, 153
ASP approach. *See* Application service provider (ASP)
Assembly language, 79
Asynchronous modes, 239
ATM devices. *See* Automatic teller machine (ATM) devices
AT&T. *See* American Telephone & Telegraph (AT&T)
Attribute, **93**, 94
Auctions, on Internet, 181
Audible Mobile Player, 152
Audio data, 5
Austin-Hayne, 269
Authentication technologies, 185
Automated TPS, 190
Automatic number identification (ANI), 135
Automatic teller machine (ATM) devices, 51
Automation. *See* Robotics
AutoNation.com, 176–177
Axiom Sports Tracker, 151
Aye.Net, 324

B

B2B e-commerce. *See* Business-to-business (B2B) e-commerce
B2C e-commerce. *See* Business-to-consumer (B2C) e-commerce
Backbone, **143**
Backing up records, 197
Backward chaining, **263**, 264
BackWeb, 156
Baldauf, Kenneth, 273
BankAmerica, software of, 304
Banking, on-line, 180–181
Bar codes, 9, 194
 scanners, 52

Bargaining power, competitive advantage and, 19, 21
BASIC, 79
Batch processing systems, **191**–192
Bellagio (hotel), 167
Bell Canada, 113
Best practices, **200**
Bidder's Edge, 269
Biometrics, **328**
Bioscrypt Enterprise, 328
Bit, 44, 92
Bits per second, as telecommunications speed, 132
Black Orifice (software), 324
Blue Mountain, 58
Boeing
 Palm computers and, 54
 quality at, 23
Boise Cascade Office Products, 120
Bonner, Glenn, 167
BOOM device, 270
Borders Books & Music, 174
Borland International. *See* Inprise (Borland International)
Borse, John, 169
Bots (knowledge robots), 269
bps, 132
Brainstorming, **232**
Brandeis, Louis, 332
Break-ins. *See* Computer crime
Britannica.com, 153
Brokerage services, natural language processing in, 255
Browser. *See* Web browser
Bugs, in software, 81–82
BuildersExpress.com, 157
Building Materials Holding Company, 157
Burlington Northern Santa Fe (BNSF), 223
Business information systems, 11–14
Business intelligence, 120–121
Business plan, expert system for developing, 278
Business resumption planning, **196**–197

Business-to-business (B2B) e-commerce, 11, **174**, 175, 183
Business-to-consumer (B2C) e-commerce, **174**, 176
Byte, **44**

C

C (language), 79
C++ (language), 80, 155
Cable Act (1992), 334
Cable modems, 136, 146
  vs. DSL, 135
CAD. *See* Computer-aided design (CAD); Computer-assisted design (CAD)
Calculator, microprocessor in, 82
Call center, artificial neural network for, 273
Caller ID, 135
Camera, digital computer, 50–51
Capacity. *See* Memory; Storage
Cards, memory, 49
Careers
  chief information officer, 27–28
  Internet-related, 28–29
  LAN administrators, 28
Carpal tunnel syndrome (CTS), **337**
Cartridges, 118
  for printers, 53
Cary®Origin2000 servers, 58
Cascading style sheets (CSS), 154
Case, Steve, 151, 157
Case(s), expert system use of, 263
CASE tools. *See* Computer-aided software engineering (CASE) tools
Catalina Marketing Corporation, data services from, 91
Catalog management software, **183**
CAVE virtual reality device, 270
CBIS. *See* Computer-based information system (CBIS)
CBS MarketWatch, 246
CD-rewritable (CD-RW) technology, 48
CD-ROM. *See* Compact disk read-only memory (CD-ROM) using, 87–88
CD-writable (CD-W) disks, 48
Celestia (ERP program), 207–208
Cell phones, Ericsson and, 88–89
Cellular transmission, 133
Centralized processing, **136**
Central processing unit (CPU), **41**, 43
CEO. *See* Chief executive officer (CEO)
Certification, in IS, **29**
CE Universe, credit card numbers and, 323
CFM International, 208
CFO. *See* Chief financial officer (CFO)
CGI. *See* Computer-generated image technology (CGI)

Chaining, backward and forward, 263–264
Chapin, John, 257
Character, **92**
Charge card, 186
Chase Manhattan Bank, 219
  Deloitte Consulting alliance with, 207
Chat room, **152**
CheckFree.com, 181
Chess games, AI and, 251
Chief executive officer (CEO), 27
Chief financial officer (CFO), 27
Chief information officer (CIO), 27–28
Children's Online Privacy Protection Act (1998) (COPPA), 334, 341–342
China, on-line access to imported music and videos, 151
Chippewa Falls, Wisconsin, computer-related school lunch mistake in, 321
Chips, 43, 54
  types of memory, 45
Choice stage, **213**–214
Cinergy Corporation, DSS and, 227
CIO. *See* Chief information officer (CIO)
Ciphertext, 161, 162
Cisco Systems, 174, 288
Claimsnet.com, 11
Clarion AutoPC, 50
Clark, George, 148
Client/server architecture, **137**–138
Clock speed, **42**–43
Closed shops, 299
CMGI, merger with Raging Bull, 20
CMOS chip, 57
CNBC, 246
Coaxial cable, 133, 134
CobaltCard, 309
COBOL, 79, 80
Coca-Cola, IT projects at, 315
Codes. *See* Programming languages
Coding. *See* Cryptography
Cognizer technology, 277
Cognos, 115
Collaboration, 244
Command-based user interface, **61**
Common carriers, **134**
Communication facilitators, 267
Communications. *See* Telecommunications
Communications Decency Act, 330
Communications protocols, **140**, 141
Communications software, **139**–140
Compact disk read-only memory (CD-ROM), **48**
Competition, electronic exchanges and, 178

Competitive advantage, **18**–21
  altering industry structure and, 20
  creating new products and services for, 20
  factors and strategies in, 21
  five-force model of, 18–19
  improving existing product lines and services for, 20
  strategic planning for, 19–21
  TPS for, 193
  using information systems for strategic purposes, 20–21
Competitive intelligence, **120**
Compiler, **81**
Complete information, 6
Complex situations, AI and, 251–252
CompUSA, 304
CompuServe, 330
Computer(s). *See also* Computer crime
  as crime-commission tool, 324
  health concerns about, 337
  manufacturers of, 304
  mistakes related to, 320–321
  as object of crime, 324–332
  waste and, 320
Computer Abuse Amendments Act (1994), 334
Computer-aided design (CAD), 56, 222, 230
Computer-aided software engineering (CASE) tools, **291**, 292, 297
Computer Associates, 279
Computer-based information system (CBIS), **9**–11
Computer crime, 322–332
  data alteration and destruction, 325–326
  illegal access and use, 324–325
  information and equipment theft as, 326–327
  international, 328
  preventing, 328–332
  scams, 327–328
  security breaches, 323–324
  software and Internet piracy, 327
Computer Crime and Security Survey, 323
Computer dealers and distributors, 304
Computer Emergency Response Team (CERT), 328
Computer Fraud and Abuse Act (1986), 328
Computer-generated image technology (CGI), 272
Computerized information system, 8–9
Computer Matching and Privacy Act (1988), 334
Computer network, **136**
Computer programs, **58**
Computer system platform, **58**

Computer systems. *See also* Hardware
  components of, 41–53
  types of, 54–58
Comshare, 115
Conclusions, expert system and, 258
Concurrency control, **105**
Consulting firms, careers with, 29
Consumer. *See also* Customer(s)
  operating systems for, 64
Consumer appliance operating systems, 67–68
Consumer goods companies, OLAP for, 115–116
Containment, in IS department, 299
Content streaming, **152**
Continuous-improvement programs, 23
Control unit, **41**
Cookie, **161**
  advertising ethics and, 162
  privacy issues and, 332
Coors Ceramics, 34
Coprocessor, **45**
Copyright, of Internet material, 331
CoreBuilder, 167
Corporations, computer crime prevention by, 328–329
Costs
  competitive, 19
  of expert system development alternatives, 266
Counterintelligence, **120**
CoverStory, 269
Covisint systems, 288
CPU. *See* Central processing unit (CPU)
Cracker, **325**
Crashes, testing and, 306
Creativity, of intelligent behavior, 252
Creativity software, 71
Credit bureaus, privacy and, 333
Credit card, 186
  fraud with, 193
  number theft and, 323
Crime, computer-related, 322–332
Criminal hacker, **325**
Crisis management, ESS and, 238
Critical success factors (CSFs), 297
CRM. *See* Customer Relationship Management (CRM)
CRT monitors, 52, 53
Cryptography, **161**–163
Cryptosystem, 162
CSFs. *See* Critical success factors (CSFs)
CSS. *See* Cascading style sheets (CSS)
Culture, moral codes and, 5
Customer(s)
  awareness and satisfaction of, 25
  bargaining power of, 19
  TPSs and, 190–191

Customer ordering database, entity-relationship (ER) diagram for, 99
Customer Relationship Management (CRM), 200
 programs, 223–224
Customer service
 artificial neural network for, 273
 wireless Net for, 317–318
Customization, of software, 69, 304
CyberCash, 186
Cybercrime. See Computer crime
Cybermall, **176**–177
Cybersecurity, 319

D

DaimlerChrysler, fuzzy logic at, 262
Daley, John V., 318
DARPA, 242
DASDs. See Direct access storage devices (DASDs)
Data, **4**
 alteration and destruction of, 325–326
 ERP for accessing, 200–201
 hierarchy of, 92–93
 information compared with, 4–6
 as input, 7
 process of transforming into information, 6
 types of, 5
Data analysis, 294–**295**
Database, **10**
 acquiring system, 305
 creating and modifying, 104–105
 data dictionary and, 104–105
 data mart and, 112
 data mining and, 112–113
 data warehouses and, 110–111, 112
 in DSS, 13
 market share of, 108
 on-line analytical processing (OLAP) and, 115–116, 117
 output of, 107
 privacy and, 333
 relational, 102–103
 software, 71, 75
 universal server, 118
 worldwide share by type, 108
Database approach to data management, **95**–97
Database design, 299
Database management system (DBMS), **92**, 95–96, 102–112, 228
 acquiring, 305
 advantages of, 97
 business intelligence and, 120–121
 concurrent users of, 109
 cost of, 110
 disadvantages of, 98
 features of, 109

integration of, 109
 performance of, 109
 popular systems, 108
 selecting, 108–112
 size of, 109
 storing and retrieving data in, 105
 user view of, 103
 vendor for, 109
Database models, 97, 99–102
DataBlades, 118
Data Broadcasting, 246
Data collection, **193**–194
 steps in, 295
 for systems analysis, 294
Data correction, **195**
Data dictionary, **104**–105
Data editing, **194**–195
Data Encryption Standard (DES) algorithm, 162–163
Data-entry operators, 26
Data-flow diagram (DFD), **295**, 296
Data glove, 16
Data havens, 138
Data integrity, **94**–95
Data item, **93**
Data management, 92–97
Data manipulation, **195**
Data manipulation language (DML), **106**–107
Data mart, **112**
Data mining, **112**–113, 122
 applications of, 115
 for e-commerce, 113
 OLAP and, 117
Data model(ing), 97, **98**–99, 295, 296
Data preparation (data conversion), **306**
Data processing
 strategies for, 136–137
 TPS activities, 196
Data redundancy, **94**
DataSage, 113
Data scrubbing, 3
Data services, from Catalina Marketing Corporation, 91
Data storage, **195**
Data tables, linking of, 101–102
Data warehouse, 57, **110**–111, 112
 at Allina Health System, 127
 OLTP database and, 116
 at Safeway, 114
Davis, Beth, 2
DB2, 102–103
dBASE (Inprise), 108, 117
DBMS. See Database management system (DBMS)
Dealtime.com, 153
Debit card, 186
 monitoring usage of, 309
Decency, Internet laws for protecting, 330
Decentralized processing, **136**
Decision
 nonprogrammed, 214
 programmed, 213

Decision making
 by expert system, 258
 problem solving and, 212–215
Decision-making phase, **212**
Decision room, **234**–235
Decision support system (DSS), 12–**13**. See also Group decision support system (GDSS)
 ad hoc, 227
 capabilities of, 227–228
 components of, 228–231
 conceptual model of, 230
 decision-making level and, 228
 elements of, 14
 e-trading ethical issues and, 231
 at Farmers Insurance Group, 3
 institutional, 227
 MIS compared with, 226, 228, 229
 overview of, 226–228
 problem structures, 228
 selected applications, 227
Decode instruction, 41
Dedicated line, **135**
Deep Blue (computer), 251
Defense Advanced Research Projects Agency, 267
Dell Computer, 187
Deloitte Consulting, 202
 Chase Manhattan Bank alliance with, 207
Delphi (language), 79
Delphi approach, **232**
DELTA (Diesel Electric Locomotive Troubleshooting Aid), 258
Demand reports, 107, **218**
Department of Defense (DOD)
 ARPANET and, 142
 Computer Emergency Response Team (CERT) and, 328
Design. See System design
Design report, **302**
Design stage, **213**
Desktop PCs, 54
Desktop publishing (DTP) software, 71
Deterrence controls, 299
Deutsche Bank, 219
Development. See also Systems analysis
 systems analyst and, 285
DFD. See Data-flow diagram (DFD)
DHTML. See Dynamic HTML (DHTML)
Dialogue manager, **228**
 for DSS, 231
Dictionary. See Data dictionary
Diffusion of technology, 18
Digital certificate, **185**
Digital computer camera, **50**–51
Digital detective, 324
Digital lecture system, 311
Digital Millennium Copyright Act (1998), 331
Digital signature, **163**

Digital subscriber line (DSL), **135**, 136, 146
Digital video disk (DVD), **48**–49
Direct access, **46**
Direct access storage devices (DASDs), **46**
Direct conversion, **306**, 307
Direct cutover. See Direct conversion
Direct Marketing Association, privacy issues and, 334
Disabled, neural networks for, 257
Disasters, business resumption planning and, 196–197
Disk
 magnetic, 47
 magneto-optical, 48
 optical, 48
Diskettes, 47
 CD-ROM and, 48
Display monitors, 52–53
Distance learning, **151**, 311
Distributed processing, **136**–137
DML. See Data manipulation language (DML)
Documentation, **58**
Document production, 196–197
Documents, on system maintenance and use, 321
Domain, **100**
Domain expert, **265**–266
Domain name, 143, 144, 189–190
Domino, 75
DOS with Windows, 64
DoubleClick, 179
Dow Chemical, 121
 e-mail monitoring and, 335
dpi (dots-per-inch), 53
Dragon Naturally Speaking Mobile, 50
DRAM. See Dynamic RAM (DRAM)
Drill down reports, **219**
Drugs, computer crime and, 328
Drugstore.com, 156
DSL. See Digital subscriber line (DSL)
DSS. See Decision support system (DSS)
Duhaime, Mark, 288
Dumpster diving, **324**
DVD. See Digital video disk (DVD)
DVD RAM, 49
Dynamic HTML (DHTML), 154
Dynamic RAM (DRAM), 44
Dynamic Web pages, 183

E

Earley, Kathleen, 283
Earning growth, 23
Ease of use, of expert systems, 258
Easi-Order, at Safeway, 114
Easy Diagnosis system, 268
eBay, 40, 153
eBillPay, 163
Echelon, 332

E-commerce, **11**. *See also*
E-commerce software
applications of, 176–181
business-to-business, 11
data mining for, 113
enterprise resource planning
and, 199–201
global, 187–188
hardware components of, 182
information and inventory
control in, 223
Internet copyright protection
and, 331
network and packet switching
for, 184–185
at Safeway, 114
strategies for, 188–190
technology components of,
182–186
trust issues and, 338
types of, 174
value chains in, 174–175
Web development and,
188–189
Weirton Steel Corporation
and, 173
E-commerce software, **183**–184
E-commerce transaction processing software, **184**
Economical information, 6
Economic feasibility, 293
Ecosystem for ASPs, 283
EDI. *See* Electronic data interchange (EDI)
EDO RAM. *See* Extended data
out (EDO) RAM
Education
distance learning and, 311
virtual reality applications
in, 272
Education Privacy Act, 334
E-ink (electronic ink), 87
EIS. *See* Executive information
system (EIS)
Electronic bill presentment, **181**
Electronic bulletin board, 150
Electronic cash, **185**
Electronic data interchange (EDI),
149–150, 174, 178
Electronic exchange, **177**–178
partnerships for, 207
Electronic Funds Transfer Act
(1979), 334
Electronic payment systems,
185–186
Electronic Postmark, 163
Electronic power business, cus-
tomer service in, 317–318
Electronic retailing (e-tailing), **176**
Electronic shopping cart, **183**, 184
Electronic Visualization
Laboratory (University of
Illinois), 270
Electronic wallet, **185**–186
Ellison, Larry, 130
E-mail, **147**
monitoring of, 335
privacy issues and, 333
Embedded computers, 54–55
AI and expert systems in
products, 268
fuzzy logic in, 261

Emerging enterprises, expert system for business plan development, 278
EMI Group, 151
Employees
expert system for perfor-
mance evaluation, 269
spying on, 59
Encryption, **161**–163
End user
development of PC-based systems by, 291
programming language pro-
totyping and, 289
End-user system development life
cycle, **291**
Enterprise application software,
76–77
Enterprise data modeling, **98**
Enterprise design projects, 298
Enterprise management
software, 279
Enterprise management
system (ERP)
integrating with other
systems, 201
risks in using one vendor, 201
Enterprise operating systems,
64, 67
Enterprise resource planning
(ERP), 199–201
in mergers and acquisitions,
207–208
at SmithKline Beecham, 316
vendors of, 199
Enterprise resource planning soft-
ware, 76, 156
Enterprise sphere of influence, 60
Enterprisewide database, 96–97
Entertainment
copyright infringement
and, 331
expert systems and, 268
Entity, **93**
Entity-relationship (ER) diagrams,
98–99, 295, 296
Environment, health issues
and, 337
E-piphany, 113
EPROM (erasable programmable
read-only memory), 45
Equipment theft, 326–327
ER diagrams. *See* Entity-
relationship (ER) diagrams
Ergonomic keyboard, 50
Ergonomics, **337**
Ericsson, 88–89
Cordless Screen Phone
HS210, 68
Ernst & Young, "Three Cs" rule
for groupware, 76
ERP. *See* Enterprise resource
planning (ERP)
ERP software. *See* Enterprise
resource planning software
Error, expert systems and, 258
ES. *See* Expert system (ES)
ESS. *See* Executive support
system (ESS)
E-tailing. *See* Electronic retailing

Ethernet, 141
Ethical and societal issues. *See
also* Privacy issues
competitive intelligence
and, 121
definition of ethics, 5
e-mail monitoring, 335
employee monitoring as, 59
e-trading and, 231
about expert systems, 259
in information systems, 320
monitoring debit card
usage, 309
neural networks for
disabled, 257
E-time. *See* Execution time
(E-time)
E-trading, ethical issues and, 231
Evaluation, of system design, 301
Event-driven review, **310**
Evolutionary Technologies
International, 119
Excel, 117
Exception reports, 107,
218–219
Exchange. *See* Electronic
exchange
Execution phase, 41–42
Execution time (E-time), **42**
Executive information system
(EIS), **236**
Executive support system (ESS),
236–238
capabilities of, 237–238
characteristics of, 236–237
layers of, 236
Expandable storage, **49**
Expert system (ES), **13**, 253,
257–269
applications of, 267–269
for business plan develop-
ment, 278
characteristics of, 258–259
components of, 260–265
development of, 265–268
human experts for, 261
legal and ethical concerns, 259
uses of, 259–260
Expert system shell, **259**
Explanation facility, 260, **264**
Extended data out (EDO)
RAM, 44
Extenders, 118
Extensible Markup Language
(XML), **154**
External data sources, 294
Extranet, 10, **158**, 159

F

Facts
as data, 4
as information, 4–5
Fair Credit Reporting Act (1970)
(FCRA), 334
Farmers Insurance Group, 3
Fax modem, 134
FBI. *See* Federal Bureau of
Investigation (FBI)
Feasibility analysis, **292**–293
Federal Bureau of Investigation
(FBI), computer crime and, 323

Federal Computer Incident
Response Capability, 161
Federal government. *See*
Government
Federal Trade Commission,
Internet privacy and, 332, 342
FedEx, UPS and, 131
Feedback, **8**
in information system, 17
Fetch instruction, 41
Fiber-optic cable, 133, 134, 138
Field, **92**
Fifth-generation programming
languages, 79
Fifth Third Bank, Internet systems
at, 128
File, **93**
File management, 63
File-oriented approach, as tradi-
tional approach, 94–95
File server architecture, **137**
File transfer protocol (FTP), **152**
Filtering software, 330
Finance, on Internet, 179–181
Financial management
software, 71
Financial MIS, **219**–221
Financial models, for DSS, **229**
Financial options, in systems
design, 300
Financial Web sites, 245–246
Fingerhut, 113
Firewall, **159**, 331
First-generation programming
languages, 78
First Union Bank, 220, 246
Five-force model, **18**–19
Flash chips, 88–89
Flexible information, 6
FLEXPERT, 267
Focus, 79
Fool.com, 180
Ford-Cisco alliance, 288
Forte, 79
FORTRAN, 79, 80
Forward chaining, **263**–264
Fourth-generation programming
languages (4GLs), 79
Fraud. *See also* Computer crime
credit card, 193
encryption and, 161–163
Freedom of Information Act
(1970), 334
computer privacy and, 333
Free speech
COPPA and, 341–342
e-mail monitoring and, 335
Fruit of the Loom, e-commerce
of, 189
FTP. *See* File transfer protocol (FTP)
Functional management informa-
tion systems, 13
Functional requirements, 297
Fuzzy logic, **261**–262
for travelers, 277

G

Games, expert systems and, 268
Garmin GPS III Plus, 151
Gartner Group, TCO model of, 35
Gavin, Michael, 157

GDSS. *See* Group decision support system (GDSS)
General Electric (GE)
　Aircraft Engine Group, 208, 269
　reapplication of Web site, 208–209
　software of, 304
General Motors (GM), 186
　distance learning at, 151
Geographic MIS, 226
Germany, on-line music in, 151
Gigabyte, 44
GIGO, 6
Glaxo Wellcome, merger with SmithKline Beecham, 316
Global e-commerce, 187–188
Globally Unique Identifier (GUID), 335–336
Global market, 187
Global positioning system (GPS), 150–151
　wireless Net and, 318
GlobalSight, 187
Globalstar Telecommunications, mistakes at, 321
GNU General Public License, 65
Goals, 59–60
Government. *See also* specific laws
　Children's Online Privacy Protection Act and, 341–342
　computer crime prevention by, 328
　privacy issues and, 332–333
GPS. *See* Global positioning system (GPS)
Graphical modeling program, **230**
Graphical user interface (GUI), **62**, 82
　Unix and, 66
Graphics software, 71, 72, 75
Greengard, Samuel, 90
Group consensus approach, **232**
Group decision support system (GDSS), 231–236, **232**
　alternatives to, 234–236
　characteristics of, 232–233
　local area decision network and, 235
　negative behavior and, 233
　parallel communication and, 233
　record keeping and, 233
　software for, 234
Group scheduling software, 76
Group videoconferencing, 148
Groupware. *See* Workgroup, application software
GTE, computer crime against, 323
GUI. *See* Graphical user interface (GUI)
GUID. *See* Globally Unique Identifier (GUID)

H
Hacker, 310, **325**
　computer crime and, 323

protecting corporate data from, 329
Handheld (palmtop) computers, 54
Haptic interface, 270
Hardesty, Christopher, 28
Hardware, **10**
　acquiring, 303–304
　backup for, 197
　career with vendor, 29
　components of, 40–42
　design, 299
　for e-commerce, 182
　functions of, 61
　global e-commerce and, 187
　independence of, 62
　input devices, 49–52
　memory characteristics and, 44–45
　multiprocessing and, 45
　output devices, 52–53
　privacy and, 334–336
　processing characteristics and functions, 42–43
　secondary storage devices, 46–49
Hartzel, Kathleen S., 202
Harvard Community Health Plan, 268
Hashing algorithm, 163
Head-mounted display (HMD), 15, 270
Healthcare companies, Kaiser Permanente and, 167–168
Health concerns, 337
Heane, Patrick, 169
Help desks, expert systems for, 268–269
Heuristics, **214**–215, 252–253
Hewlett-Packard, 184
　DSS and, 227
　hardware acquisition from, 304
Hierarchical (tree) models, **99**, 100
Hierarchy of data, 92–**93**
Highly structured problems, **228**
Hiring
　of IS personnel, 305
　predatory practices in, 341
Hits, 34
　service bottlenecks and, 160
HMD. *See* Head-mounted display (HMD)
Hoax virus, 326
Home builders, Internet sites of, 157
Home Depot, 39
Home network, 141
Home page, **153**. *See also* Internet; Web site
　Internet addresses and, 33–34
Home PNA network, 141
Hospitals, expert systems for, 268
Hotels, Mirage, 167
Household International, 186
HTML. *See* Hypertext markup language (HTML)
HTML tags, **154**, 190
http, 143, 155

Human resource MIS, **224**–225
Hunter, Mark, 210
HVAC equipment, 56
Hyperion Solutions, 115
Hypermedia, 119, **153**
　Web browser and, 154
Hypertext, 118–119
Hypertext markup language (HTML), **154**
Hypertext Transfer Protocol. *See* http

I
IBM, 103, 118
　database market share of, 108
　DecisionEdge software of, 3
　microprocessors of, 82
　virus detection and, 269
ICANN. *See* Internet Corporation for Assigned Names and Numbers (ICANN)
ICQ, 147
Idiom, 187
iDTVs, 187
IET-Intelligent Electronics, 269
If-then statements, **260**–261, 263
Illinois Power, customer service in, 317
Illinova Corporation, 317
IM. *See* Instant messaging (IM)
iMac, 54
Image data, 5
Immersive virtual reality. *See* Virtual reality system
Implementation stage, **213**
Individual design projects, 298
Infections, 325–326
Inference engine, 260, **263**
Information, **4**
　characteristics of valuable information, 6–7
　data compared with, 4–6
　sources of managerial, 216
　value of, 7
Information center, **27**
Information control, in e-commerce, 223
Information systems (IS), **4**. *See also* Systems design
　acceptance of, 18
　business, 11–14, 17–18
　computer-based, 9–11
　at Coors Ceramics, 34
　development team for, 284–285
　feedback and, 8
　input and, 7–8
　instant messaging and, 239
　manual and computerized, 8–9
　organization and, 16–18
　output and, 8
　performance-based, 21–23
　personnel for, 25–29
　personnel titles and functions in, 27–29
　processing and, 8
　responsibilities of, 26
　return on investment and value of, 23–25

social issues in, 320
　for strategic purposes, 20–21
Information systems (IS)
　department, 299
　roles and functions of, 25–27
Information systems planning, **286**
　developing Internet site for suppliers and dealers, 288
　IS plan and, 297
　steps of, 287
Information theft, 326–327
Informix, 103, 114, 118
Infrared transmission, 133
Infusion of technology, 18
In-house expert system development
　from scratch, 266
　from shell, 266–267
Initial public offerings (IPOs), Internet companies and, 28
Inprise (Borland International)
　dBASE, 108
　predatory hiring and, 341
Input, **7**–8
　devices for, 49–52
　processes for, 8
Installation, **306**
Instant messaging (IM), **147**, 239
Institutional DSS, **227**
Instructions, execution of, 42
Instruction time (I-time), **41**
Insurance companies, 3
Integrated-CASE (I-CASE) tools, 291
Integrated supply chain management software, 77
Integration, of database, 109
Integration testing, 306
Intel, 43
　chips of, 54
　flash chips and, 88–89
　microprocessor development by, 82
　privacy issues and, 334
　processor, 54
Intellectual Asset Management, 121
Intelligence, natural vs. artificial, 253
Intelligence stage, **212**–213
Intelligent behavior, **251**
　of expert system, 258
　nature of, 251–253
Interactive talk shows, on Internet, 152
Interface
　through ORDBMS, 118
　user, 61
Internal data sources, 294
Internal Revenue Service, 12
International computer crime, 328
International Computer Security Association, 326
International networks, **138**–139
Internet, **10**. *See also* Extranet; Intranet; World Wide Web (WWW, W3)
　accessing, 144–146
　careers related to, 28–29
　DSS and, 228

employee uses of, 59
Fifth Third Bank and, 128
functioning of, 143
home builder sites on, 157
job and career sites on, 28
Kaiser Permanente's strategy, 167–168
laws for libel and protection of decency, 330
management issues and, 160
marketing research on, 245
meaning of addresses, 33
new business applications for, 284
preventing crime on, 330–332
privacy issues and, 332–333, 336
routing messages over, 142
service bottlenecks on, 160–161
site for suppliers and dealers, 288
summary of users, 159
systems design and, 298
taxing sales on, 342–343
trust issues and, 338
UPS and, 131
use and functioning of, 141–146
WANs and, 236
Washington Post site, 168–169
wireless system, 317–318
Internet2 (I2), 142
Internet Activities Board (IAB), 160
Internet Corporation for Assigned Names and Numbers (ICANN), 143
Internet operators, 26
Internet piracy, **327**
Internet protocol (IP), **143**
Internet providers, of e-mail, 147
Internet Relay Chat (IRC), 152
Internet service(s)
    bidding systems, 181
    e-mail as, 147
    instant messaging as, 147
    voice mail as, 146–147
Internet service provider (ISP), **145**–146
Internet Society, 160
Internet Tax Freedom Act (1998), 343
Interpreter, **81**
Interwoven, 224
Intranet, 10, **157**–158, 159
Invasion of privacy. *See also* Privacy issues
    PocketCard as, 309
Inventory system
    at Coca-Cola, 316
    in e-commerce, 223
    JIT, 222
Investment
    on Internet, 179–181
    Web sites for, 180
IP. *See* Internet protocol (IP)
IRC. *See* Internet Relay Chat (IRC)
Iron Mountain, 196

IS. *See* Information systems (IS)
ISDN, 136, 146
ISP. *See* Internet service provider (ISP)
iTM1 (Applix), 116
IT projects, at Coca-Cola, 315

J
JAD. *See* Joint application development (JAD)
Java, 80, **155**
    applets and, 155, 156
    program maintenance and, 309
Java Studio, 79
JIT. *See* Just-in-time (JIT)
Job, Peter, 211
Job titles, of IS personnel, 27–28
John Cotton Tayloe School (Washington, North Carolina), 272
Joining, **101**
Joint application development (JAD), **290**
Judnick, Harriet, 162
Just-in-time (JIT)
    inventory approach, 222
    programs, 155

K
Kaiser Permanente, 268–269
    Web and, 167–168
Kbps, 132
Keen, Peter, 282
Kelly, James, 35–36
Key, **93**
    primary, 93–94
    secondary, 94
Keyboard, 50
Key-indicator report, **217**
Kilobyte, 44
Knowledge, defined, 5
Knowledge acquisition facility, 260, **264**
Knowledge base, 158, **260**–263
    of expert system, 258
    purpose of, 261
Knowledge engineer, 265, **266**
Knowledge management, **121**
Knowledge Query and Manipulation Language (KQML), 269
Knowledge user, 265, **266**
Kovar, Brian, 163
KPMG Peat Marwick, 10, 269
KQML. *See* Knowledge Query and Manipulation Language (KQML)
Kramer, Larry, 245–246
KWorld, 10

L
LaBarre, James E., 82
LAN. *See* Local area network (LAN)
Languages. *See* Programming languages
Language translator, **80**–81
Lanier, Jaron, 269

Law(s). *See also* specific laws
    federal privacy laws, 334
    Internet laws for libel and protection of decency, 330
Law enforcement, computer crime and, 328
Lawsuits, over privacy, 162
LCDs. *See* Liquid crystal displays (LCDs)
Learning, in intelligent behavior, 251
Learning systems, **256**
Leased line. *See* Dedicated line
Leasing, hardware equipment, 301, 304
Legal feasibility, 293
Legal profession, expert system for, 268
Lexis/Nexis, 237
Libel, Internet laws and, 330
Life cycles, systems development, 286–292
LifeTime software, 113
Line types, for telecommunications, 134–135, 136
Linking, **101**–102
Linux operating system, 65
    Mobil, 68
    Red Hat, 66
Liquid crystal displays (LCDs), 53
Loan analysis, expert system for, 269
Local access path (LAP), 105, 106
Local area network (LAN), **138**, 139
    accessing Internet via, 144–145
    administrators, 28
    GDSS, 235
    of Mirage Resorts, 167
    operators, 26
Localization, 187
Lockheed Martin, OLAP system at, 126
Logical design, **299**
Logic bomb, **326**
London Stock Exchange, crash at, 306
LOpht, 326–327
Los Alamos National Laboratory
    security issues at, 319, 323
    supercomputer at, 58
Lotus
    1-2-3, 117
    Approach, 108
    Notes, 75
Lovelace, Herbert W., 248
Lower-CASE tools, 291
Lyondell Chemical, 207

M
Machine cycle, **42**
Machine cycle time, 42
Machine language, 78
Macintosh computer. *See* Apple Computer; Mac personal computers
Mac personal computers, 65
    OS X and, 65, 67

Magnetic disk, **47**
Magnetic tape, **46**
Magneto-optical (MO) disk, **48**
Mahramas, Demetrios D., 202
Mail. *See also* E-mail; Postal Service
    SAN technology for, 47–48
    sorting, 9
Mainframe computers, 55, **56**–57
    as enterprise operating systems, 67
Mainframe relational DBMSs, 102–103
Maintenance. *See* Systems maintenance and review
Make-or-buy decisions, **304**
Management information system (MIS), **12**
    accounting, **226**
    DSS compared with, 228, 229
    financial, 219–221
    geographic, **226**
    human resource, 224–225
    inputs to, 216
    as integrated system, 220, 246
    manufacturing, 221–223
    marketing, 223–224, 245
    outputs of, 216–219
    overview of, 215–219
    for programmed decision making, 213
    reports generated by, 217–219
Management information system decision support system (MIS/DSS), 191
Manual information system, 8–9
Manuals, on system maintenance and use, 321
Manufacturing, repair, and operations (MRO) goods and services, 177
Manufacturing company, value chain of, 17
Manufacturing MIS, 215, 221–223
Manugistics, 223
Market, global, 187
Market entrants, competitive advantage and, 19, 21
Marketing
    data mining and, 113
    e-commerce and, 178–179
    ethics of, 162
    expert system for, 269
Marketing MIS, **223**–224, 245
Marketing research, on Internet, 245
Market intelligence, characteristics of valuable, 7
Marketplaces, on-line, 178
Market segmentation, **179**
Market share, 23–24
Markup languages, 154
Materials requirement planning (MRP), 222
MatheMEDics, 267
Mbps, 132
McCarthy, John, 250

McHaney, Roger, 122
McNerney, W. James, 208
Media, for telecommunications transmission, 133
Medical Manager Midwest, 244
Medicine. *See also* Health entries
expert systems for, 268
transaction processing system for, 11
virtual reality applications in, 271–272
Megabyte, 44
Melissa virus, 323, 326
Mellor, Robert, 157
Memory
characteristics and functions of, 44–45
flash chips and, 88–89
management of, 62
primary, 41
random access (RAM), 44
storage in, 41
types of, 44–45
Memory cards, 49
Memory chips, types of, 45
Menus, for search engines, 34
Mergers
ERP in, 207–208
systems development and, 317
Metadata, **119**
Metals companies, 173
MetalSite, 173
Meta tag, **190**
Metropolitan services, 150
Microprocessor, 82
Microsoft
Access, 107, 108, 117
database market share of, 108
data marts and, 112
e-mail monitoring and, 336–337
Office, 76
OLAP software of, 115
PC operating system of, 64–65
predatory hiring and, 341
software, 54
Windows, 62
Windows 2000, 109
Windows NT, 64–65, 109
Microwave transmission, 133
Midrange computer, 55, **56**
Million instructions per second (MIPS), 254
MineSet, 113
MineShare, 115
Mirage Resorts, on Net, 167
MIS. *See* Management information system (MIS)
MIS/DSS. *See* Management information systems decision support system (MIS/DSS)
Mistakes (computer-related), 320–321
preventing, 321–322
MMS. *See* Model management software (MMS)
Mobil Linux, 68
Model base, in DSS, 13, **228**–229

Model management software (MMS), in DSS, **228**–229
Modem, 134
speed of, 160
MO disk. *See* Magneto-optical (MO) disk
Moneycentral.msn.com, 153
Monitor, display, 52–53
Monitoring, of policies and procedures, 322
Monitoring stage, **213**
Monster.com, 28, 153
Moore, Gordon, 43
Moore's Law, **43**
Moral codes, 5
Morgan Stanley Dean Witter Trust, 47–48
Morris, Mike, 324
MOS Technology 6502 microprocessor, 82
Motion trackers, virtual reality and, 271
Motley Fool, The, 246
Motorola MC6800 microprocessor, 82
Mouse, 50
Mouse-controlled navigation, in 3-dimensional environment, 271
"Mousetrapping," 323
Movies, CGI for, 272
MP3, 151
MPEG-2 compression, distance learning and, 311
MRO. *See* Manufacturing, repair, and operations (MRO) goods and services
MRP. *See* Materials requirement planning (MRP)
MRPII, 222
MS-DOS, 64
Multicorporate design projects, 298
Multiplexer, 134
Multiprocessing, **45**
Multitasking, **63**
Mac OS X and, 67
Music, on-line, 151–152
MYCIN, 267
Myers Industries, 279
myhelpdesk.com, 153
MySimon, 269

N
NASA, 119
VPNs and, 159
Nathan, Susan, 162
National Audubon Society, DSS and, 227
National Fraud Information Center, 328
National Security Agency, 332
Natural language processing, **255**–256
"Nature-based models," of antivirus programs, 330
NCI Information Systems, 279
NDIS. *See* Network Driver Interface Specification (NDIS)
Netmosphere, 244
NetNanny, 330
NetPost Mailing Online, 163

Netware, 66
Network, **10**
AI software for monitoring and controlling, 279
computer, 136
for e-commerce, 184–185
home, 141
international, 138–139
local area (LAN), 138, 139
public network services, 150–151
telecommunications, 135
types of, 138–139
wide area (WAN), 138
Network computers, **55**–56
Network database model, 100
Network Driver Interface Specification (NDIS), 144
Networking
capability for, 63
at Mirage, 167
Network interface card (NIC), 138
Network management software, **140**
Network model, **99**–100
Network operating system (NOS), **140**
Neugents, 279
NeuralAct Inc., 273
Neural networks, **256**. *See also* Expert system (ES)
business applications of, 273
for disabled, 257
for virus detection, 269
New entrants. *See* Market entrants
New York (state), e-mail privacy issues in, 333
Next Generation Internet (NGI), 142
NFL GameDay, 272
NGI. *See* Next Generation Internet (NGI)
NIC. *See* Network interface card (NIC)
Niche market, competitive advantage and, 19
Nodes, 118
NOMAD mainframe reporting software, 120
Nominal group technique, **232**
Nonprogrammed decisions, **214**
NOS. *See* Network operating system (NOS)
Notebook computers, 54
Novell, Netware of, 66
Nuclear weapons testing, security and, 319

O
Object code, **80**
Object-oriented programming languages, 79–80
system maintenance and, 309
Object-relational database management system (ORDBMS), **118**–119
ODBC. *See* Open database connectivity (ODBC)

ODI. *See* Open Datalink Interface (ODI)
Off-the-shelf expert system purchase, existing packages as, 267–268
Off-the-shelf software, **68**–69, 304
compared with proprietary software, 72
OLAP. *See* On-line analytical processing (OLAP)
OLTP. *See* Online transaction processing (OLTP)
One-to-many relationship, 98–99
One-to-one relationship, 99
On-line analytical processing (OLAP), **115**–116, 117
at Lockheed Martin, 126
On-line marketplaces, 177–178
Online Privacy Alliance, 333
On-line services, 145
banking, 180–181
decency issues and, 330
information services, 71
newspapers, 168–169
stock trading, 179–180
Online transaction processing (OLTP), **191**
database, 116
Open database connectivity (ODBC), **117**
Open Datalink Interface (ODI), 144
Open shops, 299
Open systems interconnection (OSI) protocol, 141
Operating system (OS), **61**–63
Apple Computer, 65
consumer appliance, 67–68
enterprise, 67
Microsoft, 64–65
of personal computers, 64–65
workgroup, 65–67
Operation(s), of IS department, 25–26
Operational decision making, data access for, 200–201
Operational feasibility, 292–293
Optical disk, **48**
Optimization model, **214**
Oracle, 103, 114, 115, 118
ORDBMS. *See* Object-relational database management system (ORDBMS)
Order processing systems, **197**–198
Organization, **16**
ESS for, 238
information systems and, 16–18
model of, 16
Organizational goals, converting into systems requirements, 297
Organizational system, 16
OS. *See* Operating system (OS)
OSI. *See* Open systems interconnection (OSI) protocol
OS 9, 64
Osram Sylvania, 200
OS X, 64, 67

Output, **8**
    of database, 107
    devices, 10, 52–53
Outsourcing, 25
Ownership, total cost of, 35

P

Pacific Rim, linking with, 160
Packet switching technology,
    184–185
Page(s), 151
"Pagejacking," 323
Palmtop computers, 54
Palo Alto Research Center
    (PARC), 87
Parallel processing, **45**
Parallel start-up, **307**
Partnership. *See also* Strategic
    alliances
    as Internet exchanges, 207
Passive-matrix displays, 53
Password sniffer, **326**–327
Payroll transaction processing
    systems, 12, 195
PBX system, 134
PC. *See* Personal computer (PC)
PC COBOL, 80
PC-DOS, 64
PC Postage, 163
Pentium
    flash memory and, 89
    privacy issues and, 334
    processor, 82
People, in computer-based infor-
    mation systems, 10
Perceptive system, **252**
Performance, of database, 109
Performance-based information
    systems, 21–23
Performance measures, 22
Peripheral equipment manufac-
    turers, 304
Personal application software,
    69–74
    examples of, 71
Personal computer (PC),
    54–56, 55
    input devices for, 50
    Internet connections of, 10
    operating systems of, 64–65
Personal Computer Memory
    Card International Association
    (PCMCIA), 49
Personal information managers
    (PIMs), 76
Personal productivity
    software, **60**
Personify, 113
Personnel
    chief information officer
        (CIO), 27–28
    for IS, 25–29, 305
    for systems development,
        284–285
Personnel design, 299
Phase-in approach, **306**–307
Phone-answering systems, natural
    language processing in, 255

Physical access path (PAP),
    105, 106
Physical Access to Data operat-
    ing system, 62
Physical design, **299**
Piecemeal approach. *See*
    Phase-in approach
Pilot start-up, **307**
PIMs. *See* Personal information
    managers (PIMs)
Pipelining, 42
"Pipes," as transmission technolo-
    gies, 160–161
Piracy, software and Internet,
    327, 328
Pixel, **51**
    screens and, 52–53
PKI. *See* Public-key
    infrastructure (PKI)
Plaintext, 161
Planned data redundancy, **98**
Planning. *See* Strategic planning
Plant layout, expert system
    for, 268
Platinum Technology, 119
Plotters, 53
Plunge. *See* Direct conversion
PocketCard, 309
Pocket PC, 67–68
Point-of-sale (POS) devices, **51**
    transaction processing system
    and, 197
Point-to-point protocol (PPP),
    144, **145**
Polaroid Corporation, 245
Policies, to avoid waste and mis-
    takes, 321–322
Porter, Michael, 16, 19
Portfolio tracker, 180
POS devices. *See* Point-of-sale
    (POS) devices
Postal Service
    competitiveness of, 163
    mail sorting by, 9
    privacy issues and, 334
PosteCS, 163
Power, 42
PowerBuilder, 79, 290
Powerhouse, 79
PowerPC, 82
PowerPoint, 230
ppm (printed per minute), 53
PPP. *See* Point-to-point
    protocol (PPP)
Predatory hiring practices, 341
Preemptive multitasking, 67
Prevention, of computer crime,
    328–332
Priceline.com, 181
PricewaterhouseCoopers, digital
    detective and, 324
Primary key, **93**–94
Primary memory, 41
Printers, 53
Prism Solutions, 119
Privacy issues, 161–163,
    332–336
    e-mail and, 333, 335–336
    extranets, intranets, Web,
        and, 158
    fairness issues and, 336

federal privacy laws, 334
    government and, 332–333
    individual protection efforts,
        333–336
    Internet and, 336
    lawsuits over, 162
    at work, 333
Privacy Partnership, 338
Problem solving, **213**
    AI and, 252
    decision making and,
        212–215
    DSSs and, 227
Procedures, **11**
    to avoid waste and mistakes,
        321–322
Procedures and control
    design, 299
Process
    defined, 5
    subsystem, 16
    of transforming data in infor-
        mation, 5–6
Processing, **8**
    characteristics and functions
        of, 42–43
    of data, 41
    devices, 10
    operating system and, 62–63
Procter & Gamble, LAN usage
    by, 146
Procurement business, 207
Prodigy, 145
Product(s)
    improving existing, 20
    new, 20
Product configuration
    software, **183**
Product differentiation, 19
Productivity, **22**–23
Program. *See* Software
Programmable read-only mem-
    ory (PROM), 44–45
Programmed decisions,
    213–214
Programmer, 56, **285**
Programming languages, **77**–82
    attributes of, 78
Project Home Page, 244
Projecting, **101**
Project management
    collaboration and, 244
    models for DSS, **230**
    skills, 244
    software, 71
Project Oxygen, 142
PROM. *See* Programmable read-
    only memory (PROM)
Proprietary software, **68**, 304
    compared with off-the-shelf
        software, 72
Protocols
    communications, 140, 141
    FTP, 152
Prototyping, as iterative
    approach, 289
Prototyping phase, **289**, 290
Public-key encryption, 163
Public-key infrastructure (PKI), 147
Public network services,
    150–151

Purchasing equipment, advan-
    tages and disadvantages
    of, 301
Push technology, **155**–156

Q

Quality, **23**
Quicken, 73
Qvtech, 147

R

R/3. *See* SAP
RAD. *See* Rapid application
    development (RAD)
Radio, on-line, 152
Raging Bill, CMGI merger with, 20
RAID. *See* Redundant array of
    independent/inexpensive
    disks (RAID)
Random access memory
    (RAM), **44**
Rapid application development
    (RAD), **290**
RCA, DSS and, 227
Read-only memory (ROM),
    44–45
Realatrends Real Estate
    Service, 272
Real estate marketing, virtual
    reality for, 272
Record(s), **92**–94
    in DSS, 233
    off-site storage, 196
Red Brick Systems, 114
Red Hat Linux, 66
Redundancy, in e-commerce, 185
Redundant array of indepen-
    dent/inexpensive disks
    (RAID), **47**
Reel.com, 113
Regional services, 150
Registrars, 143
Relational database, 102–103
Relational models, **100**–102
Relationship management, tech-
    nology-enabled, 179
Relevant information, 6
Reliable information, 6
Remote logon, 152
Renting. *See also* Leasing
    of business software, 283
    of equipment, 301
Repair and maintenance, ACE
    expert system for, 269
Repetitive motion disorder (repet-
    itive stress injury; RSI), **337**
Reports
    design, 302
    producing, 107, 196–197
    systems analysis, 297–298
    systems investigation
        report, 293
Reports (from MIS), 217–219
    demand, 218
    drill down, **219**
    exception, 218–219
    guidelines for developing, 219
    key-indicator, 217
    scheduled, 217
Request for proposal (RFP), **300**
Requirements analysis, **296**–297

Resort Condominiums International (RCI), 116
Résumés, on Monster.com, 28
Retirement benefits, computer mistake in, 321
Return on investment (ROI), 22, **23** and value of IS, 23–25
Reusable code, 79, 80
Reuters Group, 211
Review, of policies and procedures, 322
RFP. *See* Request for proposal (RFP)
Right to Financial Privacy Act (1978), 334
Rivalry, competitive advantage and, 19, 21
Robotics, **254**–255. *See also* Bots (knowledge robots)
ROI. *See* Return on investment (ROI)
ROM. *See* Read-only memory (ROM)
Rosenfeld, John, 208
Routers, service bottlenecks and, 160
Royal Bank of Scotland, 148
Rubbermaid, 224
Rule, in expert system, **262**–263
Rules of thumb, heuristics as, 214–215

S
SABRE system, 21
Safeway, 114
St. Paul Software, 150
Samsung, 72
SAN. *See* Storage area network (SAN)
Santoro, Gerald M., 239
SAP, 156
    Customer Relationship Management (CRM) package of, 200
    predatory hiring and, 341
    R/3, 199, 207, 316
SAS (language), 79
SASDs. *See* Sequential access storage devices (SASDs)
SAS Institute, 122
Satisficing model, **214**
Scalability, **63**, 109
Scams, 327–328
Scanning, 194
    antivirus, 329
    bar code, 52
    devices, 51
Scheduled reports, 107, **217**
Schedule feasibility, 293
Scheduling, of network, 184
Schema, **103**, 104
Science fiction movies, AI and, 250
Scour, Media Agent of, 331
Screen Phone, 68
SDLC. *See* Systems development life cycle (SDLC)
SDRAM (synchronous DRAM), 44 of Samsung, 72

Search engines, 34, **154**–155
    Ask Jeeves, 249
    list of, 155
    Yahoo!, 155
Secondary keys, 94
Secondary storage, **45**–49
    access methods, 46
    devices, 46–49
Second-generation programming languages, 78–79
Secure information, 6
Security, 161–163. *See also* Computer crime
    in e-commerce, 185
    extranets, intranets, Web, and, 158–159
    at Los Alamos National Laboratory, 319
    of MetalSite, 173
    responding to incident, 325
    subschemas and, 103
    visible on screen, 185
Segmentation, market, 179
Selecting, **100**–101
Semantic processing, 249
Semistructured or unstructured problems, **228**
Sequential access, **46**
Sequential access storage devices (SASDs), **46**
Serial line protocol (SLIP), 144, **145**
Servers, 57
    accessing Internet through, 144–145
    market share of, 66
    on-line services as, 145
    Windows 2000, 66
    workgroup, 65–66
Services
    improving existing, 20
    new, 20
    telecommunications types, 134–135, 136
Shipping companies
    e-commerce transaction processing software and, 184
    Internet and, 131
Shopping cart. *See* Electronic shopping cart
SHYSTER, 267
Siebel Systems, 341
Silicon Graphics, MineSet software, 113
Simon, Herbert, 212
Simple information, 6
Single-vendor environments, 138
Site preparation, **305**–306
SLIP. *See* Serial line protocol (SLIP)
SLIP/PPP server, 145
Smalltalk, 80
Smart card, **186**
SmartMoney, 246
Smith, David, 258
Smith, Fred, 131
SmithKline Beecham, merger with Glaxo Wellcome, 316
SNA. *See* Systems network architecture (SNA)

Snail mail, 163
Snecma, 208
Social engineering, **324**
Societal issues. *See* Ethical and societal issues
Software, **10**. *See also* specific vendors and programs
    acquiring, 304
    AI for monitoring and controlling networks, 279
    application, 68–77
    bots as, 267
    bugs in, 81–82
    classified by type and sphere of influence, 60
    communications, 139–140
    for credit card fraud, 193
    customized packages, 69
    design, 299
    e-commerce, 182, 183–184
    expert systems for business plan development, 278
    filtering, 330
    for GDSS, 234
    at Home Depot, 39
    overview of, 58–60
    predatory hiring practices and, 341
    privacy and, 334–336
    quality of, 38
    rental of, 283
    sources of, 69
    systems maintenance and, 308–309
    user interface software for expert system, 264–265
    Web content management, 187
Software AG, 112
Software piracy, 5, **327**
Software Publishers Association (SPA), computer crime and, 329
Software suites, **72**–74, 76
Software vendor, career with, 29
Solaris, 66–67
Sorting system, at U.S. Post office, 9
Source code, **80**
Source data automation, **194**
Spatial data technology, 119
Specialized services, 150–151
Speed, 42–43
    mistakes based on, 321
    of modems, 160
SpeedStep chip, 54
SpendCash, 309
Sphere of influence, **59**–60
Spies. *See* Los Alamos National Laboratory
Sports simulation, CGI for, 272
Spreadsheet software, 10, 70–71, 74
Sprint, computer crime against, 323
Spying, on employees, 59
SQL. *See* Structured query language (SQL)
Staffing, ESS and, 238

Stakeholders, **284**
    user acceptance and, 308
Standardized Teleoperation System (STS), 254
Standard & Poor's (S&P), neural network system of, 256
Stanford University, 268
Start-up, 306–307
Static Web pages, 183
Statistical analysis models, for DSS, **229**–230
Stereo viewing, 271
Stock trading, on-line, 179–180
Storage, 41
    capacity, 44
    expandable, 49
    secondary, 46–49
Storage area network (SAN), **47**–48
Storefront broker, **189**
STP. *See* Straight Through Processing (STP)
Straight Through Processing (STP), 202
Strassman, Paul, 38
Strategic advantage, at UPS, 35–36
Strategic alliances, 19, 20
Strategic decision making. *See* Decision making; Decision support system (DSS)
Strategic partnerships. *See* Strategic alliances
Strategic plan
    for Ford-Cisco alliance, 288
    information systems planning and, 286
Strategic planning, **237**
    for competitive advantage, 19–21
    information systems and, 286
Strategic Processing Environment, of Deloitte Consulting, 202
Structured query language (SQL), 79, 108, 109
STS. *See* Standardized Teleoperation System (STS)
Style sheets, cascading, 154
Subnotebook computers, 54
Subschema, **103**, 104
Substitute products and services, competitive advantage and, 19, 21
Subsystems, of MIS, 215, 221
Sun Microsystems
    C++, 155
    rental software market and, 283
    Solaris, 66–67
Supercomputers, 55, **57**
    Blue Mountain, 58
Superconductivity, **43**
Suppliers, bargaining power of, 19
Supply chain management, **174**–175
Supply chain operations, on Internet, 177–178
Support, 26, 27

Support personnel, for systems development, 284–285
Surfing, 154
Swiss Reinsurance America, 74
Switched line, **134**
Sybase, 103, 112, 114
    PowerBuilder from, 290
Synchronous DRAM, 44
Syntactic processing, 249
Syntax, **78**
System(s)
    access to resources of, 63
    organization as, 16
    policies and procedures for using, 321–322
System engineers, 56
System operators, 26, 27
System programmers, 56
Systems analysis, 15, **286**–288, 293–298
    data analysis and, 294–295, 296
    data collection and, 294
    report on, 297–298
    requirements analysis and, 296–297
Systems analyst, **284**–285
Systems design, 15, **288**, 298–302
    alternatives in, 300
    evaluating and selecting, 301
    financial options in, 300
    freezing specifications for, 301, 302
    scope of, 298
Systems development, **14**–16, 26–27, 284
    at Coca-Cola, 316
    distance learning courses and, 311
    information systems planning, 286
    mergers and, 317
    participants in, 284–286
    team leaders for, 285
Systems development life cycle (SDLC), 286–292
    CASE tools and, 291
    computer-aided software engineering (CASE) tools and, 291, 292
    end-user, 291
    joint application development (JAD), 290
    prototyping, 289, 290
    rapid application development (RAD), 290
    traditional systems development, 286–289
Systems implementation, 15, **288**–289, 303–308
    database and telecommunications systems acquisition, 305
    data preparation, 306
    hardware acquisition, 303–304
    installation and, 306
    personnel hiring and training, 305
    site preparation, 305–306
    software acquisition, 304

start-up and, 306–307
    steps in, 303
    testing and, 306
    user acceptance and, 307–308
    user preparation and, 305
Systems investigation, 15, **286**, 292–293
Systems investigation report, **293**
Systems maintenance and review, 15, **289**, 308–309, 309–310
Systems network architecture (SNA), 141
Systems software, 10, **58**, 60–68
Systems testing, 306

T
T1, 136
Tags, HTML, 154, 190
TAM. See Technology acceptance model (TAM)
Tandem, 114
Tape. See Magnetic tape
Taxation, of Internet sales, 342–343
Tax preparation programs, 73
Tax Reform Act (1976), 334
TCO. See Total cost of ownership (TCO)
TCP. See Transport control protocol (TCP)
TCP/IP, 141, 143
    service bottlenecks and, 160
Teams, for systems development projects, 285
Technical feasibility, 292
Technology, 2
    e-commerce components, 182–186
    ERP for upgrading infrastructure, 201
Technology acceptance model (TAM), **18**
Technology diffusion, **18**
Technology-enabled relationship management, 179
Technology infrastructure, **9**
Technology infusion, **18**
Teenagers, PocketCard for, 309
Telecommunications, **10**, 132–135
    acquiring system, 305
    carriers and services for, 134–135
    devices for, 134
    transmission media for, 133
Telecommunications Act (1996), 330
Telecommunications design, 299
Telecommunications medium, **132**
Telecommunications networks, 135
    UPS, FedEx, and, 131
Telecommuting, **147**–148
Teleconferencing, as GDSS alternative, 235
Telephone companies, 134
    computer crime against, 323
Telephone Consumer Protection Act (1991), 334

Telephone networks, expert system for, 269
Telephone service, standard, 136
Telepresence systems, 271
Telnet, **152**
Terabyte, 44
Terminals, 51
Terminal-to-host architecture, **137**, 138
Terrorism, computer, 328
Testing, of IS systems, 306
TFT-LCDs, 72
TheStreet.com, 246
Third-generation programming languages, 79
Third-party Web service provider, for e-commerce, 182
THORASK, 267
3Com, 167
3-D Website Builder, 272
Ticket sales, automated, 4
Time-driven review, **310**
Timely information, 6
Time-share resorts, OLAP and, 116
Time-sharing, **63**
Time values, of expert system development alternatives, 266
Time Warner, AOL merger with, 20
Timken, 178
Total cost of ownership (TCO), **25**, 35
Touch-sensitive screens, 52
TPS. See Transaction processing system (TPS)
Tracker Web sites, 180
Trade secrets, 120
Traditional approach to data management, **94**–95
Traditional systems development life cycle, 286–289
Traffic data analysis software, 184, 190
Training
    of IS personnel, 305
    virtual technology for, 272
Transaction, **12**
Transaction processing cycle, **193**
Transaction processing system (TPS), 11–**12**, 190–198
    activities of, 192–197
    for competitive advantage, 193
    credit card fraud and, 193
    data processing activities common to, 196
    for document production and reports, 196–197
    in financial services, 202
    integration of firm's, 194
    MIS information from, 216
    overview of, 195
    point-of-sale, 197
    traditional methods and objectives of, 191–192
Transborder data flow, 138
Transmission media, for telecommunications, 133
Transport control protocol (TCP), **143**

Transport control protocol/Internet protocol. See (TCP/IP)
Travelers, fuzzy logic system for, 277
Treasury Department, computer crime and, 324
Tree models. See Hierarchical (tree) models
Trojan horse, **326**
TRUSTe trustmark, 338
Tunneling, **159**
TurboTax, 73
Twisted-pair wire cable, 133, 134
2001: A Space Odyssey (film), 250

U
UCAID. See University Corporation for Advanced Internet Development (UCAID)
Uncertainty, expert system and, 258
Underwriting business, 3
Unicenter TNG, 279
Uniform resource locator (URL), **143**
Uniscape.com, 187
United Distillers, 269
United Parcel Service (UPS). See UPS
U.S. Army, DSS and, 227
U.S. Forest Service, 226
U.S. Postal Service. See Postal Service
United Technologies (UTC), competitive intelligence and, 121
Unit testing, 306
Universal database server, 118
Universal Product Code (UPC), 194, 195
University Corporation for Advanced Internet Development (UCAID), 142
University of Canterbury (Christchurch, New Zealand), distance learning in, 311
Unix operating system, 65, 66, 109
Unshielded twisted-pair (UTP) wire cable, 138
Unstructured problems, **228**
UPC. See Universal Product Code (UPC)
Upper-CASE tools, 291
UPS, 131
    technology at, 35–36
    TPS of, 190–191
UPSnet, 139
URL. See Uniform resource locator (URL)
USBuild.com, 157
User(s), 10, **284**
User acceptance document, **307**–308
User interface, **61**
    of expert system, 260
    expert system software, 264–265

User preparation, in systems
implementation, 305
UTP wire cable. *See* Unshielded
twisted-pair (UTP) wire cable

V
Value-added business processes,
ESS and, 237
Value-added carriers, **134**
Value-added networks
(VANs), 184
Value-added process, 16, 17
Value chain, **16**
in e-commerce, 174–175
of manufacturing company, 17
Values, ethics and, 5
VANs. *See* Value-added net-
works (VANs)
Vargo, John, 311
Velocity, of shopper, 113
Vendors, of ERP, 199, 201
Verbind, LifeTime software
from, 113
Verifiable information, 6
Very large databases
(VLDBs), 111
Victoria's Secret, 152
Video, on-line, 151–152
Videoconferencing, **148**
Video data, 5
Video display terminal (VDT), bill
for, 337
Video Privacy Act (1988), 334
Virginia, DSS and, 227
Virtual Aircraft Maintenance
System, 272
Virtual private network (VPN),
159, 184
Virtual reality system, **13**–14,
269–272
applications of, 271–272
data glove and, 16
forms of, 271
head-mounted display and, 15
immersive, 271
interface devices for, 270
Virtual workers, 151
Virtual workgroups, **235**–236
Virtus Corporation, 272

Virus, 323, **325**–326
antivirus programs and,
329–330
e-mail monitoring and, 335
hoax (false), 326
Melissa, 323, 326
neural network for
detecting, 269
top 10, 327
Vision, ESS and, 237
Vision systems, **255**
Visteon, 226
Visual Basic, 79, 80
Visual C++, 79, 80
Visual images, AI and, 252
Visual programming
languages, 80
VLDBs. *See* Very large databases
(VLDBs)
Voice mail, **146**–147
Voice-over-IP (VOIP), **148**–149
Voice-recognition devices, **50**
VOIP. *See* Voice-over-IP (VOIP)
Volume testing, 306
VPL Research, 269
VPN. *See* Virtual private
network (VPN)

W
W3. *See* Internet; World Wide
Web (WWW, W3)
Wall Street Journal Online,
The, 246
WAN. *See* Wide area network
(WAN)
Warehouse optimization, expert
system for, 269
Washington Post, on Web,
168–169
Waste, computer-related, 320,
321–322
Weather-Tek Design Center, 157
Web. *See also* Internet; World
Wide Web (WWW, W3)
Web appliance, **56**
Web-based operations, ORDBMS
and, 118
Web browser, **154**
HTML tags in, 154

Webcasting, 155
Web content management software, 187
Webmd.com, 153
Web operators, 26, 28
Web pads, 68
Web page construction
software, **183**
Web sales, taxing, 342–343
Web server software, for
e-commerce, 183
Web site, 34. *See also* Internet;
Privacy issues
attracting customers to,
189–190
for computer security, 326
data mining and, 113
establishing, 189
financial, 245–246
of GE, 208–209
for investors, 180
listing of, 153
for suppliers and dealers, 288
theft of, 323
of U.S. companies, 188
Web site hosting
companies, **189**
Web site traffic data analysis
software, **184**, 190
Weirton Steel Corporation, 173
Welch, John F., 208
Wetherell, David, 20
WhiteLight, 115
Whiting, Rick, 2
Whitman, Meg, 40
Whyte, Ronald, 267
Wide area network (WAN),
138, 140
as GDSS alternative,
235–236
Wilder, Clinton, 172
Willamette chip, privacy issues
and, 334
Windows (Microsoft), 62
CE (Pocket PC), 67–68
development of, 82
Millennium Edition (ME), 65
Windows 95, 64
Windows 98, 64

Windows 2000, 65, 109
Windows 2000 server, 66
Wireless Net, for customer service, 317–318
Wiring, for telecommunications
transmission, 133, 134, 160
Word processing software, 70,
71, 74
Workgroup, 60
application software, 74–76
computing, 158
design projects, 298
operating systems for, 64,
65–67
sphere of influence, 60
Workplace, computer privacy
in, 333
Workstation, 55, **56**
Worldstream Communications, 152
Worldwide Retail Exchange, 178
World Wide Web (WWW, W3),
143. *See also* Internet;
Web site
browsers for, 154
business uses of, 156
ethics of advertising on, 162
Java and, 155
search engines for, 154–155
Worm, **325**
www. *See* Internet; World Wide
Web (WWW, W3)

X
X.400 standards, 141
X.500 standards, 141
Xerox, 87
XML. *See* Extensible Markup
Language (XML)

Y
Yahoo!, 34, 155
Yasunobu, Seiji, fuzzy logic and,
261–262
Yatra.net, 277

Z
ZIP codes, 9